Risk, Crisis, and Disaster Management in Small and Medium-Sized Tourism Enterprises

Diego R. Toubes
University of Vigo, Spain

Noelia Araújo-Vila
University of Vigo, Spain

A volume in the Advances in Hospitality, Tourism, and the Services Industry (AHTSI) Book Series

IGI Global
PUBLISHER of TIMELY KNOWLEDGE

Published in the United States of America by
IGI Global
Business Science Reference (an imprint of IGI Global)
701 E. Chocolate Avenue
Hershey PA, USA 17033
Tel: 717-533-8845
Fax: 717-533-8661
E-mail: cust@igi-global.com
Web site: http://www.igi-global.com

Library of Congress Cataloging-in-Publication Data

Names: Toubes, Diego R., 1966- editor. | Araujo-Vila, Noelia, 1983- editor.

Title: Risk, crisis, and disaster management in small and medium-sized tourism enterprises / Diego R. Toubes and Noelia Araujo-Vila, Editor.
Description: Hershey, PA : Business Science Reference, [2021] | Includes bibliographical references and index. | Summary: "This book connects the challenges of today's tourism industry more closely to academic research, connecting research in the field of crisis management with the risks affecting small and medium-sized tourism enterprises"-- Provided by publisher.
Identifiers: LCCN 2020055681 (print) | LCCN 2020055682 (ebook) | ISBN 9781799869962 (hardcover) | ISBN 9781799869979 (paperback) | ISBN 9781799869986 (ebook)
Subjects: LCSH: Tourism--Management. | Tourism--Risk management. | Crisis management.
Classification: LCC G155.A1 R542 2021 (print) | LCC G155.A1 (ebook) | DDC 910.68--dc23
LC record available at https://lccn.loc.gov/2020055681
LC ebook record available at https://lccn.loc.gov/2020055682

This book is published in the IGI Global book series Advances in Hospitality, Tourism, and the Services Industry (AHTSI) (ISSN: 2475-6547; eISSN: 2475-6555)

British Cataloguing in Publication Data
A Cataloguing in Publication record for this book is available from the British Library.

For electronic access to this publication, please contact: eresources@igi-global.com.

Advances in Hospitality, Tourism, and the Services Industry (AHTSI) Book Series

ISSN:2475-6547
EISSN:2475-6555

Editor-in-Chief: Maximiliano Korstanje, University of Palermo, Argentina

MISSION

Globally, the hospitality, travel, tourism, and services industries generate a significant percentage of revenue and represent a large portion of the business world. Even in tough economic times, these industries thrive as individuals continue to spend on leisure and recreation activities as well as services.

The Advances in Hospitality, Tourism, and the Services Industry (AHTSI) book series offers diverse publications relating to the management, promotion, and profitability of the leisure, recreation, and services industries. Highlighting current research pertaining to various topics within the realm of hospitality, travel, tourism, and services management, the titles found within the AHTSI book series are pertinent to the research and professional needs of managers, business practitioners, researchers, and upper-level students studying in the field.

COVERAGE

- Leisure & Business Travel
- Food and Beverage Management
- Service Training
- Customer Service Issues
- Travel Agency Management
- Service Design
- Health and Wellness Tourism
- Destination Marketing and Management
- Casino Management
- Tourism and the Environment

IGI Global is currently accepting manuscripts for publication within this series. To submit a proposal for a volume in this series, please contact our Acquisition Editors at Acquisitions@igi-global.com or visit: http://www.igi-global.com/publish/.

The Advances in Hospitality, Tourism, and the Services Industry (AHTSI) Book Series (ISSN 2475-6547) is published by IGI Global, 701 E. Chocolate Avenue, Hershey, PA 17033-1240, USA, www.igi-global.com. This series is composed of titles available for purchase individually; each title is edited to be contextually exclusive from any other title within the series. For pricing and ordering information please visit http://www.igi-global.com/book-series/advances-hospitality-tourism-services-industry/121014. Postmaster: Send all address changes to above address.

Titles in this Series

For a list of additional titles in this series, please visit: http://www.igi-global.com/book-series/

Resiliency Models and Addressing Future Risks for Family Firms in the Tourism Industry
Anita Zehrer (MCI The Entrepreneurial School, Austria) Gundula Glowka (MCI The Entrepreneurial School, Austria) Katrin Magdalena Schwaiger (MCI The Entrepreneurial School, Austria) and Victoria Ranacher-Lackner (MCI The Entrepreneurial School, Austria)
Business Science Reference • © 2021 • 361pp • H/C (ISBN: 9781799873525) • US $195.00

Strategic International Restaurant Development From Concept to Production
Angelo A. Camillo (Sonoma State University, USA)
Business Science Reference • © 2021 • 497pp • H/C (ISBN: 9781799843429) • US $195.00

ICT Tools and Applications for Accessible Tourism
Celeste Eusébio (University of Aveiro, Portugal) Leonor Teixeira (University of Aveiro, Portugal) and Maria João Carneiro (University of Aveiro, Portugal)
Business Science Reference • © 2021 • 356pp • H/C (ISBN: 9781799864288) • US $215.00

Impact of New Media in Tourism
Maria Gorete Dinis (GOVCOPP, Polytechnic Institute of Portalegre, Portugal) Luís Bonixe (ICNOVA, Polytechnic Institute of Portalegre, Portugal) Sónia Lamy (ICNOVA, Polytechnic Institute of Portalegre, Portugal) and Zélia Breda (GOVCOPP, University of Aveiro, Portugal)
Business Science Reference • © 2021 • 389pp • H/C (ISBN: 9781799870951) • US $195.00

Growth of the Medical Tourism Industry and Its Impact on Society Emerging Research and Opportunities
Manjeet Singh (Central University of Jammu, India) and Subbaraman Kumaran (Alliance University, Bangalore, India)
Business Science Reference • © 2021 • 189pp • H/C (ISBN: 9781799834274) • US $195.00

Global Development of Religious Tourism
Emilia Alaverdov (Georgian Technical University, Georgia) and Muhammad Waseem Bari (Government College University, Faisalabad, Pakistan)
Business Science Reference • © 2021 • 366pp • H/C (ISBN: 9781799857921) • US $195.00

IGI Global
PUBLISHER of TIMELY KNOWLEDGE

701 East Chocolate Avenue, Hershey, PA 17033, USA
Tel: 717-533-8845 x100 • Fax: 717-533-8661
E-Mail: cust@igi-global.com • www.igi-global.com

Editorial Advisory Board

Table of Contents

Detailed Table of Contents

during the COVID-19 pandemic. The extent of the analysis includes hotels and guest houses in the south-western part of Bulgaria that managed to keep operating despite the global pandemic situation.

This case study is intended to analyze, through a longitudinal analysis, the radical business model innovation experienced a Cantabrian tourism company that exploits a protected cultural property called "El Capricho de Gaudí." More specifically, this chapter focuses on the manager attitudes toward risk and their influence on response strategies implemented to redefine and renew this business model for competitiveness and survival. The results of the analysis contribute to generate knowledge about how a solid and well-established small experienced cultural-based tourism enterprise reformulates its business model to survive in time of crisis.

This study focuses on the effects of COVID-19 on independent restaurant operators in small cities. Proposals for recovery are assessed both in terms of business operations and service provision and in terms of the governance measures demanded by restaurateurs. Semi-structured open-ended interviews were conducted with 38 restaurateurs in Ourense (Spain) during the toughest months of the COVID-19 pandemic. This study provides key operational and governance measures for the recovery of the restaurant sector and confirms the need to generate a greater connection and empathy with the customer and to adapt to the new situation of instability and uncertainty by incorporating new business models. More effective governance measures are needed in line with increased public concern for sustainability, humanization, and environmental conservation.

Preface

Tourism destinations are traditionally dominated by small and medium-sized enterprises that provide a wide range of products to tourists, such as accommodations, travel services, transportation, recreation and entertainment, and food and beverage services, thus contributing significantly to the variety, authenticity, and quality of their tourist experience.

Small and medium-sized tourism enterprises (SMTEs) are more vulnerable to crisis impacts due to their limited capacity to reduce risk. SMTEs do not have the capacity or the financial, human and operational resources of large companies to prepare for crisis contingencies (i.e. to plan for and respond to the challenges they face). Due to their small size and limited resources and knowledge, small businesses have less capacity to identify potential threats, assemble crisis teams and draw up crisis plans. On the other hand, the imbalance and disparity in innovation levels between businesses and tourism activities is probably due to the small size of businesses in the sector. We therefore believe that this is an area of growing importance and concern, in both the public and private sectors, where specific research and more in-depth knowledge are required.

Since the end of the 20th century, the number of publications on risk, crisis and disaster management in tourism has increased. The most widely cited scientific paper in this field was written by Faulkner (2001) and published under the title "Towards a framework for disaster management in tourism". Since then, new knowledge and global risks have emerged. Recently, the COVID-19 pandemic struck all societies and economies of the world, particularly affecting the tourism sector. Many SMTEs around the world have been severely affected by the need to completely shut down for months, and medium-term recovery expectations are not optimistic. After implementing solutions and measures to overcome the crisis in the short term, most SMTEs need to plan for the medium and long term to enable them to survive the current dramatic situation and prosper. We need to better understand the impact and risks associated with global diseases and pandemics and whether we must consider rethinking the business models of the tourism industry in general and SMTEs in particular.

There is a need to connect the challenges of today's tourism industry more closely to academic research. However, despite the prevalence of small and medium enterprises in the tourism industry, there is still limited literature on SME-related planning, response and recovery. This book calls for this gap to be bridged in the field of crisis and disaster management so that, based on rigorous studies, this publication connects research in the field of crisis management with the risks affecting SMTEs.

Originally, this publication was not limited to proposals examining the threat and impact of COVID-19 on SMTEs, but to any type of risk posing a threat to tourism businesses. However, the call for chapters of this book coincided with the most critical moments of the pandemic and lockdowns in many regions of the world. The impact of the pandemic on the tourism sector, and particularly on SMEs, is reflected in that most of the contributions to this book (written by authors from different parts of the world) deal with the virus as the main threat to the different sectors analysed. In a context where the pandemic has dramatically and unexpectedly affected the tourism sector, this has been an unavoidable issue, as many companies have had to close down, some temporarily, but many permanently.

Next, we present the eleven chapters selected after going through the different stages of reviewing and editing, with a brief explanation of their content and highlighting the main topics addressed. We are confident that these contributions will help to better understand the challenges that SMEs face when confronted with crises at any of the three stages (preparedness and planning, response and recovery, and resolution and feedback) in the different business areas (strategy and planning, human resource management, operations, information and communication technology (ICT), tax solutions, and marketing). Moreover, the book provides insights and supports executives and public authorities in different types of environments and subsectors, such as rural tourism, travel and guide services, car rentals, catering, travel agencies and tour operators.

STRATEGIC MANAGEMENT OF COVID-19 CRISIS IN RURAL TOURISM SETTINGS: LESSONS LEARNED FROM SMES

During the COVID-19 pandemic, the author focuses on the crisis management of rural SMTEs. Through semi-structured interviews carried out with owners, general managers and professionals in the rural tourism segment, the research addresses the three traditional phases of crisis management: the measures taken to alleviate the crisis, the lessons learned during the period of coexistence with the pandemic, and the opportunities for rural tourism development in the post-virus scenario. Concerning the final stage, a number of key topics are proposed, such as safety and hygiene, risk of rejection of tourism, destination governance, marketing policies, compatibility between short-term emergency measures and long-term vision, creativity and innovation, and people-centred organizations. Finally, the author offers a wide range of practical recommendations, including business strategies for customer base diversification and building loyalty among new rural tourism clients, as well as measures to strengthen governance structures and partnership agreements with business associations for the technological and business skills update of rural SMTEs. In the academic field, this study contributes to filling the gap in the scientific literature on crisis management in rural SMTEs.

ICTs AND RURAL TOURISM: EXPERIENCES OF THE COVID-19 PANDEMIC IN CENTRAL MEXICO

This study continues the analysis of the rural tourism sector and the consequences of the pandemic, focusing on the role that ICTs have played in the process of adaptation, learning and tourism reactivation in rural areas. The authors develop a case study of nine community-based rural tourism enterprises located in indigenous territories of the Central Mexican Highlands, collecting the experiences accumulated by companies throughout the pandemic. This study highlights the role of ICTs as elements that allowed for maintaining social cohesion and articulation between groups with specific interests, such as those involved in rural tourism. The increase in the use of ICTs as a result of the outbreak of the COVID-19 pandemic has an ambivalent meaning that includes the exacerbation of inequalities and generates opportunities for small rural tourism enterprises. The authors consider that the still incipient use of ICTs in rural areas should lead to the creation of mechanisms to bridge the technology gap.

THE IMPACT OF COVID-19 ON THE TOURISM INDUSTRY: THE CASE OF MICRO, SMALL AND MEDIUM-SIZED ENTERPRISES (MSMEs) IN BARCELONA

This chapter analyses the impact of COVID-19 on the tourism activity of an important urban tourist destination, Barcelona. After exploring the effects and impact of the pandemic on Barcelona's tourism sector, a number of problems faced by SMEs and the self-employed are identified. As a second part of this research, a questionnaire was administered to about 200 travel agencies (small tour operators and tour guides). Nearly 90 per cent of the companies interviewed do not offer services that they used to offer before COVID-19; that is, they have had to close the business wholly or partially. These findings indicate that most of the participating enterprises are struggling to survive. Innovation and development of the Internet are seen as key factors in improving the quality of services in the tourism industry and seizing the opportunities that might arise. However, this study shows that not all companies are working towards that holistic and innovative mentality. Only 34 per cent of organizations offer new tourism products, so the results show a lack of innovation and resilience to adapt to a new way of life and business. Moreover, it should be noted that conventional policy measures might not be able to overcome this crisis, as companies are not adapting their activities to the new situation. The bailouts have highlighted the growing reliance of the market on the state, and the dependence of companies on subsidies to sustain themselves. In short, the authors argue that companies should offer new tourism products aimed at local visitors and look for formulas that enable flexible cancellations and refunds. Governments should plan a new tourism strategy based on domestic tourism for the immediate future.

PUBLIC HEALTH RISK ENVIRONMENT FOR BULGARIAN SMES IN TOURISM DURING COVID-19 PANDEMIC

This chapter focuses on examining the public health risk environment for Bulgarian SMTEs during the COVID-19 pandemic. An analysis of the hotels and guesthouses network in south-western Bulgaria is carried out. After identifying the main characteristics of SMTEs and their survival conditions in times of threat to public health, the study delves into the legislative and regulatory aspects. The authors conducted exploratory research through stationary observations of guesthouses and family hotels to analyse the main direction of tourism activity of SMEs at the regional level and to identify specific risks in the public health environment. In addition, seven in-depth interviews were conducted with hotel and tourism professionals. The study identifies six main risks attributed to the public health environment of the SMEs in the south-west Bulgarian tourism. In this area, the authors recommend strengthening three basic principles: protection of human life and health, openness and transparency to the public in understanding the nature of risks, and truthful information on risks and precautionary measures.

SMALL AND MEDIUM TOURISM ENTERPRISE SURVIVAL IN TIMES OF CRISIS: "EL CAPRICHO DE GAUDI"

This chapter focuses on the risk attitudes of tourism business managers and their influence on response strategies implemented to redefine and renew the business model to be competitive and even survive. Through an exploratory descriptive method, the authors perform an in-depth analysis of the management of a protected cultural property called "El Capricho" by the artist Gaudí. This study argues that experience-based tourism enterprises, such as the one analysed in this case, may experience faster growth resulting from the implementation of different business model innovations. The results of the analysis contribute to generating knowledge about how a solid and well-established small cultural experience-based tourism enterprise reformulates its business model to survive in a time of crisis. In fact, "El Capricho" was the first building to recover activity after quarantine due to the COVID-19 health crisis, thus demonstrating again the innovative endeavour of this SMTE to survive in a time of crisis.

EFFECTS OF COVID-19 ON SMALL BUSINESSES IN THE CATERING SECTOR: MEASURES FOR RECOVERY

The study of the impact of the pandemic on cafés, bars and restaurants is the aim of this chapter. The research focuses on the operational problems of independent restaurant operators in a small city. Recovery proposals are evaluated from the point of view of both commercial operations and service provision and the governance measures demanded by restaurateurs. Through semi-structured open-ended interviews conducted with 38 micro-enterprises in Spain during the toughest months of the COVID-19 pandemic, this study provides key operational and governance measures for the recovery of the restaurant sector. The main measures preferred by businesses include aid to cover fixed costs during the closure of establishments, Social Security exemptions for employees on furlough, and extension of the furlough period. On the supply side, the benefit of adapting to the needs of customers is observed, for example, by increasing takeaway or home delivery options and making it easier for customers to consume in open spaces. Due to the increased public concern for sustainability, humanization and environmental conservation, the chapter also opens an interesting discussion on the responsibility of local governments for promoting open and innovative urban planning.

THE EFFECTS OF COVID-19 IN THE RENT-A-CAR INDUSTRY: A STUDY CASE BASED ON ARGENTINA

This chapter describes the impact of COVID-19 not only in terms of the statistics and sales of a major rent-a-car organization based in Argentina, but also the socio-psychological impact of uncertainty on staff. In this context, this work represents a novel contribution given the under-representation of the car rental subsector in academic studies. The author proffers an interesting discussion how global risks, which range from terrorism to virus outbreaks, are gradually changing the nature of tourism towards more globalized forms. The effects of viruses in the tourism industry, as well as the future of the industry in a post-COVID-19 context, are explored. The last section is a snapshot of the economic effects of the pandemic in a renowned rent-a-car company that covers rentals throughout Argentina. The outcomes confirm not only that the studied company was seriously affected by the lockdown, but also that tourist locations were more prone to experience a rapid downturn in comparison with industrial ones. The author argues that, in the absence of firm commercial aviation industry, the rent-a-car subsector presents a promising option of economic recovery to boost local tourism.

A TERRORISM-INFLICTED CASUALTY IN KASHMIR: PERSPECTIVES FROM TRAVEL AGENCIES AND TOUR OPERATORS

The last three chapters focus on the management of different threats by travel agents and tour operators. In this chapter, the authors discuss the harmful effects of terrorism on SMEs operating in Kashmir, India. Data are collected through semi-structured interviews with 12 travel agency owners. The findings show how terrorism has detrimental effects on this tourist destination and, consequently, on travel agencies and tour operators operating in the region. In this chapter, the authors also address the changes in Kashmir's legal and political status, how they affected tourism and its relationship to the terrorist threat. Interviewees hope that political stability and proper promotion will lead to a positive destination image and better management. The authors suggest that the media and the government – both regional and national – should play a greater role in promoting the region as a safe destination through the use of necessary measures such as the implementation of a disaster management plan promoted by regional and national governments.

DIFFERENT STRATEGIES FOR DIFFERENT CRISES TO RESCUE MICRO, SMALL AND MEDIUM TRAVEL AGENCIES

In the second chapter dealing with travel agencies, the author examines the different crises that travel agencies have faced since the 1990s. Travel agencies have implemented similar strategies in response to different crises. Practices such as management support, customer focus and process management appear to be effective in small service companies. The author points out that lack of resources, expertise and access to finances limit the scope of action of SMEs, so strategic planning can provide an appropriate advantage in dealing with a crisis. The proactive approach helps agencies adapt to the new situation and speed up recovery times; however, they recognize that implementing a proactive approach is not always easy. Among the recommendations for travel agencies in crisis situations cited in the chapter are the need to specialize and intensify the digitization of business and to improve the quality of customer relations, for example by focusing on strengthening the professionalization of customer-contact employees. Finally, the author presents strategies that have already been adopted in practice and others that could be implemented to face the recent health and economic crisis caused by the COVID-19 pandemic.

TRAVEL AGENCIES: A SECTOR HIT BY THE COVID-19 CRISIS

The final chapter discusses how the COVID-19 pandemic has affected travel agency operations. Many of these businesses have not only stopped booking new trips, but also had to cancel bookings made before 2020. The authors conduct a content analysis of academic studies published in the Scopus database containing the keywords "covid" and "travel agency". The analysis suggests that, during the pandemic, the use of new technologies and online platforms by travel agencies has increased substantially. The authors argue there has been a growing positive attitude towards technology, which may be the path for recovering this sector. The increased use of online tools and platforms can help with quick recovery once travel restrictions are eased. At the same time, it is noted that the reinvention of the sector includes a shift towards more sustainable, resilient and inclusive forms of tourism.

In conclusion, the chapters presented in this book propose practical and theoretical approaches that can be used by tourism professionals and entrepreneurs, researchers, postgraduate students and educators working in the field of crisis management. Crisis management in the tourism sector as a discipline, hitherto almost absent from the curricula of academic institutions, has acquired special prominence due to the events of the past year.

The COVID-19 crisis presents a unique opportunity to undertake a transformation of the tourism system more aligned with the Sustainable Development Goals. Specifically, at the business level, sustainable economic growth integrated into the local economy (objectives 8.9 and 12b) can be promoted, and innovation and technology can be fostered to improve competitiveness and resilience (objective 9). This is an opportunity to critically rethink the growth trajectory of tourism and question the rationale of focusing exclusively on more arrivals, as it is supposed to imply greater benefits. The transformation process requires a more complete understanding of the link between the nature of tourism consumption in the new setting and the resilience of tourism systems. We hope that the insights provided in this book will help build more resilient business models for the future.

We hope you enjoy this book.

Acknowledgment

We acknowledge and thank all the professionals who participated in the double-blind peer review process of this publication. Their generous dedication of time and experience contributed to the quality of this book.

Chapter 1
Strategic Management of the COVID-19 Crisis in Rural Tourism Settings:
Lessons Learned From SMEs

Alfonso Vargas-Sánchez
https://orcid.org/0000-0003-0588-8654
University of Huelva, Spain

ABSTRACT

This chapter is intended to contribute to filling the gap in the scientific literature on crisis management in small and medium-sized rural tourism enterprises in the occasion of the COVID-19 pandemic. Taking as the object of this study the Spanish province of Huelva, and under a qualitative methodological approach, the research aim was threefold: the measures taken to alleviate the crisis, the lessons learnt during the period of coexistence with the pandemic, and the opportunities for the development of rural tourism in the post-viral scenario. The latter ones were related to key topics such as safety and hygiene, risk of rejection of tourism, destinations governance, marketing policies, compatibility between short-term emergency measures and long-term vision, creativity and innovation, and people-centred organizations. Finally, some practical implications were concluded on this strategic issue.

DOI: 10.4018/978-1-7998-6996-2.ch001

INTRODUCTION

As stated by Mikusová (2013), research on crisis management has not devoted much attention to small and medium-sized enterprises (SMEs). This neglect is especially striking in the tourism industry because the vast majority of firms engaged in it are SMEs. This represents an evident research gap and, therefore, an opportunity.

The abovementioned author concludes that this kind of organisations "still put a relatively small emphasis on crisis planning. The only impulse for concern seems to be a real crisis occurrence. Organisations which have not gone through crises yet are dealing with prevention rarely" (p. 62). This unconcern should be corrected, at least to some extent, after the traumatic experience of the COVID-19 pandemic. Negative events of a sudden or unforeseen nature, and particularly those which have little potential of being controlled (such as the case of this health disaster), are the ones most prone to causing a crisis which can be further aggravated by incorrect management (Vargas-Sánchez, 2018). It is worth noting that the strategy and actions to manage this effectively will rely on the nature of the disaster (Hall, 2010), In this case it is global, long-lasting and with no precedents that could be taken as a benchmark.

Authors such as Vargas-Sánchez (2020b) have expressed their conviction that the COVID-19 pandemic and its legacy, together with other factors, such as the ongoing exponential tech revolution and the pressures in favour of sustainability coming from governments' plans for the mitigation of climate change and the application of circular economy principles, may transform tourism. In this regard, Jones and Comfort (2020) echo some voices that have seen the COVID-19 crisis as a test for climate change and a strong focus on sustainability, taking tourism into a new era. As stated by Zenker and Kock (2020, p. 3): "Sentences like 'things will never be the way they used to be' are often heard after disasters and crises, but mostly proved wrong as we go back to our normal routines. However, this time some aspects of our behaviour might be affected by true paradigm-shifts".

In general, the tourism sector is being severely beaten by this disaster, although not all segments are being equally affected. In relative terms, losers and winners could be identified, with rural tourism as part of the latter group (Vargas-Sánchez, 2020a), with this market segment being widely dominated by SMEs. Thereby, the results of a recent study at the Spanish State level published in a specialised portal[1], titled 'The impact of COVID-19 on rural tourism', support that assertion. As a token, during the 2020 summer, almost half of rural accommodations experienced an increase in demand compared to pre-pandemic figures.

This is also the case of the area chosen for this study, the Spanish province of Huelva, with a renowned natural heritage and where rural tourism was already gaining traction before the pandemic. In this geographical area, which could serve as a starting point to expand this research to other rural destinations, the study has three aims:

- to find out how the aftermath of this unprecedented crisis is being managed;
- to know what lessons have been learnt from it and, finally, when heading towards the future,
- to identify opportunities at a time of rethinking the rural tourism model/offer to secure its sustainability (economic, social and environmental) in the post-COVID-19 scenario.

In short, to stimulate the development of rural areas and alleviate the chronic problems of their depopulation, the progressive ageing of these communities, their lack of infrastructures, environmental degradation, etc., which affect society as a whole, tourism has been extensively encouraged to diversify their economic base and generate new opportunities for their residents. Accordingly, this diversification has been proposed as an anti-crisis development strategy for rural territories (Ilnitsky and Chikurova, 2015), and as an engine to increase their resilience (Ibanescu et al., 2020) and improve their territorial dynamics (Sánchez-Zamora et al., 2014).

In this vein, several references to rural environment/development can be found in the statements, key data and targets of the United Nations' 2030 Agenda for Sustainable Development. Certainly, these matters are closely connected to this Agenda, so that some of its Sustainable Development Goals (SDGs) refer to the problems mentioned above. In this regard, tourism has been viewed as an activity with the potential "to truncate or minimise" these undesirable trends (Jurado-Almonte and Pazos-García, 2016).

More recently, the Framework Convention on Tourism Ethics (WTO, 2020) made some specific references related to this type of tourism, namely: "Nature tourism and ecotourism are recognized as being particularly conducive to enriching and enhancing the standing of tourism, provided they respect the natural heritage and local populations and are in keeping with the carrying capacity of the sites" (article 6.5); "Special attention should be paid to … vulnerable rural or mountain regions, for which tourism often represents a rare opportunity for development in the face of the decline of traditional economic activities" (article 8.3).

Within this framework, and under the recognition of the importance of rural tourism for the economic life of these communities, this research work aims to explore how rural tourism SMEs have been dealing with the unparalleled consequences of the pandemic caused by the COVID-19 pandemic, which lessons they have already learnt and how they are preparing to deal with the post-coronavirus scenario.

The rest of this chapter is structured in the following sections: literature review, methodology, results, discussion and conclusions.

LITERATURE REVIEW

Crisis/Disaster Management and Resilience in Rural Tourism SMEs

Carter (1991) defines a disaster as an event, natural or man-made, sudden or progressive, which impacts with such severity that the community affected has to respond by taking exceptional measures. The investigation of such measures taken by SMEs in the rural tourism segment inspires this work. The aim is to find out to what extent they have been able to mitigate the generalised business crisis brought about by the pandemic disaster, which is seriously detrimental to profitability and even puts survival at risk.

Disaster and crisis are interrelated terms, but not synonymous. As stated by Hao et al. (2020, 2), "based on the distinction of Faulkner (2001), an organization faces a crisis due to internal failures, while it confronts a disaster due to uncontrollable external factors. In this regard, COVID-19 should be defined as a disaster". Evidently, after this disaster, the tourism industry in general (and beyond) is being hit by a crisis with consequences that are still impossible to fully calibrate. In this respect, Stafford et al. (2002) note that all such crises share three common characteristics which certainly fit the current situation: suddenness, uncertainty and time compression. Quarantelli (1988, 373-374), in his general approach to disaster management, presents the main problems to tackle: those related to "the communication process, the exercise of authority, and the development of co-ordination". At the same time, he warns that prior planning can limit these management difficulties but cannot eliminate them. This is because of the usual "big gap between what was planned and what actually happens in a major disaster crisis", which requires, respectively, the handling of strategy and tactics principles.

Therefore, following the approach of Irvine and Anderson (2005), the interplay between disaster, crisis and action forms the basis of our research question, since disaster/crisis and its impact can be modelled as a linear continuum between catastrophe and the control of its effects. The specific position within it is determined by the set of implemented actions. In this sense, although tourism organisations regulate all their functional components through their specific mechanisms of crisis management (Mazilu et al., 2019), Irvine and Anderson (2005) concluded in their research that smaller firms responded more effectively than larger ones. Yet it is not clear how their planning operated. Conversely, Zenker and Kock (2020) underline that innovative capabilities play a key role in crisis recovery, while tourism businesses often suffer deficiencies in this domain, particularly small operators.

In the same line, Phillipson et al. (2003) found that rural micro-firms are very resilient to external events, although, for the same crisis (the Foot and Mouth

Epidemic in the UK), "a downturn in the business performance of these rural tourism enterprises…, comparative to many general businesses, has, potentially, significantly greater impacts on the financial standing of the business given their limited margins and flexibility" (Leslie and Black, 2005, p. 44). Concerning this latter case, Williams and Fergusson (2005) highlight the importance of an integrated vision of rural economies to avoid the domino effect that occurred as a result of a myopic policy that put aside the implications of a crisis in the farming industry for other activities, causing the devastation of the rural tourism and leisure industries in the areas affected. Nevertheless, the Organisation for Economic Co-operation and Development (OECD. 2020, p. 7) states that SMEs "are generally less resilient and lack the capacity and resources to cope with the changes and costs that such shocks entail".

Within this field of study, the topic of resilience has gained momentum in the last years, although not much attention has been paid to the specificities of rural areas and the impact of tourism activities on them (Sánchez-Zamora et al, 2014). Nevertheless, the findings of Ibanescu et al. (2020) show "a positive effect of tourism activities upon economic and demographic resilience performance in highly accessible rural areas, while in peripheral areas, the impact was insignificant". Since the tourism sector consists mostly of SMEs, its resilience is based on a complex set of networks (Scott & Laws, 2005). R.-Toubes et al. (2020) point out that social networking and community engagement contribute to self-organisation, a typical element of resilient systems.

However, the mere reactive approach to a crisis has been emphasised in the literature as insufficient. Thereby, for a better way of limiting the impacts of crises, Ritchie (2004) warns that those responsible for tourism policy and management should move on from a reactive focus to one that is both proactive and holistic, with prevention and crisis/disaster plans as the first step in crisis management strategies, before they occur. Besides, "there is a need for cooperation between a wide number of stakeholders both internal and external to the organisation to effectively plan and manage crises and disasters" (Ritchie, 2004, pp. 680-681), so leadership is essential to bring them together.

It is also relevant to note that, although the intersection between biosecurity management and ecotourism has been a topic that has led to some research efforts, the "literature on biosecurity threats and tourism is underdeveloped" (Ivanov et al., 2020, 3). Despite that, to avert major impacts from biosecurity threats, the importance of having preventive plans in place is paramount.

Similarly, Chen et al. (2020, p. 9) conclude that "researchers lack attention to the risk of tourism diseases" and, in addition to the COVID-19 outbreak, speculate about "the disease risks associated with new ways of travelling, including medical, cruise, and wildlife tourism" as a hotspot for the research on this topic in the next few years.

The COVID-19 Pandemic and Rural Tourism: Looking Ahead

As part of a post-viral strategy for rural tourism transformation, Santos Lacueva (2020) proposes an agenda with seven key topics: safety and hygiene; risk of rejection of tourism; new models of tourism destination governance; flexibility and transparency in marketing policies; how to build a compatible short-term emergency and long-term vision; the consolidation of creativity and innovation beyond the crisis; and the relevance of people-centred organisations.

Concerning safety and hygiene, although it is clear that they have become more important when choosing a destination or tourist service, for Santos Lacueva (2020) this topic raises several questions: (a) safety or false perception of safety? That is to say, to what extent a destination or establishment can be labelled as safe; (b) more waste means greater safety? That is, will safety measures reverse the efforts made to reducing waste (plastic, disposable utensils, etc.)?; (c) how to make a place safe both for visitors and residents in rural areas, with communities characterised by an ageing population and limited and/or distant availability of health services?; (d) by definition, is there no overcrowding in rural areas? Are there no enclaves that suffer congestion as a result of the intense inflow of daily visitors, beyond those who stay overnight? The rural sites equal safety and urban territories equal unsafety dichotomy is too simplistic; (e) are there enough resources (technical, human and financial) and effective control mechanisms to guarantee safety?; (f) who is responsible for ensuring safety in these circumstances, tourist services providers, their users…?; (g) is the limitation of access/use of public spaces democratic? Is it based on the principle of guaranteeing their limited enjoyment under equal conditions?

About the potential risk of tourism rejection, the pandemic has brought with it some episodes of an adverse reaction, for example upon the arrival of people at their second homes in rural areas, with some visitors being less welcomed by locals due to the perceived risk of contagion. Rejection is usually generated by fear or ignorance, and both are fought with transparency and information, particularly about the safety measures implemented. Rural residents should be informed about the implications of tourism in their quality of life, and also channels for their effective participation in decisions regarding the tourist activity in their municipality ought to be articulated. Likewise, visitors must be involved in caring for the local population and the environment.

Tourism governance is another key topic given that public-private cooperation is essential to face a challenge of the magnitude of this pandemic. This crisis has shown the need to strengthen collaboration among all the stakeholders and governance in tourist destinations. The challenge is to consolidate these relationships beyond the crisis.

Commercial policies have also been affected. Potential clients have required, to mitigate the risk of booking in a scenario with such a high uncertainty, more measures of flexibility and transparency, having been advised to encourage reservations without reducing prices: flexible rates, free cancellations, refundable reservations through vouchers, gift vouchers, discount vouchers, special packages, etc.[2]

A trade-off between short-term emergency and long-term vision is critical as well. The immediate reaction to the pandemic is the struggle for survival, but, beyond that obvious aim, fundamental issues such as sustainability, climate change, the circular economy, a lack of basic infrastructures, or rural depopulation have to be part of the strategic priorities to thrive in the post-COVID-19 competitive landscape.

Another core issue raised by this crisis is the necessity to turn creativity and innovation into everyday practice so that businesses and tourist destinations can adapt and be resilient to new scenarios of change in the so-called new normality. And rural areas cannot be an exception.

And last but not least important, the people topic: the ability to set up people-centric organisations that fit a new period, which means empathy (to understand the new patterns of customers' behaviours), customisation (to captivate them), emotions (to provide memorable experiences), rethinking and evolving (to accommodate the offer of tourist services to the people's new values and needs).

METHODOLOGY

From a methodological point of view, the qualitative approach followed by Cioccio and Michael (2007, p. 4) was followed here, because, in this inductive kind of research, "it was deemed to be important to allow the participants to narrate and describe their own responses to the event in their own terms". Seven semi-structured interviews were carried out with owners, general managers and professionals active in the tourism segment under study, with open-ended questions to guide the interview and also stimulate the conversation, as detailed below. More informants were not selected because it was felt that data saturation (Fusch & Ness, 2015; Saunders et al., 2018) had already been reached.

The area chosen for this study was the Spanish province of Huelva, with a renowned natural heritage at a national and international level (with 'Doñana Natural Park' as the most iconic). Despite the pre-eminence of coastal tourism, the rural segment had already been gaining traction before the pandemic and, using data provided by the National Statistics Institute of Spain, it has performed better than 'sun and beach' tourism during the 2020 summer season (the peak months of July and August were considered). More specifically:

*While hotels on the coast (an area designated as ‘Costa de la Luz’) have lost 43.5% of the number of overnight stays with respect to July/August 2019, rural tourism has suffered a fall of only 6.5% -in July the number of overnight stays increased 9.3% (-50.8% on the coast).

*Concerning the average stay, during the summer of 2020 this indicator fell 21.5% (from 4.88 to 3.83 days) in coastal hotels, but rose 14.6% in the rural tourism segment and particularly, within this, in the most popular area for this kind of tourism, in the mountains, known as ‘Sierra de Aracena y Picos de Aroche’ Natural Park, with a remarkable increase of 56.1% (from 2.64 to 4.12 days).

*In short, rural tourism is being one of the winners during this exceptional period where people are looking for uncrowded areas, open spaces and getting back in touch with nature after a lengthy lockdown. As the OECD points out (2020, p. 10), this is not an exception in the context of the “new preferences for low tourist density destinations and outdoor activities away from big cities and in regions with a safer image in terms of COVID-19 circulation”.

It is also relevant to note that in the same period of the year 2019, the international market represented 12.5% of the total number of overnight stays in rural tourism establishments (17.2% in the ‘Sierra de Aracena y Picos de Aroche’ Natural Park). In 2020 those percentages have fallen to almost insignificant values. This severe fall of international tourism has been similar on the coast.

The interviewees chosen operate in the above-mentioned natural parks (‘Doñana’ and ‘Sierra de Aracena y Picos de Aroche’), three in each of them, in addition to a tourism professional, as follows:

- In the ‘Doñana’ district, the President of the Doñana Nature Tourism Business Association and Secretary of the Spanish Ecotourism Association, the entity that manages the Ecotourism Club in Spain, who at the same time is the general manager of ‘Wild Doñana’, an active tourism services company in that district (birdwatching, trekking, etc.); the general manager of the rural tourist complex ‘Cortijo de los Mimbrales’, with 100 beds in 6 houses and 38 rooms; and the general manager of a group of four rural houses (‘El Cazadero Real’, ‘Doñana 51’, ‘Arenas de El Rocío’ and ‘Arenas de El Rocío II’, with a total of 34 rooms and 68 beds.
- In the ‘Sierra de Aracena y Picos de Aroche’ district, the President of the Tourism Companies Association in that district, at the same time owner and general manager of “La Media Legua” rural tourism apartments, with a total of 41 beds in 12 apartments; the founder and general manager of “Babel Tourism”, a company that offers active tourism services (mountain biking,

climbing, canoeing, hiking, etc.) together with ecotourism and environmental education activities (mycology, astronomy, ethnobotany, birdwatching, etc.); and the co-owner of "La Posada de San Marcos" rural house, a recognised environmentally-friendly establishment with 6 double rooms.
- Finally, the director of promotion at the Tourist Board of the province of Huelva and Head of the Technical Office in this district.

The average length of the interviews was approximately one hour. All of them (except in one case) were done remotely and recorded with due authorisation. They took place during September and October 2020.

Consistent with the study's goals, the interviews were divided into three sections: measures taken to alleviate the crisis (in domains such as health risk prevention, commercial policies, etc.), lessons learnt in this period of coexistence with the pandemic, and opportunities facing the new post-coronavirus scenario. Concerning the latter, the seven key topics proposed by Santos Lacueva (2020), already mentioned in the previous section, were used to articulate the conversation: (1) safety and hygiene; (2) risk of rejection of tourism; (3) new models of tourism destination governance; (4) flexibility and transparency in marketing policies; (5) how to make compatible a short-term emergency with a long-term vision; (6) the consolidation of creativity and innovation beyond the crisis; (7) the relevance of people-centred organisations in the design of the post-COVID-19 rural tourism strategy.

In addition to the abovementioned topics, the seven types of gains that may arise from a crisis, according to Burnett (1998), are explored in this case: (1) heroes are born; (2) changes are accelerated; (3) latent problems are faced; (4) people can be changed; (5) new strategies evolve; (6) early warning systems develop; (7) new competitive edges appear.

The findings gathered as a result of crossing topics and opportunities represent an extra element of novelty in this research. Focusing on each topic, the author intends to identify if new heroes have been born, which changes have been accelerated, the underlying problems companies have faced, to what extent people have changed, how strategies have evolved, if early alert systems have been developed and new competitive limits/advantages identified. Burnett (1998) also suggests five types of potential losses (severe impact on corporate profits; unwanted public and government scrutiny; damage to corporate integrity and name; unproductive diversion of employees, time, and capital; and negative employee morale). However, the present work focuses on the first and positive angle.

RESULTS

The findings obtained are presented next, following the same order of the sections covered in the interviews.

1. Measures That Have Been Taken to Alleviate the Crisis

Not surprisingly, the safety of guests and workers has been the first concern during this pandemic, turning health risk prevention into the priority in this new scenario. The approach to face this challenge has been purely reactive, with the implementation of the anti-COVID-19 protocols and contingency plans following the rules dictated by public authorities for each type of tourism establishment and the advice of insurance companies. The recommendations for cleaning/disinfection and risk management included in the regional government's 'Andalucía Segura' label have been also widely executed, according to the statement of responsibility submitted by each firm. Whereas bigger companies and chains have enough resources to implement certified labels or even their singular safety programmes under the umbrella of a reputed commercial brand, this distinctive has special relevance for SMEs as guidance in this field, and with no extra costs.

In line with minimising health risks, informing customers about the applicable protection protocol has been considered essential to recover their trust. In general, and although these safety measures are not always available in the establishment's website, the corresponding information has been sent to the client before arrival, whether this has been requested or not[3].

The services offered have been revisited to favour social distancing, avoid physical contact and utensil sharing. Solutions include serving breakfast in the rooms (or providing assisted service instead of self-service) and conducting outdoor activities with smaller groups. Online reservations and payments have also been encouraged.

It is worth noting that anti-COVID-19 sanitary measures have significantly increased waste generation, especially of plastics (packaging of single food portions, single-use utensils, face masks...), in a sort of involution of the pre-pandemic trend. This is something that should be tackled due to the importance of environmental sustainability in the protected natural settings where rural tourism activities take place.

In the commercial domain, flexibility and agility have been the dominant features with relation to bookings (last minute reservations[4], shorter periods for no penalty cancellations/postponements, vouchers), sales channels (reinforcing the online and direct sales), pricing (more segmented and adaptable to market circumstances at a very short notice, also bearing in mind seasonality), etc. In some cases, prices rose to make up for the higher costs borne by companies because, in one informant's words, "safety delivers value to customers and the company".

Concerning the tourists' origins, the near absence of international tourists due to travel restrictions has forced companies mainly focused on that public to re-orientate their marketing efforts to attract tourists at local, regional and national levels.

A remarkable commercial change in rural houses during the summer season has been the possibility of booking the whole house, instead of individual rooms. This offer adaptation has been the response to a noticeable change in tourist demand, with groups of family members or friends that, after the lockdown, wanted to book the full house to avoid the risk of contact with strangers. As a matter of fact, full rental accommodations have been those most requested after the state of alarm in Spain[5]. This new pattern motivated the offering of customised products for these groups.

Regarding procurement, because of its relatively small volume, there have been establishments that have chosen to make the purchases themselves directly and avoid interaction with suppliers in the delivery process.

2. Lessons Learnt in This Period of Coexistence With the Pandemic

In this regard, the issues pointed out by the informants concern the importance of:

- Innovation to adapt their businesses to the new circumstances.
- Keeping a good reputation about health and safety measures, together with a profile unconnected with mass destinations.
- Preserving a smooth cohabitation between tourists and locals to avoid outbreaks of tourism-phobia in fragile rural settings because of their ageing population and weak health services.

Additionally, the reflection of one of the interviewees is of particular interest: "We should have read the situation better. We have looked inward perhaps too much and have not been able to connect enough with our public during the lockdown. We have not had enough dialogue with them to make them our best commercial allies, and we have not asked them what they expected from us after the lockdown in its different stages". Another said: "It is time to resist and innovate. We have had much time to reflect".

In general, the exploration of new market trends is hard for rural tourism SMEs. This is a factor that has contributed to a better understanding of the relevance of mutual help through their collaborative ties and to strengthening their business associations for favouring commercial agreements (promotion campaigns and others). In a VUCA[6] scenario, lobbying public administrations and other players in the industry, as well as providing reliable information and advice to their members, became particularly important.

3. Opportunities for Rural Tourism in the Post-coronavirus Scenario

In this section, respondents/interviewees were asked about several key topics, based on which ideas are organised as follows.

- Safety and hygiene.

In this respect, opportunities are in those attributes that have demonstrated their positive impact on clients' trust, such as no massification and the mental link between rural/natural settings and a healthy environment. Taking advantage of these opportunities requires tight cooperation among the stakeholders in a destination and their commitment to making a value proposition of this kind real, with consistent behaviours at all levels.

In general, people are now more careful and prudent about hygiene and safety. Thus, the perception of quality is more influenced by this dimension of the tourism offer. In this context, ecotourism is seen as a winning option. The umbrella provided by protected natural parks delivers an extra value to reinforcing the customers' perception of a healthy environment.

In short, it is believed that these opportunities can be exploited if the necessary changes are introduced in the configuration of products (strengthening the above-mentioned attributes) and the delivery of some services (such as breakfast or cleaning). However, specific training is necessary to overcome the barrier of knowledge about these aspects. Besides, new competitive advantages have come up, such as the availability of large open spaces and roomy areas that ensure social distancing. It is also expected that this pandemic can change people's attitudes and sense of responsibility, as well as the recognition of the importance of implementing early warnings (a temperature control of all travellers is an example).

- Risk of rejection of tourism.

Particularly during a disaster that affects people's health, tourism has to gain the acceptance of the host communities, whose involvement in the tourist experience can make the difference. This industry must continue to be respected as a means to earn a living because of its trade-off between economic recovery and the safety of locals, as well as because of its commitment to striking a balance between tourism growth and environmental sustainability. The framework for new opportunities (such as the exploration of new customer segments with potential) is in this intersection.

The joint effort of public authorities and business representatives is fundamental to avoid a new type of tourism-phobia and cause damage to a place's image as a tourist destination. It is advisable to work with residents to show them what tourism firms are doing to take care of them, as well as the importance of tourism for the local economy and wellbeing. Nevertheless, the excess of second residences is a latent problem that could cause a feeling of rejection.

- New models of tourism destination governance.

Beyond promotion, the development of rural tourism depends on the creation of an effective mechanism of destination governance, which is currently missing. An ample room for improvement is felt in this domain since it is considered a facilitator for the materialisation of opportunities. Strengthening public-private partnerships is required, but with a more professional and less political, base.

The fact that business associations are experiencing a significant increase in their members, as well as going into a more intense collaboration dynamic with public administrations to mitigate the crisis, is seen as the prelude to the opportunities that can arise from business cooperation and public-private partnerships. Nevertheless, a lack of leadership at the political level is felt (with powers scattered among several public bodies), a gap which, to a certain extent, is being occupied by business associations.

- Flexibility and transparency in marketing policies.

To seize opportunities in a business environment described above as VUCA, extreme agility and flexibility are a must in all tourism firm operations, not in the marketing function only. These opportunities are envisaged in the creation of new products (such as slow tourism) that allow the enjoyment of nature in its broadest sense and the generation of higher added value. The increased tourist loyalty as a result of the pandemic, together with new customers that have been attracted, are the foundations of this new momentum.

Transparency in booking conditions (regarding cancellation policies, services offered during this time of sanitary exceptionality, etc.) have also been improved to respond to a hostile scenario.

Besides, the existence of greater coordination at the destination level on the part of tourism companies, particularly hoteliers, has been signalled, so as not to fall into an excessive price drop.

- How to make compatible short-term emergency and long-term vision.

A short-term stimulus with promotional actions based on attributes that have worked during the pandemic has to be compatible with a long-term plan, elaborated in partnership with public administrations. The idea is to consolidate this new impulse and, at the same time, prepare companies to face the challenges derived from the new tech developments, the mitigation of climate change, rural depopulation, etc. Technological and sustainability vectors are other sources of opportunities that cannot go unnoticed for rural tourism SMEs.

The traction of renowned hotel chains that are beginning to back this market segment is another factor which will in this respect push forward its professionalisation.

- The consolidation of creativity and innovation beyond the crisis.

Crises favour the generation of new ideas and the co-creation of new tourism experiences. In this sense, it is felt that the sector needs to reflect and move away from the traditional approach to rural tourism based on relaxation and leisure. Other domains such as remote work, training activities, events on a relatively small scale, etc., are potential sources of opportunities.

Accelerated digitisation is pointed out as another source of opportunities for these SMEs. Nevertheless, this process faces a significant barrier in the poor availability of quality digital infrastructure in rural areas.

Digitisation brings people closer and facilitates their collaborative work. However, the digital gap between generations is also a factor to take into account, requiring a training effort to combat digital illiteracy.

It is also recognised that the consolidation of creativity and innovation needs a type of leadership and organisational culture that promotes staff participation. As stated by one of the informants: “it is important to support employees’ self-esteem and help them to overcome the embarrassment or fear of proposing their ideas”. Or, when talking about innovation in small establishments, “it is more about copying and improving than trying to invent”.

- The relevance of people-centred organisations in the design of the post-coronavirus rural tourism strategy.

In a fast-changing market, companies are under increasing pressure to remain customer-centric and in this matter people’s ideas and skills are critical. An example is the need to improve the online experience. The higher the value offered to the customer, the more the inertia of lowering prices to compete can be offset.

The diversification of well-known hotel chains to target the rural market can generate new opportunities for SMEs in this segment. These destinations can gain visibility (together with the increasing attention of mass media), although the implementation of early signals to detect a potential excessive growth that can surpass their carrying capacity and hurt their environmental sustainability is advisable.

Table 1 summarises the main interactions identified when crossing topics and opportunities, limited to the two most significant ones.

More specifically, for each key topic:

1. A higher concern for safety and hygiene has given rise to new standards that, to some extent, will remain in the future and will serve as a benchmark of this dimension of tourism quality. Besides, this factor has arisen as a new source of competitive advantages for those companies and segments (such as rural tourism) with the ability to tune in to what tourists are looking for in this new period.
2. The risk of tourism rejection is a latent problem that must be addressed in the current scenario and beyond. Once tourism players become aware of this risk, which is only just beginning in the case observed, monitoring the local community regularly will be the origin of early warnings to counterbalance potential unfavourable reactions.
3. The lack of effective tourism governance is felt as another latent problem, having serious long-term implications. The new "heroes" in this field will be those rural destinations which, having suitable governance structures, can consolidate the more fluent dynamics of public-private cooperation sparked off by the pandemic crisis.
4. Marketing policies have evolved during this crisis in line with the requirements of a more complex and hostile market. Some changes (such as digitalisation) have been accelerated by the pandemic, and some marketing tools are beginning to be used more professionally (revenue management, customer relationship management).
5. Although what has been referred to in the previous point are actions motivated by a short-term pressure for survival, there is also an awareness of several profound challenges (climate change, the circular economy, new technologies) for the sector that should be tackled with a long-term plan in close collaboration with public authorities. Early warnings on progress made are essential through a proper set of indicators for each challenge.
6. The new scenario as a result of the pandemic has reinforced the importance of creativity and innovation in traditional business environments such as rural tourism. In this regard, in addition to the technological factor, people-centric leadership and culture are required to inject a permanent bottom-up flow of ideas able to improve the firm's competitiveness.
7. This people-centric focus embraces employees, locals and, of course, tourists. People can make a difference if they are properly managed; that is, under a balanced strategic approach and with the support of managerial and technological resources. As an example, the online tourist experience must be particularly looked after in the current technological era.

Table 1. Cross matrix

	Types of Opportunities (Burnett, 1998)						
Key topics (Santos Lacueva, 2020)	**O1: Heroes Are Born**	**O2: Changes Are Accelerated**	**O3: Latent Problems Are Faced**	**O4: People Can Be Changed**	**O5: New Strategies Evolve**	**O6: Early Warning Systems Develop**	**O7: New Competitive Edges Appear**
T1: safety and hygiene	X						X
T2: risk of rejection of tourism			X			X	
T3: new models of tourism destination governance	X		X				
T4: flexibility and transparency in marketing policies		X			X		
T5: how to make compatible the short-term emergency and the long-term vision					X	X	
T6: the consolidation of creativity and innovation beyond the current crisis		X		X			
T7: the relevance of people-centred organisations in the design of the post-coronavirus rural tourism strategy				X			X

Source: own elaboration.

DISCUSSION AND CONCLUSION

Wilson et al. (2001) documented that successful tourism development in rural settings implies much more than the involvement of individual businesspersons. In other words, they found that the importance of a community approach is essential. More specifically, ten key factors/conditions were identified: (1) a complete tourism package, (2) good community leadership, (3) the support and participation of local government, (4) sufficient funds for tourism development, (5) strategic planning, (6) coordination and cooperation between businesspersons and local leadership, (7) coordination and cooperation between rural tourism entrepreneurs, (8) information and technical assistance for tourism development and promotion, (9) good convention and visitors bureaus, and (10) widespread community support for tourism. This list, some of whose elements have also been revealed by this study, could serve to identify the strengths and weaknesses of rural tourism in a certain territory.

Concerning general trends reshaping the tourism industry (and the rural segment is not an exception), before the COVID-19 pandemic companies and destinations were already facing increasing pressure to become more customer-centric, digital and sustainable. With this disaster, changes are being accelerated in these directions to improve their competitiveness. As examples, it can be expected that:

- Increased tourist concern for hygiene, health, safety and sustainability will continue after the pandemic.
- Tourists' preference for contactless services supported by digital and smart technologies will be renewed.
- The spread of these technologies will pave the way to process automation and the improvement of operational efficiency that will benefit both tourism services providers and their customers. Authors such as Ivanov et al. (2020) present COVID-19 as a driver of automation and robotisation in post-viral tourism, considering these technologies as a critical element in the reorganisation of the industry.
- The digital customer relationships management will gain momentum as part of business models in the new normality.

A question mark is if the "war" against SARS-COV-2 is going to leave rural tourism SMEs with enough financial resources to face the disruptive changes, some of them still latent, associated with the tech revolution that is underway and that advances faster than many people usually think. The author advocates public support programmes in this direction which will help them to not be left out of the new technological normality. The results of this research lever this proposal because no sign of technological adaptation has been found. Changes have been limited to compliance with the safety protocols dictated by the health authorities (with measures to improve hygiene standards and food and beverages offers) and to incorporating as much flexibility as possible regarding commercial policies. An effort to respond with agility and creativity to a VUCA environment has also been noticeable.

In an uncertain environment, the effects of the loyalty gained in the past, the trust conveyed by a known place and the umbrella provided by the 'Natural Park' brand have been factors to attract what can be designated as crisis-resistant tourists. This is particularly important for rural tourism and related segments (ecotourism, nature tourism), because "COVID-19 may provide an impetus for individuals to transform their travel behaviours; however, transformation of the tourism system is extremely difficult" (Hall et al., 2020, p. 584). Hence, the risk is that, according to Jones and Comfort (2020, p. 343), "within any new normality, the environmental gains produced by the COVID-19 crisis may be lost, social and economic recovery may be slow and corporate sustainability programmes … may be put on hold". To avoid it they "stress the importance of governments and businesses fully integrating sustainable development, crisis management and resilience planning into their corporate and political structures and policy-making" (p. 3046). For instance, the European Green Deal, launched in 2019, goes in this direction[7].

Concerning the identification of opportunities for rural tourism in the post-coronavirus scenario, following the typology proposed by Burnett (1998), the main conclusions are synthesised as follows:

1. Crises create heroes, that is, new points of reference able to illuminate others. In this case, for SME rural tourism, safety and hygiene standards and new advances of destination governance seem to be the most outstanding. In this sense, opportunities are, on the one hand, in the ability to match the post-viral tourist's expectations in the health domain (physical and emotional), consolidating the perception of rural areas as safe/healthy destinations immersed in a natural environment. And, on the other hand, the opportunities of the new post-virus landscape, with its new patterns of behaviour, will only be taken advantage of to the fullest extent if a suitable destination governance structure is put into place, understood as effective mechanisms of the cooperation, coordination, planning and support of tourism players. Those destinations able to consolidate the momentum gained during the pandemic with solid public-private partnerships will be on the winning side of the industry reconfiguration.
2. Opportunities can also be found in those areas where changes have been accelerated, particularly in marketing policies and the emphasis on creativity and innovation. Both areas are related to technological changes in the fields of digitisation and automation. Because of their limited resources rural tourism

SMEs need some support programmes to strengthen their professionalisation and stay on the right track.

3. Opportunities are also in the determination to face latent problems. By definition, these are pre-existing ones which, for some reason, have remained unresolved. If the pandemic provides the necessary extra motivation to address them, this will be a source of improvement. In this study, two main latent problems have arisen: the risk of tourism rejection and the deficiencies in terms of destination governance. The interconnection between the two could be discussed, in the sense of advocating an alliance with local communities to respond to their concerns about how tourism development could comprehensively affect their lives (health, economy, etc.).
4. People in general (including tourists) can change in a context of crisis, and this is another source of opportunities if these changes are well understood and managed. In this facet, people's role in the process of consolidating innovation-driven companies in rural tourism is critical. Concerning the formulation and implementation of strategies in the post-coronavirus scenario, those SMEs with the ability to move faster and more thoroughly in this direction, will be in a more competitive position. Coming from a traditional business environment, this is a big change which will require in many cases an effort to re-skill staff and owners.
5. Putting into motion new strategies represents another source of opportunities. This has been noticeable in how commercial policies have evolved in the short term to respond to new (and adverse) market circumstances and, at the same time, in the awareness that a long-range vision is also needed to prepare tourism firms for a new competitive landscape dominated by sustainability issues and exponential technological shifts. While environmental matters are felt as being close to rural tourism SMEs, the technological challenge is worrying, not only because of their limited resources, but mainly due to the poor infrastructure available in rural areas (broadband, quality of Internet connections, etc.).
6. The implementation of early warning systems helps detect problems regarding which an early action has a fundamental repercussion, representing a clear opportunity for better management. As reflected in the present study, this is the case of a potential adverse reaction of local communities towards an increase of tourism flows that locals could perceive as having more negative than positive impacts in their quality of life. These systems are also indispensable for tracking company and destination progress towards sustainability (and other) goals (the increase in the waste generation as a result of the pandemic has been particularly pointed out in this research). These warnings would be part of an information system made up of carefully chosen indicators for more solid decision making at the firm and destination levels. This data-driven logic

is aligned with the smart tourism approach which has gained momentum in this new period.

7. Finally, new opportunities are linked to the emergence of new sources of competitive advantages. This has been the case of the reinforced safe and healthy conditions for tourism activities, which have favoured segments such as rural tourism and companies with a certain type of facilities (terraces, open spaces, roomy places…). People-centric organisations have, in the broadest sense, reinvigorated their role and gained a better position in the new tourism market landscape. Yet this is not the only factor to be considered to compete successfully, rather their combination with appropriate managerial and technological resources is to be contemplated.

These findings lead to some theoretical and practical implications on this strategic issue.

From a theoretical point of view, the right management of this kind of disasters/crisis requires a contingent approach that considers the particularities of each market segment (such as rural tourism) and type of companies (such as SMEs), among other variables. A generic approach has been revealed as insufficient. As complexity theory teaches us, for a correct understanding of a complex problem, such as the COVID-19 pandemic and its impact on tourism, that problem cannot be separated from its specific context and dynamics (as happens in each country, for example). But, conversely, this must be tackled in its entirety, including, therefore, its contextual factors. Accordingly, each segment and type of companies has its specificities, which must be taken into account as endogenous factors, leading to a conclusion that the corresponding models of management and actions (in our case focused on seven key topics) should be adapted based on these specificities.

As practical implications, our findings give rise to some recommendations, mainly: to provide more online visibility to safety and health protocols; deliberate commercial strategies to build loyalty among those tourists who have moved towards rural tourism during the pandemic, and to increase the resilience ability through a more diversified customer base; a renewed effort to consolidate effective destination governance structures in natural parks able to provide leadership and balance short-term measures with a long-term vision and planning; to articulate a new alliance with local communities to support tourism development and make locals feel safe and comfortable with its growth in places with acute fragilities in terms of health services, infrastructures of all kinds, etc.; the implementation by public authorities, in collaboration with businesses associations, of a programme to support the technological and business skills update of rural tourism SMEs (digitisation, automation, online marketing, the circular economy and sustainability).

The focus on the drivers of business opportunities for the future of rural tourism emerges as a recommendation as well, namely: effective destination governance structures, new technology endowments (digitisation, automation, etc.) and the corresponding human and managerial skills, creativity and innovation capabilities, diversification and resilience.

Finally, this study is not without limitations, mainly linked to its geographical scope and the type of informants (other stakeholders' views could be added for triangulation purposes). The qualitative analysis carried out could be followed by quantitative research at the national level to submit the findings to an additional scrutiny and check possible differences between regions. Besides, a more formal prospective study on the opportunities for rural tourism could be implemented, with the support of techniques such as the Delphi method.

ACKNOWLEDGMENT

The author wants to thank the kind attention and time devoted by the following tourism entrepreneurs and professionals (in alphabetical order): Mª Cinta Aguilar Coello, Lucy Arkwright, Daniel Calleja Salido, Juan Carlos Díaz González, Manuel Iglesias Fernández, Manuel Mojarro Soler and Joaquin Niza Contreras.

REFERENCES

Burnett, J. J. (1998). A strategic approach to managing crises. *Public Relations Review*, *24*(4), 475–488. doi:10.1016/S0363-8111(99)80112-X

Carter, W. N. (1991). *Disaster management: A disaster manager's handbook*. Asian Development Bank.

Chen, S., Law, R., & Zhang, M. (2020). Review of research on tourism-related diseases. *Asia Pacific Journal of Tourism Research*, 1-15.

Cioccio, L., & Michael, E. J. (2007). Hazard or disaster: Tourism management for the inevitable in Northeast Victoria. *Tourism Management*, *28*(1), 1–11. doi:10.1016/j.tourman.2005.07.015

Faulkner, B. (2001). Towards a framework for tourism disaster management. *Tourism Management*, *22*(2), 135–147. doi:10.1016/S0261-5177(00)00048-0

Fusch, P. I., & Ness, L. R. (2015). Are we there yet? Data saturation in qualitative research. *Qualitative Report*, *20*(9), 1408–1416.

Hall, C. M., Scott, D., & Gössling, S. (2020). Pandemics, transformations and tourism: Be careful what you wish for. *Tourism Geographies*, *22*(3), 577–598. doi:10.1080/14616688.2020.1759131

Hall, M. (2010). Crisis events in tourism: Subjects of crisis in tourism. *Current Issues in Tourism*, *13*(5), 401–417. doi:10.1080/13683500.2010.491900

Hao, F., Xiao, Q., & Chon, K. (2020). COVID-19 and China's Hotel Industry: Impacts, a Disaster Management Framework, and Post-Pandemic Agenda. *International Journal of Hospitality Management*, *90*, 102636. doi:10.1016/j.ijhm.2020.102636 PMID:32834356

Ibanescu, B.-C., Eva, M., & Gheorghiu, A. (2020). Questioning the Role of Tourism as an Engine for Resilience: The Role of Accessibility and Economic Performance. *Sustainability*, *12*(14), 5527. doi:10.3390u12145527

Ilnitsky, A., & Chikurova, A. (2015). Diversification as a tool of anti-crisis strategy of development of rural territories in Ukraine. *Baltic Journal of Economic Studies*, *1*(2), 51–55. doi:10.30525/2256-0742/2015-1-2-51-55

Irvine, W., & Anderson, A. R. (2005). The Impacts of Foot and Mouth Disease on a Peripheral Tourism Area: The Role and Effect of Crisis Management. *Journal of Travel & Tourism Marketing*, *19*(2-3), 47–60. doi:10.1300/J073v19n02_05

Ivanov, S.H., Webster, C., Stoilova, E., & Slobodskoy, D. (2020). Biosecurity, crisis management, automation technologies and economic performance of travel, tourism and hospitality companies – A conceptual framework. *Tourism Economics*, 1-24.

Jones, P., & Comfort, D. (2020). The COVID-19 crisis and sustainability in the hospitality industry. *International Journal of Contemporary Hospitality Management*, *32*(10), 3037–3050. doi:10.1108/IJCHM-04-2020-0357

Jurado-Almonte, J. M., & Pazos-García, F. J. (2016). Population and rural tourism in low demographic density territories in Spain. *Boletín de la Asociación de Geógrafos Españoles*, *71*, 511–514.

Leslie, D., & Black, L. (2005). Tourism and the Impact of the Foot and Mouth Epidemic in the UK. *Journal of Travel & Tourism Marketing*, *19*(2-3), 35–46. doi:10.1300/J073v19n02_04

Mazilu, M., Marinescu, R., Bălă, D., & Dragomir, L. (2019). Crisis Management in Tourism. *Quaestus Multidisciplinary Research Journal*, *14*(April), 289–298.

Mikusová, M. (2013). Do Small Organizations Have an Effort to Survive? Survey from Small Czech Organizations. *Economic Research-Ekonomska Istraživanja, 26*(4), 59–76. doi:10.1080/1331677X.2013.11517630

OECD. (2020). *Rebuilding tourism for the future: COVID-19 policy response and recovery*. https://read.oecd-ilibrary.org/view/?ref=137_137392-qsvjt75vnh&title=Rebuilding-tourism-for-the-future-COVID-19-policy-response-and-recovery

Phillipson, J., Bennett, K., Lowe, P., & Raley, M. (2003). Adaptive responses and asset strategies: The experience of rural micro-firms and Foot and Mouth Disease. *Journal of Rural Studies, 20*(2), 227–243. doi:10.1016/j.jrurstud.2003.08.006

Quarantelli, E. L. (1988). Disaster Crisis Management: A Summary of Research Findings. *Journal of Management Studies, 25*(4), 373–385. doi:10.1111/j.1467-6486.1988.tb00043.x

Ritchie, B. W. (2004). Chaos, crises and disasters: A strategic approach to crisis management in the tourism industry. *Tourism Management, 25*(6), 669–683. doi:10.1016/j.tourman.2003.09.004

Sánchez-Zamora, P., Gallardo-Cobos, R., & Ceña-Delgado, F. (2014). Rural areas face the economic crisis: Analyzing the determinants of successful territorial dynamics. *Journal of Rural Studies, 35*, 11–25. doi:10.1016/j.jrurstud.2014.03.007

Santos Lacueva, R. (2020). Estrategia de transformación del turismo rural frente a la COVID-19. In *M. Simancas Cruz, R. Hernández Martín, & N. Padrón Fumero (Coord.), Turismo pos-COVID-19. Reflexiones, retos y oportunidades* (pp. 523–534). Cátedra de Turismo Caja Canarias-Ashotel de la Universidad de La Laguna.

Saunders, B., Sim, J., Kingstone, T., Baker, S., Waterfield, J., Bartlam, B., Burroughs, H., & Jinks, C. (2018). Saturation in qualitative research: Exploring its conceptualization and operationalization. *Quality & Quantity-International Journal of Methodology, 52*(4), 1893–1907. doi:10.100711135-017-0574-8 PMID:29937585

Scott, N., & Laws, E. (2005). Tourism Crises and Disasters: Enhancing Understanding of System Effects. *Journal of Travel & Tourism Marketing, 19*(2-3), 149–158. doi:10.1300/J073v19n02_12

Stafford, G., Yu, L., & Armoo, A. K. (2002). Crisis management and recovery. *The Cornell Hotel and Restaurant Administration Quarterly, 43*(5), 27–40.

Toubes, D., Araújo-Vila, N., & Fraiz-Brea, J.A. (2020). Be water my friend: Building a liquid destination through collaborative networks. *Tourism Management Perspectives*, *33*, 100619. doi:10.1016/j.tmp.2019.100619

Vargas-Sánchez, A. (2018). Crises in tourist destinations: how can they be managed? *Enlightening Tourism. A Pathmaking Journal*, *8*(1), 47-69.

Vargas-Sánchez, A. (2020a). Entender el turismo post-coronavirus: posibles escenarios. In F.J. Bouzá Martorell, & F.J. Melgosa Arcos (Dir.), *El turismo después de la pandemia global: análisis, perspectivas y vías de recuperación*. Documento de Trabajo de AECIT. Retrieved from https://aecit.org/documento-de-trabajo-el-impacto-de-la-crisis-sanitaria-en-el-sector-t/news/193/

Vargas-Sánchez, A. (2020b). Estrategias de destinos y empresas en el mercado turístico pos-COVID-19. In M. Simancas Cruz, R. Hernández Martín, & N. Padrón Fumero (Eds.), *Turismo pos-COVID-19. Reflexiones, retos y oportunidades* (pp. 147–160). Cátedra de Turismo Caja Canarias-Ashotel de la Universidad de La Laguna.

Williams, C., & Ferguson, M. (2005). Recovering from crisis: Strategic alternatives for leisure and tourism providers based within a rural economy. *International Journal of Public Sector Management*, *18*(4), 350–366. doi:10.1108/09513550510599265

Wilson, S., Fesenmaier, D. R., Fesenmaier, J., & Van Es, J. C. (2001). Factors for Success in Rural Tourism Development. *Journal of Travel Research*, *40*(2), 132–138. doi:10.1177/004728750104000203

World Tourism Organization. (2020). *Framework Convention on Tourism Ethics*. UNWTO. doi:10.18111/9789284421671

Zenker, S., & Kock, F. (2020). The coronavirus pandemic – A critical discussion of a tourism research agenda. *Tourism Management*, *81*, 104164. doi:10.1016/j.tourman.2020.104164 PMID:32518437

ADDITIONAL READING

Antonacopoulou, E. P., & Sheaffer, Z. (2014). Learning in Crisis: Rethinking the Relationship Between Organizational Learning and Crisis Management. *Journal of Management Inquiry*, *23*(1), 5–21. doi:10.1177/1056492612472730

Blackman, D., & Ritchie, B. W. (2008). Tourism crisis management and organizational learning: The role of reflection in developing effective DMO crisis strategies. *Journal of Travel & Tourism Marketing*, *23*(2–4), 45–57. doi:10.1300/J073v23n02_04

Cavlek, N. (2002). Tour operators and destination safety. *Annals of Tourism Research*, *29*(2), 478–496. doi:10.1016/S0160-7383(01)00067-6

Deverell, E. (2009). Crises as Learning Triggers: Exploring a Conceptual Framework of Crisis-Induced Learning. *Journal of Contingencies and Crisis Management*, *17*(3), 179–188. doi:10.1111/j.1468-5973.2009.00578.x

Hall, C. M. (2010). Crisis events in tourism: Subjects of crisis in tourism. *Current Issues in Tourism*, *13*(5), 401–417. doi:10.1080/13683500.2010.491900

Johnson Tew, P., Lu, Z., Tolomiczenko, G., & Gellatly, J. (2008). SARS: Lessons in strategic planning for hoteliers and destination marketers. *International Journal of Contemporary Hospitality Management*, *20*(3), 332–346. doi:10.1108/09596110810866145

Miller, G. A., & Ritchie, B. W. (2003). A Farming Crisis or a Tourism Disaster? An Analysis of the Foot and Mouth Disease in the UK. *Current Issues in Tourism*, *6*(2), 150–171. doi:10.1080/13683500308667949

Shaw, D., Hall, M., Edwards, J. S., & Baker, B. (2007). Responding to crisis through strategic knowledge management. *Journal of Organizational Change Management*, *20*(4), 559–578. doi:10.1108/09534810710760081

Smith, D., & Elliott, D. (2007). Exploring the barriers to learning from crisis - Organizational learning and crisis. *Management Learning*, *38*(5), 519–538. doi:10.1177/1350507607083205

Toubes, D. R., Gössling, S., Hall, C. M., & Scott, D. (2017). Vulnerability of Coastal Beach Tourism to Flooding: A Case Study of Galicia, Spain. *Environments*, *4*(83), 1–23. doi:10.3390/environments4040083

KEY TERMS AND DEFINITIONS

COVID-19: Name of the disease caused by a novel coronavirus designated as SARS-Cov-2 (severe acute respiratory syndrome coronavirus 2). This coronavirus disease started in late 2019 and on 11 March 2020, the World Health Organization announced its outbreak as a pandemic.

Crisis: Severe risk suffered by organizations due to internal failures.

Disaster: Severe risk suffered by organizations due to uncontrollable external factors.

Resilience: Ability of individuals or organizations to better withstand stressful or adverse situations and recover more quickly from them.

Rural Tourism: This refers to tourism activities carried out in rural areas. It encompasses accommodation and all kind of activities in rural (non-urban) settings.

Small and Medium-Sized Enterprise (SME): This is a concept related to the size of a company. Usually, the business size is measured through variables such as staff headcount, turnover or balance sheet total. When those variables are lower than certain thresholds, the enterprise is categorized as an SME. For example, at the EU level, those ceilings can be found at https://ec.europa.eu/growth/smes/sme-definition_en.

Strategic Management: Management approach guided to reach the greatest possible coupling between the company and its environment. Such a fit is reached through the formulation, implementation and monitoring of strategies.

ENDNOTES

1 https://www.escapadarural.com/blog/impacto-covid-19-turismo-rural/ (last access: October 29, 2020).

2 74.9% of owners of rural establishments state not having changed prices due to the health crisis. Source: https://www.escapadarural.com/blog/impacto-covid-19-turismo-rural/ (last access: October 29, 2020).

3 At the national level, 97.3% of travellers claim to have stayed in rural establishments where they have been informed of hygiene and safety measure and where such health precautions have been complied with. Source: https://www.escapadarural.com/blog/impacto-covid-19-turismo-rural/ (last access: October 29, 2020).

4 85.2% of owners of rural establishments say that during the summer of 2020 their customers have booked less in advance than in the summer of 2019. Source: https://www.escapadarural.com/blog/impacto-covid-19-turismo-rural/ (last access: October 29, 2020).

5 Source: https://www.escapadarural.com/blog/impacto-covid-19-turismo-rural/ (last access: October 29, 2020).

6 Volatile, Uncertain, Complex, Ambiguous.

7 More details available at https://ec.europa.eu/info/strategy/priorities-2019-2024/european-green-deal_en (last access: October 29, 2020).

Chapter 2
ICTs and Rural Tourism:
Experiences of the COVID-19 Pandemic in Central Mexico

Humberto Thomé-Ortiz
Universidad Autónoma del Estado de México, Mexico

Luis Felipe García-Rodea
Universidad Autónoma del Estado de México, Mexico

Irais González Domínguez
Universidad Autónoma del Estado de México, Mexico

Elizabeth López-Carré
Universidad Autónoma del Estado de México, Mexico

ABSTRACT

This chapter aims to analyze the role that information and communication technologies (ICTs) have played in the processes of adaptation, learning, and tourist reactivation in rural areas, in the context of the pandemic caused by the spread of the zoonotic virus SARS-Cov-2. For this, a case study of the rural tourism offer was developed that includes the experiences of nine community tourism enterprises, settled in indigenous territories of the Central Mexican Highlands. The research approach was qualitative, and it was proposed to systematize the experiences that the enterprises have accumulated throughout the pandemic and during the process of social isolation. It is concluded that the use and exploitation of ICTs in rural areas is still incipient, in terms of their potential as tools for the promotion, marketing, and positioning of rural tourist destinations. It is necessary to overcome the technological gap in the context of the new socioeconomic characteristics that rural tourism will face, many of which will remain in force in the medium and long terms.

DOI: 10.4018/978-1-7998-6996-2.ch002

INTRODUCTION

The advance of globalization in the different spheres of human life has caused adverse effects, especially in rural areas. Rural citizens had to develop strategies to mitigate negative impacts, including the implementation of activities such as tourism, based on the territory's multifunctionality (Gerritsen, 2014). The development of these new activities has led to economic, social, cultural, and political changes in rural areas. These changes have given rise to rural societies in transition (Djanibekov et al., 2014), where the logics of negotiation and consensus are the organizing principle of the different interests that derive from a multifunctional perspective of the territory (Reig, 2002).

In this way, the rural territories have gone from an exclusive model of primary production to a mixed one, which incorporates secondary activities and services, diversifies family income (Schneider, 2009), and develops its economic potential. Rural tourism has been studied from very different perspectives, among which its conceptualization, limitations, importance, and links with other productive activities predominate. From its conception as an activity-oriented to the urban middle class, rural tourism is presented as a development tool (Pratt, 2014).

In countries like Mexico, the conceptualization of rural tourism as a development tool (Zapata et al., 2011) is due to the very nature of their rural territories, where traditional forms of organization and collective land tenure predominate, a motive for which they must be approached from the perspective of social enterprise and rural development. However, it is also essential to consider that rural tourism is a market-oriented product (Briedenhann & Butts, 2006).

Given that a large part of rural tourism definitions implies a broad conception like any tourist practice developed in rural areas, the conception of the specific spatial dimension where this modality of tourism occurs is a fundamental question. In developing countries, rural tourism plays an essential role in rural development (Reddy & Shilpa, 2016) based on the incentive of micro, small and medium-sized enterprises that contribute to a rural region's socio-economic development.

Rural tourism can generate multiple benefits, among which it stands out: the generation of new jobs, the creation of alternative sources of income, stimulating regional development, promoting social inclusion, stimulating territorial innovation, and diversifying the national tourism product. However, the consolidation of rural tourism implies that rural citizens move from a model of agricultural production to another that incorporates the provision of rural tourism services. These adaptation processes imply constant negotiation processes, commitments, and changes (Xue & Kerstetter, 2019).

Meanwhile, the global health crisis associated with COVID-19 has highlighted the tremendous technological lag in rural areas and generated various economic and social impacts on the tourism sector (Romagosa, 2020). Given the untimely nature with which the spread of the virus escalated to the level of a pandemic, the study of local governance processes was privileged since the crisis associated with COVID-19 leads to thinking about the need to reactivate local economies as a response to the economic slowdown and social uncertainty (Cadena et al., 2020; Thomé-Ortiz, 2020).

According to Gaffney & Eeckels (2020), the tourism and hospitality sectors have been among the main affected by the pandemic, generating significant adverse effects on the national economies of countries like Mexico, whose Gross Domestic Product depends, to a large extent, on tourism activity (Organisation for Economic Co-operation and Development [OECD], 2017). In this sense, it can be pointed out that rural tourism has also been affected by the pandemic, but it is precisely this proximity and non-crowded tourism from which the tourist reactivation can begin in the post-pandemic period (Seraphin & Dosquet, 2020).

As has been observed, the worldwide spread of COVID-19 produced a critical disruption in the socio-economic dynamics that had been characterizing conspicuous and hedonic consumption of post-industrial societies (Rundshagen, 2013). The health emergency had severe implications in terms of human mobility, social isolation, and the global economic slowdown, which generated strong effects in the service sector. In particular, the tourism sector suffered a historical decline given confinement conditions, border closures, and various restrictive policies aimed at containing the virus (Reza et al., 2020). However, the reactivation of tourist activity is expected to start from rural tourism expressions, which are not very crowded, and which allow mobilization by their means, in small social groups called "bubble communities" (Sigala, 2020).

Even though rural tourism can be considered an alternative for the reactivation of tourism, the risks implied by the spread of the virus to small rural communities that do not have health services and marginalization and poverty conditions are also recognized (Rascombe, 2020). For this reason, the role of rural tourism, during and after the pandemic, is ambivalent, debating between the opportunity to boost rural economies and an eminent health risk in vulnerable contexts.

In this sense, it is imperative to underline the importance of Information and Communication Technologies (ICTs) during the pandemic, with messages and communication tools being important intermediary agents in the social construction of risk perception (Malecki et al., 2020). On the other hand, it is essential to understand that communication supports have been, in recent months, an essential commercial nexus that has allowed maintaining a large part of the market dynamics, which is why ICTs have intensified their presence in social life, starting from a new dynamism that is expressed between the flow of information and the unpredictable behavior of the virus (Yu et al., 2020).

In the face of the crisis that affected the tourism sector, it is essential to promote resilient and multifunctional territories capable of facing this and future crises. This aspect requires social articulation processes that, in the context of physical distancing, have only been possible through the remote communication tools currently available. The vital role of social networks as a means of linking between social actors is highlighted, as well as e-commerce platforms that have made it possible to commercialize not only the services of rural areas but also their food products (Leone et al., 2020), which are essential for social groups confined to their domestic spaces. In the same way, one can mention educational platforms and videoconferences that have been substantive to generate reflections on the pandemic and generate horizontal and democratic learning processes that point towards a rapid adaptation to what has been called a "new normal."

Technology can be considered a tool for social transformation, which has had a profound influence on different economic sectors during the last century. In the tourism sector, it has generated profound transformations in the last decade. Undoubtedly, technology has transformed tourism into an interactive and efficient industry whose main applications are: i) travel management, ii) digital payments, iii) security for tourists through traceability and GPS, iv) personalization of tourism activities oriented to market niches, v) database management, and vi) access to remote markets at a relatively low price (Kumar & Shekhar, 2020). Likewise, ICTs play a fundamental role in the social perception of risk about the COVID-19 pandemic and tourists' dynamics and attitudes (Yu, et al., 2020). However, there are information gaps regarding the role that technology can play for rural tourism to increase its function as a development strategy. In this sense, it is essential to highlight public policies' role in the adoption of technologies as development strategies.

Although the urban-rural digital divide is a widely documented phenomenon (Salemink, et al., 2017), the disruption of COVID-19 in the face of productive activities such as rural tourism, makes it necessary to study three areas: i) access to quality internet, ii) the development of capacities for the adoption of ITCs and iii) access to electronic devices. All of them are aspects that, in turn, depend on contextual variables such as income level, educational level, and the type of economic activity carried out (Whitacre & Mills, 2007).

This chapter aims to analyze the role that ICTs have played in the processes of adaptation, learning, and tourist reactivation in rural areas, in the context of the pandemic caused by COVID-19. For this, the manuscript studies three critical moments of the pandemic: i) the disruption that the virus produced in the world with particular emphasis on rural tourism activities, ii) the long period of social distancing, and iii) the moment of reopening of economic activities. For this, the manuscript is divided into five parts. After this introduction, a methodological section is presented. The research results are developed below, based on a diagnosis of the use of ICTs in the cases of rural tourism studied, as well as their experience in managing the pandemic in the three moments indicated above. Subsequently, a research agenda was outlined on the intersections between rural tourism, COVID-19, and ICTs. Finally, the conclusions are presented.

METHODOLOGICAL APPROACH

This chapter is based on empirical exploratory research. From a random convenience sampling, nine indigenous rural tourism companies were selected. These companies are distributed in various central Mexico areas (Figure 1), whose activity had shown broad dynamism before the pandemic period and which have a good level of consolidation within the regional tourism offer (Table 1).

Table 1. Rural tourism companies studied

Company	Location	Ethnic Group	Offer of Activities
Parque Ecoturístico El Borbollón	Temascalcingo	Otomí	Accommodation Thermalism
Parque Corral de Piedra	Amanalco de Becerra	Otomí	Accommodation Trekking Kayak
Parque Bosque Esmeralda	Amecameca	Nahua	Accommodation Trekking Environmental workshops
Parque Bosque Detiña	Acambay	Otomí	Accommodation Zipline
Parque Ha Maaloma	Villa del Carbón	Otomí	Accommodation Zipline
Parque La Mesita	Acambay	Otomí	Accommodation Temazcal (Pre-Hispanic steam bath)
Parque Matawi	Temascaltepec	Matlatzinca	Accommodation Trekking
Parque Presa del Llano	Villa del Carbón	Otomí	Accommodation Zipline Kayak
Parque Tiacaque	Jocotitlán	Mazahua	Accommodation Camping Zipline Kayak

Source: Own elaboration.

Different tools of the ethnographic method (Bernard, 2006) were used to develop a deep understanding of a conjunctural phenomenon (COVID-19) that needed to be contextualized within the framework of the rural economy. The collection of field data referring to the use of ICTs in rural tourism was carried out *in situ* during the period between mid-March 2020 and November 2020. A characterization of the companies studied, an inventory of the digital resources used in the tourist activity and informal conversations was carried out. The data were recorded in a field diary.

Specialized bibliographic sources, hemerographic sources, websites, social networks (Facebook, Instagram, and Whatsapp), as well as audiovisual documents from Youtube were used to characterize the use of ICTs in rural tourism. Given the restrictive conditions declared by the government of Mexico in mid-March 2020, remote forms of research were privileged, which is why the work protocol adhered to the ethical guidelines for research during health emergencies (World Health Organization [WHO], 2020). The empirical references were recovered from testimonies, data, videos, and images, mostly obtained from internet sites and social networks.

Semi-structured interviews (n = 9) were applied to the managers of small tourism companies, taking into account the following categories: i) configuration of their socio-digital networks; ii) pre-pandemic, pandemic and post-pandemic strategies; iii) qualitative assessment of the importance of ICTs during the pandemic; iv) socio-digital gap and; v) future prospects. The interviews were conducted in person, respecting biosafety protocols to guarantee the health of the interviewees and the researchers. Likewise, several subsequent telephone calls were held, at different times in which it was required to expand some information obtained within the face-to-face interviews. Pseudonyms replaced the names of the informants to guarantee the confidentiality of their testimonies.

Study Area

The rural area of central Mexico is characterized by wooded ecosystems with a temperate climate that extends along with its Central Highlands. Tourism activity in these rural societies in transition has brought critical economic transformations, including the transition from an exclusively agro-forestry-pastoral model to one that incorporates secondary and tertiary activities, with particular emphasis on tourism in those territories that present a well-preserved natural and cultural heritage.

These relatively new activities have generated new opportunities for some social actors while also redefined the meanings of quality of life and have generated new tensions around endogenous resources, based on their potential economic valorization mediated by various technological platforms for social communication.

Due to the recent and unpredictable nature of the pandemic, the relationship between communication technologies and the impacts of COVID-19 have only been studied tangentially, which is why this study incorporates recent data (December 2020) that can provide valuable elements, not only for the companies under study but also for other rural areas that are going through similar situations.

Figure 1. Study area
Source: Own elaboration.

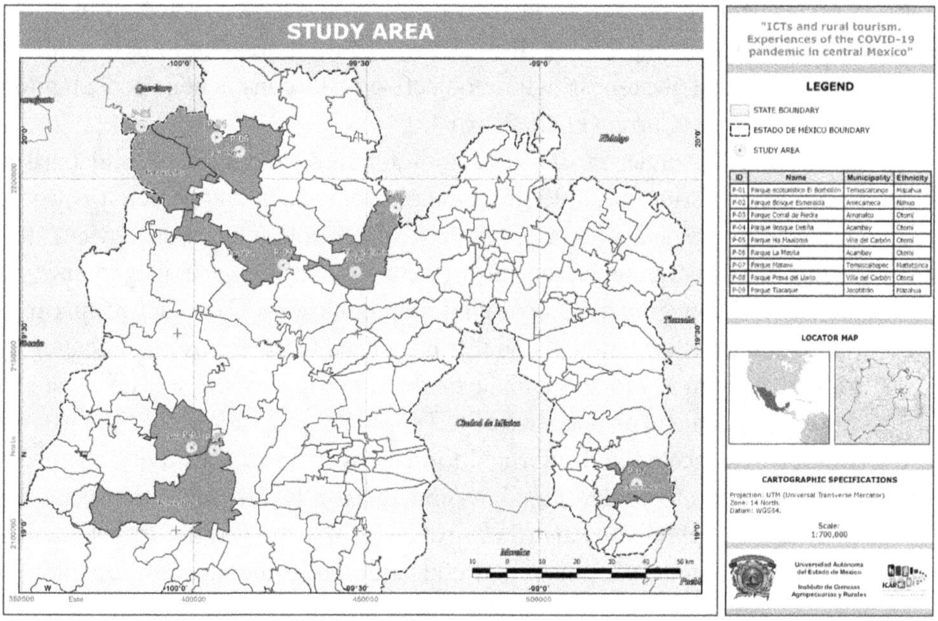

RESULTS

The technological and educational lag in rural areas is evident and is exacerbated by the critical conditions that emerged from the pandemic, which to date is not clear for how long they will persist or if they will even extend beyond the sanitary emergency is declared over (Phillipson et al., 2020). From the diagnosis made, it is possible to affirm that the use and adoption of ICTs are limited, despite being critical tools for tourism planning, education, communication, market positioning, and information dissemination. However, there is awareness regarding the importance of ICTs in rural tourism:

We are investing more in social networks, there is a Facebook page and a boy from the community is advertising the cabins, the events that are sometimes organized and also the CDI (Commission for the Development of Indigenous Peoples) supports us in that [publish] (Carlos, personal communication, August 23, 2020).

ICTs play an essential role in the development of tourist activities. However, according to Bethapudi (2015), the use of ICTs in rural tourism represents a field of opportunities and threats. From the point of view of opportunities, they allow tourists to identify, personalize, and buy tourism products through communication tools, which openly offer worldwide. While the threat is overcrowding, with its consequent ecological impact, the displacement to rural areas of tourists without cultural or ecological awareness, whose objective is the conspicuous consumption of nature and culture (Correia et al., 2016).

The problem of the underuse of ICTs as tools for the development of rural tourism is an aspect that underlies multiple structural phenomena, among which it is worth highlighting technological illiteracy, the lack of telecommunication devices, the poor quality of the communication infrastructures, and the general ignorance of the strategies and functionalities of digital communication. This fact results in a simplistic digital transformation vision that is reduced to the exposure of content to open markets, without any focus on niche markets. For this reason, it is necessary to democratize the good and planned use of ICTs, which would allow connecting with those profiles of educated and informed tourists, who show sensitivity about the recreational activities to be developed in rural areas. It is relevant to transcend the simplified view of supply-demand in the use of ICTs, in such a way that its potential as tools for tourism planning, administration, education, and evaluation can be explored.

The data presented (Table 2) is descriptive and deals with each of the rural tourism companies studied promptly. It is essential to focus on the political meanings that configurations of these socio-digital networks have: the absence or presence of each of the items analyzed in terms of the support provided to the companies studied in the pandemic context. The use of specific media for specific purposes denotes territorialization forms that extend to exogenous horizons through digital worlds. Territorialization being a form of resilience against the threats posed by the pandemic to rural societies.

ICTs management is quite conservative, and it focuses on conventional forms of communication, which can be anachronistic to establish contact with contemporary tourists. Telephone communication stands out, although it is essential to say that its consistency is not maintained since different failures constantly interrupt communication. Internet signal coverage, in general, has poor quality, an aspect that represents a limitation in terms of connection not only for establishments but also as a basic need for most tourists.

Table 2. Use of ICTs in indigenous rural tourism companies in central Mexico

Indicator Company	Phone Signal	Internet Connectivity	Telephone Reservations	Reservations Through ICTs	Dissemination Through ICTs	Intermediaries	Training in the Use of ICTs	Training Through ICTs	They Make Use of ICTs in Administrative Processes
EL BORBOLLON		-	-		-		-		-
BOSQUE ESMERALDA									
CORRAL DE PIEDRA		-							
BOSQUE DETIÑA				-	-	-	-		-
HA MAALOMA				-		-	-		-
LA MESITA			-	-	-	-	-		-
MATAWI		-							
PRESA DEL LLANO				-	-		-		-
TIACAQUE		-		-	-		-		-

Table 3. Symbology

	The company has adequate cell phone signal reception.
	Has internet connectivity
	Use of telephony to attend reservations
	They have reservation mechanisms through the use of ICTs.
	Use of digital platforms for the dissemination and commercialization of services (email, Facebook, Instagram, website)
	Use of intermediaries to link with the market (travel agencies, schools, tourist guides.)
	They have or have had support from bodies (public or private) for training in the use of ICTs.
	They have received training through ICTs.
	Use of ICTs in administrative processes

Source: Own elaboration.

There is limited use of ICTs as tools to establish communication with customers and make reservations, which has opened a virtual space for the entry of intermediaries (tour operators), an aspect that must be analyzed from the fair trade point of view. There is a limited training process for local actors regarding the use and management of ICTs, which translates into a weak use of technological resources for training, administration, and marketing aspects. Social isolation and government regulations focused on the closure of non-essential activities led small rural tourism ventures to experience great uncertainty regarding their operation, profitability, and marketing.

Telephone signal is a primary communication tool; not having it for coordination with health institutions for the transfer of injured people or with symptoms of COVID-19 contagion generates difficulties in managing the crisis. The same happens with reservations by phone or communication with suppliers and potential customers. It is essential to point out that the telephone signal is the only stable means of communication in many rural settings, and on some occasions, it is not even available. In this regard, an interviewee points out:

COVID-19 affected us because we were closed for five months and we did not work, during that time we did not have contact with customers because we do not have a mobile phone (Gerardo, personal communication, August 15, 2020).

Connectivity is a communication limitation in several rural spaces (Puschita et al., 2014). For tourists represent a moment of disconnection with the world that allows them a better coexistence with family or friends and enjoying greater communion with nature, which is why they seek this kind of tourist destination. For other people, given their profession or social configuration, it represents an obstacle that prevents them from going to places without an internet connection (Dickinson et al., 2016). The possibility that exists on the Internet to develop innovative tools in the tourism sector and the provision of training courses by public institutions or NGOs can be diminished by a poor internet connection or the lack of knowledge about using these technological tools.

ICTs represent an opportunity to link with specific tourist profiles, from the attention to market niches on which it is more profitable to focus the marketing strategies. These media have acquired the importance for reservations, given that from the pandemic's confinement, people seek to move to nearby places, a little overcrowded, and that allow direct contact with nature. In this regard, it was observed that the rural tourism companies that have communication with their clients and are already in operation (September 2020) present a greater demand than that existing before quarantine, which is why the mandatory establishment has been necessary of reservations to receive tourists. Rural tourism companies that lack a digital link with their customers have not been able to notify them that they are already in operation and have less demand than before the quarantine, an aspect that has worsened their economic situation.

Since we reopened the cabins, we have not stopped receiving people, we are operating at 50% of our capacity to maintain a healthy distance, but we always have full within that occupancy limit(Elva, personal communication, October 2, 2020).

In terms of capacity development, during the quarantine, there has been an excellent availability of seminars, conferences, and training courses on tourism, as well as the implementation of strategies by public institutions to create collaboration networks between rural tourism companies, to which the members of the companies who have basic communication knowledge and the technological devices to take advantage of them have access.

The "new normal" will imply a rethinking of the dynamics of the markets and rural tourism establishments, where ICTs are positioned as a critical tool that will have to be strengthened soon, incorporating the use of platforms, apps, web pages, and social networks. Some key questions are: Who is responsible for promoting innovation processes in rural tourism through ICTs? Where will they provide the necessary resources for the technological transition, considering the conditions of marginalization and poverty in rural Mexico?

The Pandemic and Its Disruptive Nature

The first stage of the pandemic, known as the disruptive stage, implied raising awareness of a crisis that had already been announced: the Anthropocene (Steffen et al., 2015), an aspect to which the necessary attention had not been paid. The irruption of the virus reveals the technological, social, and emotional lag with which it managed the civilizational crisis in which humanity was already immersed before the pandemic.

In this phase, ICTs were critical supports to generate information processes about the virus, but at the same time, they constituted relevant spaces to generate the first reflections and debates about the impact that the pandemic was generating in rural tourism. These put in evidence the complicated situation faced by some rural territories and revealed some countryside paradoxes as a space-oriented tertiary sector. The pandemic's unequal face was included in the political debate, which affects differently those territories that present less technological development. The key aspects were the flow of information on this emerging phenomenon and the need to generate reflections around the disruptive nature of COVID-19.

The measures taken by the Mexican government to contain the spread of the virus had essential effects on rural economies. The possibility that these effects may extend over the medium and long term is remarkable, given that many of the economic and social consequences of the pandemic have not yet manifested themselves. Given the characteristics of the rural space, referring to the atomization of its population in relatively large spaces, as well as its greater autonomy concerning the tasks aimed at food production and self-consumption, it is possible to infer that there were a greater resilience and adaptation capacity, even among those social actors that were strongly oriented to tourism.

Although we had to close the rural tourism business for a few months, we never lacked anything to eat; we always have enough corn for our tortillas and beans to eat a taco, although we have tourism, we have never stopped planting or raising our animals. (Elena, personal communication, August 2, 2020).

However, it cannot be ignored that due to the restriction of non-essential activities such as tourism, the economic flow produced by visitors decreased, generating effects on the living conditions of rural families (Steiner & Atterton, 2015). However, those territories that developed rural tourism as a complementary economic activity, without putting aside their traditional agricultural activities, were the ones who were most resilient in the face of the pandemic.

It is possible to infer that the magnitude of the impacts related to the pandemic on rural economies would depend on each territory's productive configurations, in this case, the low demand of goods and services that could be consumed outside the home (Phillipson et al., 2020). It is necessary to reorient productive activities towards schemes focused on meeting basic needs and entertainment within the domestic space, emerging as a complex scenario for rural tourism.

At this time, there was no possibility of operating rural tourism companies, but it was necessary to maintain some link with the tourists, who express the need for contact with nature and culture, enormously intensified by the period of confinement. Many of these changes were not carried out from the beginning of the pandemic.

We had no idea how long this would last, the first closure of activities lasted five months, and they say that it is very likely that by the end of the year [2020], it will close again due to a possible re-outbreak. All we did was wait for things to return to normal, thinking that they would be resolved quickly. (Daniel, personal communication August 16, 2020).

The Limbo of Social Distancing

The second moment identified as a period of social distancing, implied a total stoppage of tourist activities, even in which many academic and technical webinars proliferate, but at the same time, they generated a confusing and oversaturated communicative environment. It was presented an opportunity to systematize and share the necessary knowledge for the adaptation and redefinition of rural tourism in the post-pandemic period.

The prolonged period of confinement offered social actors the opportunity to acquire the knowledge and skills necessary to face the challenges of tourism that would now arise under different conditions. However, the experience was not the same for everyone since, once again, the conditions of connectivity and management of technologies reveal the different conditions in which various social actors face the pandemic (De Sousa Santos, 2020). The key aspect during this process was the flow of knowledge. However, some enterprises took a passive and waiting attitude; most of them understood that this stage, of apparent inactivity, should be used to learn and develop new capacities that would allow them to adapt to the new conditions in which tourism activity will develop the future.

One of the aspects that have been capitalized within the organization processes developed in the light of the pandemic was the strengthening of pre-existing community-based social capital in the rural communities of Central Mexico. This social capital translates into the emergence of social enterprise models, based on multiple expressions of collectivity, mutual help, and empathy, which are considered aspects that could play a central role in rural resilience in the face of the crisis (Steinerowski & Steinerowska, 2012). COVID-19 has intensely stimulated social capital construction, highlighting the importance of ICTs in strengthening various forms of collective action.

Here during the tough months of the pandemic, no one died of hunger, there is no shortage of someone who gives you a taco because where two people eat, three eat...... the WhatsApp group has been of great help because that way we have more communication and we stay informed, the health of the people and the opportunities that arise, today more than ever we must be together. (Andrea, personal communication, July 3, 2020).

As shown in this second phase of the COVID-19 pandemic, socio-digitality had increased considerably, highlighting some benefits that the digitalization of many areas of social life could bring. Many of these dynamics are here to stay; however, technological dependence also reflects severe gaps between rural and urban areas. Undoubtedly, digital social networks will be a crucial aspect of partial reactivation of rural tourism, which could appear intermittently, alternated with moments of partial closure. It will be essential to maintain communication inside and outside the rural communities.

At this stage of the analysis, the formation of virtual communities was evident, through which various forms of solidarity were detonated. COVID-19 has generated significant uncertainty and instability, these forums being spaces to share experiences, making it clear that different social actors faced similar situations (Haywood, 2020).

Reactivate From the Small and the Local

The third moment, called reactivation, was developed after the governments authorized the reopening of non-essential economic activities, including tourism, which would take place entirely different conditions than they had been previously developing. In this process, ICTs have played a fundamental role in the digitalization of rural tourism, generating a direct and quality link between hosts and tourists. The necessary aspect for the projection of users' physical and psychological safety is one of the most significant demands of tourists.

How rural communities experienced the COVID-19 pandemic shows the ability to adapt to these types of disruptive phenomena, which vary significantly from one community to another and within them. Likewise, the different productive sectors show essential differences in adapting to the fast changes and adaptations required by the new conditions of sociability, market, and biosafety. In rural tourism, it is essential to point out that given its role as a complementary economic activity, it should not be the only source of income for the communities, but only the reduction of this economic input in a context of crisis. In other words, the productive diversification of rural areas is one of how the countryside has historically expressed its capacity for adaptation and resilience.

Given the non-essential nature of tourist activity, this sector was one of the most affected, and it immediately experienced the effects of the health crisis (Anestis et al., 2021). As mentioned above, the non-mass and proximity forms of tourism, from which it is possible to think about the sector's reactivation in short and medium-term.

As previously mentioned, domestic tourism will have the best recovery opportunities in the short term. For that reason, a key aspect in the reactivation process of rural tourism is the reorientation of the activity towards a direct relationship with the clients, a highly personalized service focused entirely on the needs of each tourist. This customization of rural tourism is a considerable challenge, which from the technical and financial perspective, it is not always possible to face small rural tourism companies, such as those analyzed in this study.

In addition to the limited capacities of small businesses, it is crucial to consider the technological and infrastructure gap that rural spaces have. In the specific case of ICTs, it can be observed that in the countryside, there are shortcomings regarding the speed and quality of broadband, as well as the mobile phone signal, aspects that do not allow facing social isolation and the need to establish new communication bridges in the same way that large companies do. Likewise, it should be thought that these changes imply a redefinition of the tourist offer since it needs to be inserted into digital platforms through semiotic matrices of production and consumption of tourists.

It is essential to take into account that the process of reactivation of rural tourism does not necessarily reflect a consensus between local actors; on the contrary, it is possible to identify social tensions within the communities regarding the idea of receiving tourists, given the possibilities for the virus to spread in communities that have a weak health system (Brouder et al., 2020). In this sense, ICTs could also play a relevant role in terms of clearly informing rural communities about the risks and opportunities that rural tourism represents in the face of the COVID-19 pandemic.

How different tourism ventures have resisted and adapted to the pandemic has a close relationship with their rural capitals' possession and use (Garrod et al., 2006), which are collective resources accumulated historically. These rural capitals include various forms of assets (natural, cultural, physical, social, human, technological, financial, and political), through which it is possible to cope with the disruption caused by COVID-19. Therefore, small rural businesses, where self-employment predominates and has limited resources, experience the most significant difficulties in coping with the crisis and develop organizational strategies and collective action.

It can be said that the emergence of COVID-19 has stimulated territorial innovation through the creative response of local businesses. However, innovation processes towards rural spaces have not always been inclusive and require the deployment of financial resources that are currently scarce. In the case of rural tourism, we refer specifically to innovation in terms of processes and services, which involve biosafety and digitalization protocols.

We tried several things, during the quarantine we tried to sell food and crafts in nearby cities or online, we also wanted to sell tourist packages at a good price to use in the future, but none of these ideas have worked because sometimes we do not have the means to develop them or we cannot find interested clients. (Rebeca, personal communication, November 18, 2020).

Many of these innovations are conditioned on the availability of financial resources and the development of capacities, two areas in which the Mexican countryside has lagged behind the urban space. Overcoming these gaps requires more than the goodwill of rural communities but rather a structural policy guided by the State, in terms of rural development. In other words, equitable distribution is necessary for the management and access of ICTs as territorial innovation tools.

The adaptation of rural tourism to the disruptive process of COVID-19 leads us to think about society's configuration and the identities present in the "new rurality," where it is very likely that emerging actors come into play that leads to social and productive restructuring, marked by digital intermediation. It is possible to identify new rural socio-economic configurations, which will question the basis of the identities of the ethnic groups studied.

The use of the WhatsApp social network stands out, from which organizational and collective action strategies have been developed. In this sense, this kind of network has helped maintain contact between social actors, share training processes, reactivation strategies for tourism activity, and deploy networks of solidarity collaboration.

Thanks to the WhatsApp group it was that we learned about the course on biosafety protocols for the reactivation of tourism that the Mexican Rural Tourism Association gave us; in that group, they always put interesting things, and it is the way to stay together and up to date with what it happens in the community. (Iván, personal communication, October 24, 2020).

One aspect observed in these beginnings of the economic reactivation is that beyond all the possibilities of commercial, technological intermediation, small companies have relied on social networks. Mainly Facebook, WhatsApp, and Instagram have served to maintain social ties during the quarantine time. However, they have also been significant in the reactivation process; for example, tourists have rediscovered rural enterprises from their social networks, and the first reservations made by small businesses have been made through WhatsApp. It is important to note that all these platforms were part of the personal communication networks of almost anyone with an internet connection and an electronic device (laptop, mobile phone, or tablet), which is why it has been easy and intuitive to use them for commercial purposes.

FUTURE RESEARCH DIRECTIONS

The uncertainty about the pandemic's duration and its collateral effects need to reflect these aspects from a long-term perspective. Therefore, it is essential to outline a research agenda on the relationship between COVID-19, ICTs, and rural tourism. It is crucial to reflect on the management of risk and the crisis as two variables present in the immediate future. In this sense, it is essential to consider that according to Kumar & Shekhar (2020), the main research trends on tourism and technologies are concentrated in three main areas: marketing, travel management, and other applications of technology.

In the first place, it would be essential to think about the possible persistence of digital mediations in cultural, commercial, and educational practices, which once the pandemic is over can replace face-to-face relationships. In this sense, it is vital to think about how rural tourism enterprises will carry out their marketing, promotion, and commercialization tasks, with social networks being the main tools that will be used to generate this type of small-scale commercial intermediation. This opportunity is due to these platforms' popularity, which are widely used and whose handling is easy and intuitive. However, for this to work, it is crucial to achieving digital democratization and universal connectivity as an immediate project to propose more resilient and equitable societies in the Post-COVID context (Pardo & Ladeiras, 2020) since the precarious socio-digital conditions of rural spaces put pieces of evidence the problematic conditions that its inhabitants have experienced during the pandemic.

It will be essential to understand the new uses of the Internet within the commercial dynamics of small companies, which, beyond being a matter of will, translates into a series of structural problems that must be solved through various technological and social innovation forms. It is crucial to consider closer proximity between tourists and service providers, through ICTs, for effective dissemination of information, reduction of distances between supply and demand, increased transparency in tourism management (Shanker, 2008), consolidation of the digital economy (Hojeghan & Esfangareh, 2011), as well as the increase in productivity and competitiveness of rural tourism ventures (Armendia & Ollo, 2013).

The digital transition in rural tourism has a solid material base that requires investment processes in telecommunications infrastructure and access to electronic devices such as smartphones, tablets, laptops, and desktop computers and the investment necessary for the development of technological capabilities. The need to guide these actions is observed from a State policy that could include a fifth helix model in which concerted actions can be developed between the government, universities, companies, society, and the environment.

Indeed, it will be necessary to explore alternative organizational models where new and diverse actors converge in the objective of overcoming the rural digital divide and allow open access to the challenges and opportunities offered by the new digitalization scenarios of the economic and social dynamics that have taken place as a result of COVID-19. One of the most significant challenges in this regard is to harmonize the interests of small-scale social economies, such as those that guide these forms of rural tourism, with the market's interests. Thus, it is necessary to explore the opportunities offered by technology to link with market niches for which it is necessary to develop tourism products with high added value. Technological tools have had an important impact on tourist activities' personalization; simultaneously, they make it possible to generate actual databases that allow travelers' needs and desires to be identified (Kumar & Shekhar, 2020). In these terms, creativity will be a fundamental input to face the crisis, which could be analyzed from design thinking (Liedtka, 2018).

The use of GPS systems is essential to manage tourists' mobility, who will become more independent every day. In contrast, digital payments' dissemination is also a pending issue to achieve self-managed community rural tourism projects. Mipti (2014) points out that the diversity of technologies available for rural tourism is essential to achieve the communities' development objectives. The preceding is based on the evidence that ICTs are redefining how tourism services are provided in the context of the pandemic (Kansakar et al., 2017).

On the other hand, it is essential to reflect on the role that rurality will play as a future scenario for tourist leisure, to the extent that these areas are projected as non-crowded spaces, where it is possible to develop tourism practices with less exposure to the virus. Paradoxically, this contrasts with the possibility of turning rural tourism into a vector of transmission of the virus from the cities to the countryside and thereby threatening the security of rural societies whose health coverage is precarious and insufficient. Evidence shows that tourism activities can be an essential factor in the spread of epidemics (Mason, et al., 2005). At the same time, the prevalence of zoonotic viral threats will continue to be present.

Therefore, the relationship between rural tourism, ICTs, and COVID-19 can present a high ambivalence degree. Undoubtedly, it will be necessary to think about how the pandemic will drive technological innovation in the new urban-rural articulations, specifically about rural tourism, in terms of the use of digital technology and the transformation of its socio-economic dynamics.

It is necessary to think rural areas are spaces with limitations of infrastructure and technological knowledge, so it is essential to find a way to bring technology closer to people, through investments in equipment and fundamentally in capacity development, to take advantage of the potential that rural tourism has to promote the development of poor areas of countries with emerging economies.

Given that urban-rural digital divide, it is a problem that affects all rural territories around the world, including the most developed countries (Davies, 2021). It is necessary to think about alternatives such as centralizing communication, promotion, and commercialization of rural tourism in areas where there are better connectivity conditions. In this regard, it is suggested to work on the construction of regional rural tourism networks.

It is considered that the study of the crisis and disruption that COVID-19 presents to rural tourism cannot be anchored to stagnant models of technological adoption but must be based on a future-back approach (Johnson & Suskewicz, 2020), in which the compilation of the empirical data of the present will have crucial importance in the approach of the strategies to be followed. There are not enough elements to make recommendations about a health crisis that is still ongoing, so it is considered very important to document small rural tourism companies' survival strategies to have greater clarity on the starting points to face the change. It is important to remember that the current moment presents an essential change in values and behaviors so that rural tourism businesses must be flexible and adaptable.

In general terms, the academy must generate profound changes to support small rural tourism companies. In this sense, it is crucial to generate creative approaches to problem-solving based on exchanging information and knowledge from a horizontal perspective. The pandemic represents an excellent opportunity for students and researchers to get involved in community tourism projects seeking solutions. This fact responds to the urgent need expressed by small tourism businesses for advice on the business skills necessary to survive in this volatile context. Many of the changes to be developed must be implemented from the universities where social laboratories for rural tourism can be articulated, whose proposals for experimentation and transformation can have a safe exit to the markets.

CONCLUSION

Tourism has been one of the economic sectors most affected by COVID-19. The pandemic's negative impacts have had substantial implications for the rural space, which has influenced people dedicated to rural tourism to return to traditional productive activities (mainly agricultural) as a subsistence strategy. Many of the food produced on a small scale, in the pandemic context, covered the food needs of millions of city dwellers while the food was guaranteed for rural families through self-consumption.

The global scope of the pandemic has produced changes not only in economic activities but also in the way people think in the strategies developed by small rural tourism ventures; in such a way that the reactivation of rural tourism is an opportunity for the reconfiguration of the productive activities of the territory, with significant participation of the technological component.

At this historical moment where ICTs are positioned as a tool for connectivity between people, products, and territories, it is necessary to reflect on how the rural space will be inserted into the logic of virtual interaction. However, for this, it is necessary to close the gaps in existing technologies. It can then be affirmed that there is a pre-existing or acquired "technological capital" that plays a crucial role in the adaptation and resilience capacity of small rural tourism companies in the face of the crisis caused by COVID-19.

ICTs have been valuable tools to face the economic and social crisis generated by the outbreak of the COVID-19 pandemic. However, it is recognized that there is still much to do in terms of learning and taking advantage of available technological resources, which not only refers to a solo effort on the part of rural communities but also as a social process that must be oriented from a State policy, related to welfare and the democratization of technological innovation.

In this sense, the role of ICTs is recognized as elements that have allowed maintaining social cohesion and articulation between groups with specific interests such as those involved in rural tourism. From these technological applications, it has been possible the timely circulation of information, the construction of spaces for reflection, to have a vehicle for the flow of consciousness about the historical moment we are living, to share and obtain knowledge, develop new capacities and adapt to the new conditions of life.

However, it was also found that these technologies reveal the contexts of inequality in which the health crisis is experienced, which mainly affect rural areas, where there are not always communication networks, infrastructures, equipment, and knowledge for management of these technologies. The increase in the use of ICTs as a result of the outbreak of the COVID-19 pandemic has an ambivalent meaning that includes the exacerbation of inequalities and generates opportunities for small rural tourism enterprises.

Therefore, one of the most critical aspects to make visible is the need to generate inclusive and democratic technological societies, from which it is possible to establish ourselves as a civilization capable of facing this and future challenges. It is important to remember that rural well-being is a critical aspect in determining society's quality of life as a whole.

There are elements to think that tourism will be reactivated from short trips, which allow better control of security measures and low cost, in congruence with the pandemic's economic crisis. However, it is necessary to rethink a new perspective of tourist leisure from which value is created for tourists, based on the development of experiences conscious and transformative while for the host societies, it is a tool to improve their quality of life comprehensively.

There is a real technological influence that is producing transformations in the territory. However, the depth of the Covid-19 crisis in the face of tourist activities is so strong that a principle of adaptation towards the future implies the social capacity to propose alternative and complementary economic systems to tourist activity. It is not limited to thinking about productive diversification processes in the territory but proposing profound social changes to overcome the crisis, which implies the alternative search for income.

The current moment of social distancing, marked by the second moment of sanitary restrictions (December 2020), is presented as a moment to propose different initiatives, reflecting on the future scenarios of touristified rural environments. In this sense, it is essential to develop creative solutions that include the digitalization of local productive structures and their commercial networks, but this requires new actors to serve as bridges between the digital and rural areas.

The critical state that rural tourism presents in the face of the COVID-19 pandemic, to a large extent, is part of an intellectual lag in terms of designing more conscious tourism alternatives with social value. For this reason, it is necessary to promote economic, scientific, organizational, technological, and cultural advances that lead to sustainable business models and that present stability in the face of an uncertain future. The joint action of various sectors of society is necessary to achieve this change.

REFERENCES

Anestis, F., Stathis, P., & Tzung-Cheng, H. (2021). The good, the bad, and the ugly on COVID-19 tourism recovery. *Annals of Tourism Research*, *87*, 103117. doi:10.1016/j.annals.2020.103117 PMID:33518847

Aramendia, E., & Ollo, A. (2013). ICT impact on tourism industry. *International Journal of Management Cases*, *15*(2), 87–98.

Bernard, H. (2006). *Research methods in anthropology: Qualitative and quantitative approaches*. Alta Mira Press.

Bethapudi, A. (2015). Role of ICT in promoting rural tourism product. *Journal of Tourism & Hospitality (Los Angeles, Calif.)*, *4*(3), 1–3.

Briedenhann, J., & Butts, S. (2006). Application of the Delphi technique to rural tourism project evaluation. *Current Issues in Tourism*, *9*(2), 171–190. doi:10.1080/13683500608668246

Brouder, P., Teoh, S., Salazar, N., Mostafanezhad, M., Mei Pung, J., Lapointe, D., Higgins, F., Haywood, M., Hall, M., & Balslev, H. (2020). Reflections and discussions: Tourism matters in the new normal post-COVID-19. *Tourism Geographies*, *22*(3), 735–746. doi:10.1080/14616688.2020.1770325

Cadena, A., Child, F., Craven, M., Ferrari, F., Fine, D., Franco, J., & Wilson, M. (2020). *How to restart national economies during the coronavirus crisis*. McKinsey. https://tec.mx/sites/default/files/inline-files/How-to-restart-national-economies-during-the-coronavirus-crisis-vF_1.pdf

Correia, A., Kozak, M., & Reis, H. (2016). Conspicuous Consumption of the Elite: Social and Self-Congruity in Tourism Choices. *Journal of Travel Research*, *55*(6), 738–750. doi:10.1177/0047287514563337

Davies, A. (2021). COVID-19 and ICT-Supported RemoteWorking: Opportunities for Rural Economies. *WORLD (Oakland, Calif.)*, *2*, 139–152.

De Sousa Santos, B. (2020). *La cruel pedagogía del vírus* [The cruel pedagogy of the virus]. CLACSO. https://www.clacso.org/la-cruel-pedagogia-del-virus/

Dickinson, J., Hibbert, J., & Filimonau, V. (2016). Mobile technology and the tourist experience: (Dis)connection at the campsite. *Tourism Management*, *57*, 193–201. doi:10.1016/j.tourman.2016.06.005

Djanibekov, N., Herzfeld, T., & Weingarten, P. (2014). Rural areas in transition: An introduction. *Appl Agric Forestry Res*, *64*, 125–126.

Gaffney, C., & Eeckels, B. (2020). Covid-19 and tourism risk in the Americas. *Journal of Latin American Geography*, *19*(3), 308–313. doi:10.1353/lag.2020.0054

Garrod, B., Wornell, R., & Youell, R. (2006). Re-conceptualising rural resources as countryside capital: The case of rural tourism. *Journal of Rural Studies*, *22*(1), 117–128. doi:10.1016/j.jrurstud.2005.08.001

Gerritsen, P. (2014). Working with indigenous women on multifunctionality and sustainable rural tourism in western Mexico. *Journal of Rural and Community Development*, *9*(3), 243–257.

Haywood, K. (2020). A post-COVID future: Tourism community reimagined and enabled. *Tourism Geographies*, *22*(3), 599–609. doi:10.1080/14616688.2020.1762120

Hojeghan, S., & Esfangareh, N. (2011). Digital economy and tourism impacts, influences and challenges. *Procedia: Social and Behavioral Sciences*, *19*, 308–316. doi:10.1016/j.sbspro.2011.05.136

Johnson, M., & Suskewicz, J. (2020). *Lead from the future*. HBR Press.

Kansakar, P., Munir, A., & Shabani, N. (2017). Technology in hospitality industry: Prospects and challenges. *IEEE Consum Electron Mag*, *8*(3), 60–65. doi:10.1109/MCE.2019.2892245

Kumar, S., & Shekhar. (2020). Technology and innovation: Changing concept of rural tourism - A systematic review. *Open Geosciences*, *12*(1), 737–752. doi:10.1515/geo-2020-0183

Leone, L., Fleischhacker, S., Anderson-Steeves, B., Harper, K., Winkler, M., Racine, E., Baquero, B., & Gittelsohn, J. (2020). Healthy Food Retail during the COVID-19 Pandemic: Challenges and Future Directions. *International Journal of Environmental Research and Public Health*, *17*(20), 1–14. doi:10.3390/ijerph17207397 PMID:33050600

Liedtka, J. (2021, April 3). *Why design thinking works*. Harvard Business Review. https://hbr.org/2018/09/why-design-thinking-works

Malecki, K., Keating, J., & Safdar, N. (2020). Crisis Communication and Public Perception of COVID-19 Risk in the Era of Social Media. *Clinical Infectious Diseases*, ciaa758.

Mason, P., Grabowski, P., & Du, W. (2005). Severe acute respiratory syndrome, tourism, and the media. *International Journal of Tourism Research*, *7*(1), 11–21. doi:10.1002/jtr.519

Mpiti, K. (2014). The potential use of ICT to enhance agritourism in Lesotho. In Technology and innovation: Changing concept of rural tourism (pp. 129-136). 751 Editors.

Organisation for Economic Co-operation and Development. (2017). *Tourism Policy Review of Mexico*. Organisation for Economic Co-operation and Development. https://read.oecd-ilibrary.org/industry-and-services/tourism-policy-review-of-mexico_9789264266575-en#page2

Pardo, C., & Ladeiras, A. (2020). Covid-19 "tourism in flight mode": A lost opportunity to rethink tourism – towards a more sustainable and inclusive society. *Worldwide Hospitality and Tourism Themes*, *12*(6), 671–678. doi:10.1108/WHATT-07-2020-0064

Phillipson, J., Gorton, M., Turner, R., Shucksmith, M., Aitken-McDermott, K., Areal, F., Cowie, P., Hubbard, C., Maioli, S., McAreavey, R., Souza-Monteiro, D., Newbery, R., Panzone, L., Rowe, F., & Shortall, S. (2020). The COVID-19 Pandemic and Its Implications for Rural Economies. *Sustainability*, *12*(10), 1–9. doi:10.3390u12103973

Pratt, S. (2014). Potential economic contribution of regional tourism development in China: A comparative analysis. *International Journal of Tourism Research*, *17*(3), 303–312. doi:10.1002/jtr.1990

Puschita, E., Constantinescu, A., Dobrab, R., Colda, I., Vermesana, A., & Tudor, M. (2014). Challenges for a broadband service strategy in rural areas: A Romanian case study. *Telecommunications Policy*, *38*(2), 147–156. doi:10.1016/j.telpol.2013.08.001

Rascombe, P. (2020). Rural areas at risk during the COVID-19 pandemic. *Lancet*, *20*(5), 545. doi:10.1016/S1473-3099(20)30301-7 PMID:32311327

Reddy, V., & Shilpa, R. (2016). Rural tourism – a catalyst for rural economic growth. International. *Journal of Human Social Sciences and Humanities Invention*, *5*(5), 14–19.

Reig, E. (2002). La multifuncionalidad del mundo rural [The multifunctionality of the rural world]. *ICE Revista de Economía*, *803*, 33–44.

Reza, M., Gholipour, H., Feizi, M., Nunkoo, R., & Andargoli, A. (2020). International Tourism and Outbreak of Coronavirus (COVID-19): A Cross-Country Analysis. *Journal of Travel Research*, *60*(3), 1–6.

Romagosa, F. (2020). The COVID-19 crisis: Opportunities for sustainable and proximity tourism. *Tourism Geographies*, *20*(3), 690–694. doi:10.1080/14616688.2020.1763447

Rundshagen, V. M. (2013). Post Industrial Society. In S. Idowu, N. Capaldi, L. Zu, & A. Gupta (Eds.), *Encyclopedia of Corporate Social Responsibility* (p. 68). Springer. doi:10.1007/978-3-642-28036-8_175

Salemink, K., Strijker, D., & Bosworth, G. (2017). Rural development in the digital age: A systematic literature review on unequal ICT availability, adoption, and use in rural areas. *Journal of Rural Studies*, *54*, 360–371. doi:10.1016/j.jrurstud.2015.09.001

Seraphin, H., & Dosquet, F. (2020). Mountain tourism and second home tourism as post-COVID 19 lockdown placebo? *Worldwide Hospitality and Tourism Themes*, *12*(4), 485–500. doi:10.1108/WHATT-05-2020-0027

Shanker, D. (2008). ICT and tourism: Challenges and opportunities. *J Manage Res Anal*, *5*(1), 276–283.

Sigala, M. (2020). Tourism and COVID-19: Impacts and implications for advancing and resetting industry and research. *Journal of Business Research*, *117*, 312–321. doi:10.1016/j.jbusres.2020.06.015 PMID:32546875

Steffen, W., Broadgate, W., Deutsch, L., Gaffney, O., & Ludwig, C. (2015). The trajectory of the Anthropocene: The Great Acceleration. *The Anthropocene Review*, *2*(1), 81–98. doi:10.1177/2053019614564785

Steiner, A., & Atterton, J. (2015). Exploring the contribution of rural enterprises to local resilience. *Journal of Rural Studies*, *40*, 30–45. doi:10.1016/j.jrurstud.2015.05.004

Steinerowski, A., & Steinerowska-Streb, I. (2012). Can social enterprise contribute to creating sustainable rural communities? Using the lens of structuration theory to analyze the emergence of rural social enterprise. *Local Economy*, *27*(2), 167–182. doi:10.1177/0269094211429650

Thomé-Ortiz, H. (2020). O coronavírus reescreverá o turismo rural? Reinvenção, adaptação e ação no contexto latino-americano [Will the coronavirus rewrite rural tourism? Reinvention, adaptation and action in the Latin American context]. *Cenário*, *8*, 55–72.

Whitacre, B., & Mills, B. (2007). Infrastructure and the rural urban divide in high-speed residential internet access. *International Regional Science Review*, *2007*(30), 249–273. doi:10.1177/0160017607301606

World Health Organization. (2020). *Ethical standards for research during public health emergencies: Distilling existing guidance to support COVID-19 R&D*. WHO. https://apps.who.int/iris/handle/10665/331507

Xue, L., & Kerstetter, D. (2019). Rural Tourism and Livelihood Change: An Emic Perspective. *Journal of Hospitality & Tourism Research (Washington, D.C.)*, *43*(3), 416–437. doi:10.1177/1096348018807289

Yu, M., Li, Z., Yu, Z., He, J., & Zhou, J. (2020). Communication-related health crisis on social media: A case of COVID-19 outbreak. *Current Issues in Tourism*, 1–7. doi:10.1080/13683500.2020.1752632

Zapata, M., Hall, M., Lindo, P., & Vanderschaeghe, M. (2011). Can community-based tourism contribute to development and poverty alleviation? Lessons from Nicaragua. *Current Issues in Tourism*, *14*(8), 725–749. doi:10.1080/13683500.2011.559200

ADDITIONAL READING

Ajzen, I. (1985). From intention to actions: A theory of planned behavior. In J. Kuhl & J. Beckmann (Eds.), *Action-control: From cognition to behavior* (pp. 11–39). Springer. doi:10.1007/978-3-642-69746-3_2

Ajzen, I. (1991). The theory of planned behavior. *Organizational Behavior and Human Decision Processes*, *50*(2), 179–211. doi:10.1016/0749-5978(91)90020-T

Davis, F., Bagozzi, R., & Warshaw, P. (1989). User acceptance of computer technology: A comparison of two theoretical models. *Management Science*, *35*(8), 982–1003. doi:10.1287/mnsc.35.8.982

Fishbein, M., & Ajzen, I. (1975). *Belief, Attitude, Intention and Behavior: An Introduction to Theory and Research*. Addison-Wesley.

Rogers, E. (1995). *Diffusion of Innovations*. The Free Press.

Taylor, S., & Todd, P. (1995). Understanding information technology usage: A test of competing models. *Information Systems Research*, *6*(2), 144–176. doi:10.1287/isre.6.2.144

Taylor, S., & Todd, P. (1995a). Assessing IT Usage: The Role of Prior Experiences. *Management Information Systems Quarterly*, *19*(3), 561–570. doi:10.2307/249633

Venkatesh, V., & Bala, H. (2008). Technology Acceptance Model 3 and a Research Agenda on Interventions. *Decision Sciences*, *39*(2), 273–315. doi:10.1111/j.1540-5915.2008.00192.x

Venkatesh, V., & Davis, F. (2000). A Theoretical Extension of the Technology Acceptance Model: Four Longitudinal Field Studies. *Management Science*, *46*(2), 186–204. doi:10.1287/mnsc.46.2.186.11926

KEY TERMS AND DEFINITIONS

Central Mexican Highlands: It is a natural region that makes up the central part of Mexico. It is found between the Western and Eastern Sierras Madre and the south with the Neovolcanic Axis.

ICTs: Information and communication technologies (ICTs) are tools related to the transmission, processing, and digitized storage of information.

Resilience: Ability to adapt slightly with positive results in the face of adverse situations.

Rural Societies in Transition: It refers to rural societies that have new commercial, cultural, and political articulations with the urban space, based on models of multifunctionality and pluriactivity. They are characterized by the transition towards productive schemes that incorporate tertiary activities such as tourism.

Rural Tourism: Tourist activity developed in a rural or natural space, usually in small towns or outside the urban area in larger towns.

Social Economy: It refers to the relations of production and distribution organized by the principle of solidarity and does not pursue profit. It is a sector of the economy that would be halfway between the private and public sectors.

Socio-Digital Networks: Socio-digital networks are structures formed on the Internet by people or organizations that connect based on shared interests or values. Through them, relationships between individuals or companies are created quickly, without hierarchy or physical limits.

Chapter 3
The Impact of COVID-19 on the Tourism Industry:
The Case of MSMEs in Barcelona

Najat Tort
Universitat de Girona, Spain

Elena Puiggros
https://orcid.org/0000-0003-2196-4581
EU Mediterrani-Universitat de Girona, Spain

ABSTRACT

The purpose of the present research is to analyze the impact of COVID-19 in touristic activity enterprises, which offer several experiences such as routes, visits, among others, in Barcelona. Many of these businesses are micro, small, and medium-sized enterprises (MSMEs). An online questionnaire has been designed in order to collect data. The study provides timely and insightful implications for tourism practitioners who will be expected to prepare the post-corona field for a new normal after the experience of restricted living during an unprecedented pandemic.

INTRODUCTION

This study focuses on tourist guide micro, small and medium-sized enterprises (MSMEs) in Barcelona. Although it could be considered that it is premature to carry out an analysis of the impacts of the pandemic, the impact has been so excessive that it requires immediate action. Publications and studies about covid-19 consequences in tourism have been already published in April 2020. The contribution of this study

DOI: 10.4018/978-1-7998-6996-2.ch003

lies more in the research rather than a topic of discussion. From particular studies, issues can be detected and be a contribution to MSMEs. Although each destination has its own particularities, concrete examples could shed light on this matter. The first part of this study is focused on risks in tourism from a general perspective. Secondly, we explain the impact of covid-19 on the tourism sector, particularly in Spain, Catalonia and Barcelona. After this theoretical framework, the study presents methodology and findings, which are focused on touristic MSMEs in Barcelona. Finally, like any other traditional study, there is a section dedicated to conclusions.

Since the covid-19 outbreak in China in January 2020, the coronavirus has spread rapidly around the world with social, economic, environmental and cultural impacts. Covid-19 is perceived as a crisis and tourism is a risk-sensitive and vulnerable industry and this has evoked considerable research that has focused on impacts, changing the perception of travel, and destination image (Li & Nguyen, 2020; Reddy et al. 2020; Ritchie & Jiang, 2019). In tourism research, the scholars studied both natural and human-induced disasters from different perspectives.

Table 1. Perspectives of research focused on crisis impact

Perspective	Scholars
Crisis knowledge and risk intelligent	Paraskevas, Altinay, McLean, & Cooper, 2013; Paraskevas & Quek, 2019
Forecasting	Page, Yeoman, Munro, Connell, & Walker, 2006; Prideaux, Laws, & Faulkner, 2003
Destination image	Avraham, 2015; Ketter, 2016; Zenker, von Wallpach, Braun, & Vallaster, 2019
Perceptions of risk and management	Liu-Lastres, Schroeder & Pennington-Gray, 2019; Tsai & Chen, 2010; Sembada & Kalantari, 2020; Li et al., 2020; Wolff et al., 2019; Ritchie & Jian, 2019; Richard et al., 2020.
Crisis indicators	De Sausmarez, 2007
Post-crisis recovery	Campiranon & Scott, 2014; Scott, Laws & Prideaux, 2007; Reddy et al, 2020; Qiu et al, 2020; Li et al., 2020; Romão, 2020; Romagosa, 2020; Shafi et al., 2020; Lew, 2020; Bhuiyan, et al., 2020.
Crisis detection	Paraskevas & Altinay, 2013
Stakeholder collaboration	Hystad & Keller, 2008; Jiang & Ritchie, 2017
Crisis communication strategies	Henderson, 2003; Liu-Lastres et al., 2019; Wang & Ritchie, 2012
Terrorism prevention	Kılıçlar, Uşaklı, & Tayfun, 2018; Paraskevas & Arendell, 2007

Source: (Adapted from: Li et al, 2020; Reddy et al. 2020)

Table 1 shows different perspectives of research focused on crisis impacts on tourism. There are many scholars, who studied crises in tourism from different perspectives. Perception of risk and management and post-crisis recovery are the most studied perspectives by academics. Specifically, crises in tourism fall into two main contexts: risk perception at the individual level or demand side, and crisis management at the collective level or supply side (Richard et al., 2020.). Touristic studies of covid-19 have focused on perceptions on risk and management and post-crisis recovery. Perception of risks studies are focused on natural disasters like tsunamis or humans' disasters like terrorism o financial, or physical, social, psychological or health. Objective or absolute risk is defined as the probability of negative outcomes weighted by their severity (Wolf et al., 2019). However, the concept of perceived risk has been defined in many different ways. Indeed, the fuzziness of the concept has also been pointed at by various scholars within the field (Larsen, et al., 2017). Scholars seem to agree, perceived risk is related to the possibility or perception of the visitor of losing something of value. Therefore covid-19 has an objective risk because it has affected day to day life, businesses and disrupted world trade and movement (Haleem, et al. 2020). Also, covid-19 is perceived as a subjectively perceived risk and that influence the tourist industry. Existing studies focused on post-crisis recovery related to covid-19. Scholars propose strategies of growth after pandemic (Romao, 2020), some based on sustainable tourism (Romagosa, 2020; Bhuiyan, et al., 2020; Lew, 2020). Some other studies explore the consequences and future impacts of coronavirus in tourism (Richard, et al.; Li et al., 2020; Shafi, et al., 2020; Qiu, et al., 2020).

While a number of studies have investigated the pandemic broadly on tourism, less attention has been paid to how covid-19 affects a specific destination. This research analyses the effects of the pandemic in the tourism industry broadly and specifically in Spain, as well as the economic impacts of tourist MSMEs, especially tour guides of Barcelona during the pandemic. For this proposal a survey and descriptive analysis is adopted to explore changes, difficulties MSMEs are facing. This research provides a contribution to existing knowledge by providing data from a well-known touristic city (Barcelona).

COVID – 19 AND TOURISM SECTOR

Tourism provides livelihoods for millions of people and allows billions more to appreciate their own and different cultures, as well as the natural world. For some countries, tourism represents (as of May 2020) over 20% of their GDP and, overall, it is the third largest export sector of the global economy. The world is facing an unprecedented crisis with the covid-19 pandemic. Travel and tourism sector are among the most affected sectors: 100% of UNWTO countries impose restrictions on travelling and 72% (as of May 2020) have completely closed their borders to international tourism (UNWTO, 2020a). As a consequence of the restrictions and measures, tourism is among the hardest hit sectors due to many tourists cancel their trips because some destinations had been closed by the local governments. According to UNWTO (2020b), international tourism arrivals worldwide this year could fall by as much as 30%, with a corresponding loss of international tourism receipts up to 450 billion US dollars. Especially for countries where the national economy relies heavily on the tourism industry, such as Spain in which tourism generates 11,4% of its GDP (as of May 2020) (EUROSTAT, 2020).

Figure 1. Percentage of GDP contributed by tourism.
Source: (OECD, 2020a)

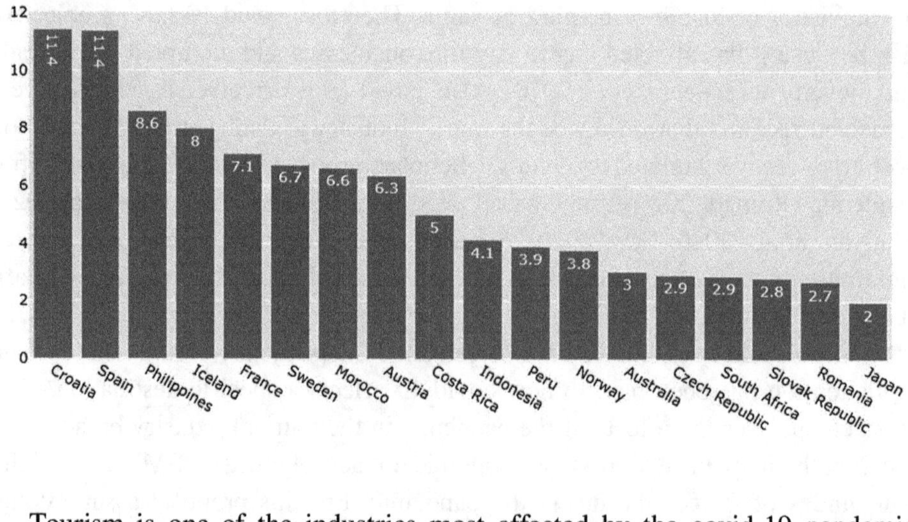

Tourism is one of the industries most affected by the covid-19 pandemic, affecting economies, daily life, public services and opportunities on all continents. No country has escaped the decimation of its tourism sector. Tourism is also one of the most labour-intensive sectors of our economies. This crisis is a major shock for developed economies and an emergency for the most vulnerable people and developing countries. Tourism generated 7 per cent of global trade, employed one in every ten people globally and supports 330 million jobs worldwide, which are at risk in the current situation (UNWTO, 2020b; UN, 2020). To create a better understanding of the number of people working in the tourism industry, the European Data Portal created a visual (see figure 2) based on 2019 data from Eurostat.

Figure 2. European overview of percentage of population employed in tourism-related sectors
Source: (EUROSTAT. 2020)

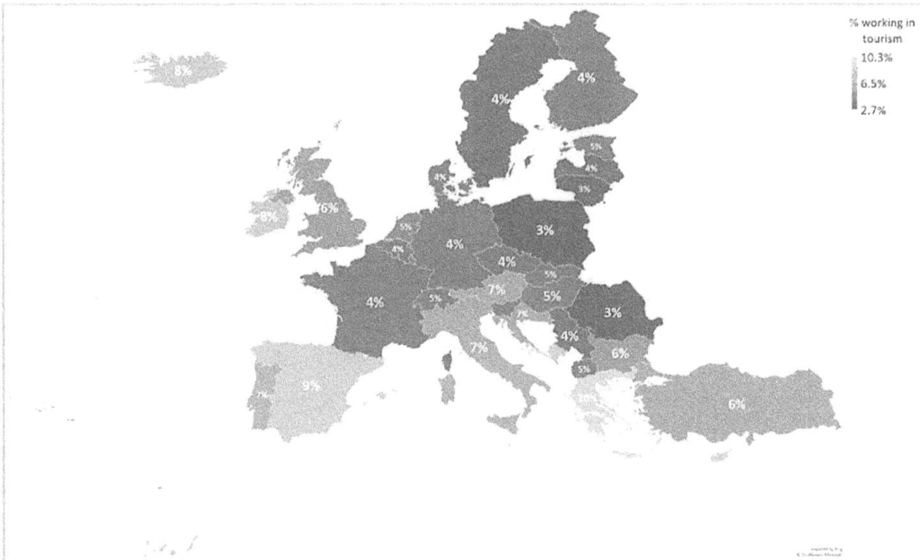

Due to the pandemic, 100 to 120 million direct tourism jobs are at stake at risk. It is also expected to lose $910 billion to $1.2 trillion in exports from tourism – international visitors' spending (UN, 2020;). Additionally, covid-19 involves the loss of 1.5% to 2.8% of global GDP (UNCTAD, 2020; UN, 2020). The UNWTO published a summary with special focus on the impact of covid-19 on the tourism industry. This document explores three scenarios (figure 4) depending on the course of the virus and the duration of the travel restrictions and estimated a possible decline in international tourist arrivals between 58% and 78% in 2020. In the first quarter of 2020, international tourism already decreased by 22%, with a decline of 57% in March. Creating a forecast for the rest of 2020 is complicated due to the rapidly changing circumstances, the speed of containment of the pandemic, the duration of travel restrictions and the gradual re-opening of borders that has now begun but remains uncertain. UNWTO estimated a possible drop in international tourists' arrivals between 58% and 78% in 2020. Consequently, this is the worst crisis by far that international tourism has faced so far. The global economy is projected to contract sharply by 4.9 per cent in 2020, though the outlook is expected to pick up in 2021, according to the International Monetary Fund. (UNWTO 2020a; EUROSTAT, 2020).

Figure 3. Three scenarios created by the UNWTO about the effect on international tourism.
Source: (UNWTOa, 2020)

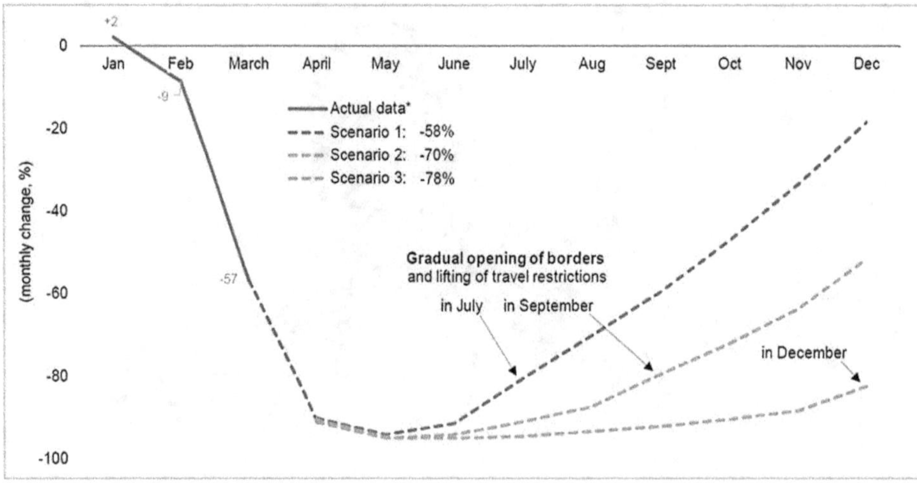

Pandemic not only has an impact on the economy but also has environmental, biodiversity and cultural consequences. The tourism sector has a high climate and environmental footprint and places stress on land and maritime systems. However, covid-19 and the subsequent closure of natural protected areas have had a destructive impact on wildlife and the communities that protect it. In some protected parks and natural areas there has already been a recent rise in poaching and looting, partly due to the reduced presence of tourists and staff (UN, 2020; UNTWO, 2020a). Protected areas and other conserved landscapes serve as important reservoirs for wildlife and other biodiversity resources and form the centrepiece of the continent's nature-based tourism.

The impact of covid-19 on tourism also places pressure on heritage conservation in the cultural sector as well as on the cultural and social fabric of communities, particularly for indigenous people and ethnic groups. Many intangible cultural heritage practices such as traditional festivals and gatherings have also been halted or postponed, with important consequences for the social and cultural lives of communities everywhere (UN, 2020). Besides, Cultural World Heritage sites and museums, also rely highly on tourism revenues to carry out instrumental monitoring, conservation and archaeological work. With more than 80% of UNESCO World Heritage properties have closed down, the livelihoods of millions of cultural professionals have been seriously jeopardized (UNTWO, 2020a). The reduction in visitors has had a direct negative impact on these operational budgets.

Nevertheless, covid-19 bringing global tourism to a standstill, millions of people in quarantine have been seeking out cultural and travel experiences from their homes. Culture has proven indispensable during this period, and the demand for virtual access to museums, heritage sites, theatres and performances has reached unprecedented levels. Therefore, Tourism biodiversity and culture sectors are facing an opportunity to create new partnerships and collaboration. They are bound to jointly reinvent and diversify the offer, attract new audiences, develop new skills and support the world's transition to the new conditions.

THE SITUATION OF COVID-19 IN SPAIN AND BARCELONA CITY AS TOURIST DESTINATIONS

Spanish borders were closed for two months and then reopened. Different measures have been implemented in Spain in response to covid-19 such as limitations on travel rights, closure of commercial establishments, and the prohibition on holding events. Teleworking was also encouraged, and distance learning was promoted. Consumers and companies had to adjust their behaviour. Data reflect the impact of this crisis in the tourism industry (Table 2).

Although countries and international organizations have implemented a variety of measures to alleviate the socio-economic impacts of covid-19 and to stimulate the recovery of tourism, the significance of the crisis requires extra efforts and extended support. Governments have in general been swift to mitigate the impacts on the sector and good examples of support packages already exist.

Table 2. Spanish tourism indicators (Year to year percentage change)

Indicators	Jan	Feb	Mar	Apr	May	Jun	Jul	Aug	Sep	Oct	Nov	Des	YTD
International Tourist Arrivals	-1	-1	-64	-100	-100	-98	-75	-76	-87	-87	-90	-85	-77
Occupancy Rates	61	67	24	9	10	19	35	40	27	35	34	26	32
Hotel Searches	5	-6	-64	-89	-87	-69	-52	-50	-75	-50	-52	-50	-56
Hotel Booking	3	3	-73	-97	-95	-78	-72	-80	-86	-80	-80	-78	-66
Short-term rentals	24	34	2	-36	-39	-37	-25	-29	-39	-39	-39	-29	-14

Source: (UNWTO,2020c; INE, 2020)

Table number 2 provides an overview of monthly change of international tourist arrivals, occupancy rates, hotel searches, hotel booking and short-term rentals in Spain during 2020. Data express year to year percentage chance. Table 2 illustrates the dramatic decrease in most indicators all year. The majority of the indicators are relatively low compared to 2019. It can be seen that all indicators fell between March and June, especially the number of international tourist arrivals. Thus, in 2019 83,509,151 international tourists visited Spain and in 2020 a total of 18,952,508, representing a reduction of 77%. Overall, we can see a clear decreased trend all year 2020.

Due to this data and the consequences of the lockdown on tourist sector, Spanish government presented an aid package that resolves around four main guidelines: easing the conditions of the temporary collective layoffs (known as ERTEs), supporting workers a business affected by the slump in activity, guaranteeing liquidity for businesses, and supporting research to find a coronavirus vaccine.

The aid package aims to:

- Provide relief of business, with a special focus on micro, small and medium-sized enterprises (MSMEs) and self-employers and reduce their operational cost.
- Facilitate access to finance and payment of credits.
- Provide fiscal stimulus by deferring and introducing flexibility mechanisms in tax payment with special focus on MSMEs and self-employers.
- Support jobs reinforced protection especially to permanent workers with discontinuous contracts.
- Protect the most vulnerable groups.

More measures to complement the aid package have been approved to mitigate the economic and social impact of covid-19 in the tourism industry. These measures are (Ministry of Industry, Trade and Tourism, 2020a; UNWTO, 2020c; UNWTO 2020d):

- Special credit line provided for the Thomas Cook crisis is enlarged and extended to all the companies and self-employed workers included in the economic sectors defined by law.
- The budget line of the Ministry of Industry, Commerce and Tourism to guarantee 50% of the credits disbursed from the ICO line, is expanded from the initial 100 million to EUR 200 million to cover the expanded financing line of up to 400 million Euros.

Measure to support the extension of the period of activity of workers with discontinuous permanent contracts in the tourism and commerce and hospitality sectors linked to tourism were approved:

- 50% bonus in Social Security contributions for tourism companies and tourism related businesses (commerce and hospitality) who start or keep the occupation of workers with discontinuous fixed-term contracts between February and June.
- Delay up 30k tax payment during six months for MSMEs and self-employed and delay in mortgage payments, easing social security contributions and allowances for employees caring for dependent relatives.

Further from fiscal and monetary policy, the Secretary of State of Tourism, in coordination with the Ministry of Health, has prepared the specification guides for the reduction of contagion by the covid-19. The guides contain service, cleaning and disinfection, maintenance and risk management specification for the different subsectors of the tourism sector. Finally, to promote domestic tourism, Spain has launched a campaign in social media: #DescubreLoIncreible (Discover the Unbelieavable) (UNTWO, 2020d).

Although EU countries will start vaccinating covid-19 on December 27, the pandemic continues to cause restrictions, including Spain. Since November 23, Spain has required a negative PCR test for all passengers coming from European countries that exceed the limits set by EU member states in October: those exceeding 150 cases per 100,000 inhabitants or the 50 cases and a positivity rate of more than 4% and several Autonomous Communities in Spain continue to apply confinements perimeters (Observatori de turisme, 2020).

Catalonia is an autonomous community of the north-eastern corner of Spain, designated as a nationality by its Statute of Autonomy. Catalonia consists of four provinces: Barcelona, Girona, Lleida, and Tarragona. The capital and largest city, Barcelona is the second-most populated municipality in Spain and one of the most important tourist destinations.

Pandemic has a deep impact in Catalonia as well, that's because the Catalan government approved some complementary aids. In Catalonia since November 23, the plan for the progressive opening of section activities has been applied. Currently (as in February 2021), the following apply restrictions affecting the tourism sector are:

- Partial opening of the restaurant: with a capacity of 30% indoors. Opening hours: 6 a.m. to 6 p.m. Only dinner is served from 6 pm to 9.30pm. Home delivery until 11 pm and collection until 10 pm.
- From 21/12 until 11/01, 30% capacity in both interiors and terraces. Opening hours: 7.30 am - 9.30 am to serve breakfast and from 1 pm to 3.30 pm to serve lunch. From 7 pm to 10 pm, it can only be used for taking away and home delivery will be allowed until 11 p.m.
- Fairs, congresses and conventions are cancelled

- Night curfew: between 10 pm and 6 am
- Perimeter confinement: from all over Catalonia.
- Mobility restrictions

Barcelona is a city on the coast of north-eastern Spain. It is the capital and largest city of the autonomous community of Catalonia, as well as the second most populous municipality of Spain. Barcelona is one of the world's most popular tourist destinations, attracting millions upon millions of visitors each year. Barcelona has a rich cultural heritage and is today an important cultural centre and a major tourist destination. Besides Barcelona has easy access, favourable weather conditions, attractive beaches and surrounding mountains, buzzing nightlife and tasty local cuisine.

Some characteristics of Barcelona as a tourist destination before and covid-19 (2019) (Observatori de Turisme, 2019):

- Accommodation
  - Barcelona city has 442 hotels, 290 guesthouses, 9.584 homes for tourist use and 128 youth hostels.
  - The city offers almost 150.000 beds,
  - Accommodation is mainly located in the following districts: Ciutat Vella, Eixample and Sants Montjuïc.
  - The city received 13.953.209 of tourist
  - 33.807.811 overnights
  - 66,7% of occupancy
  - 53,9% of tourist prefer 4 stars hotels
  - 2,5 average length of stay
- Transport
  - 52.686.314 passengers at Barcelona Airport, mainly from Europe commercial flights
  - 3.137.918 passengers at Barcelona Port
  - More than 4.300.000 passengers from hight-speed train Barcelona-Madrid
  - 850.343 international travellers by train
- Tourist profile
  - 51% of tourist 1st visit
  - 58,5% men and 41,5 women
  - Tourist come from the following nationalities: Spain (16,9%); United States (11,6%), United Kingdom (8%), France (7,7%), Italy (6,2%), Germany (5,5%)
  - The main purpose of the visit is Leisure (70%)
  - Most visited places are: La Rambla, Plaça de Catalunya, el Barri Gòtic, Sagrada Família, La Barcelona, el Park Güell and Passeig de Gràcia

- Evaluation (on a scale of 0 to 10):
    - Architecture: 9,24
    - Culture: 8,94
    - Entertainment: 8,51
    - Public transport: 8,49
    - Character and kindness of the local people: 8, 52
    - Bars and restaurants: 8,42
    - Shopping: 8,36
    - Accommodation: 8,24
    - Safety in the city: 8,09
    - Beaches: 7,97
    - General cleanliness: 7.86
    - Noise: 7.03

Data reflect the impact of this crisis in tourism industry in Barcelona tourist destination (Observatori de Turisme, 2020):

- Accommodation
    - On November 14 there were only 113 hotels open in Barcelona, which means 330 are closed. Today the number of hotels currently operating is lower than in early August.
    - The currently open hotel offers 25.030 beds, which is -67% compared to January 2020.
    - Opened hotels are located in Eixample and Ciutat Vella districts.
- Cultural attractions:
    - 68% of cultural attractions are open
    - Some of the most touristic cultural attractions such as Sagrada Família, Casa Batlló, or Tibidabo are closed.
- Events:
    - Fair, events and congress are cancelled since 2021.
    - The next congress scheduled for Destination Barcelona, ITWordEdu 2021, will be held on 18 and 19 February 2021 with 350 delegates.
    - There are currently 110 confirmed conferences for the coming years
- Barcelona online searches
    - Research in European markets remains well below that of the domestic market.
    - Most searches of the city of Barcelona are made from Catalonia

The tourism sector has grown significantly during the last decades and is today one of the most dynamic and fastest developing economic sectors globally. Tourism can help provide diverse employment opportunities, not only in cities but also in rural and coastal areas, where alternative opportunities are limited. The pandemic has been disruptive across branches of the tourism sectors, firms and destinations, with some parts of the sector more affected than o others, now and in the longer term. This has obvious connections with MSMEs as most firms in the tourism sectors are small in scale. In Spain, the share of tourism employment represents 13.5%. More specifically, in Barcelona

As the OECD (2020a) signals, there is a risk that otherwise solvent firms, particularly MSMEs, could go bankrupt while containment measures are in force. Some tourism companies that were viable before the pandemic, now are vulnerable, that's because SEMs may have less resilience to cope. Costs associated with prevention and changes in work processes, such as the adoption of digital tools and implementation of new operating protocols, may also be relatively higher for MSMEs. Despite the economic measures adopted by Spanish and Catalan governments in order to help tourism MSMEs the proportion of MSMEs that use bank loans rose from 6 to 37% between October 2019 and March 2020 (OECD, 2020a).

In addition to being characterised by a very large group of small and micro-businesses, the tourism sector is also highly fragmented and diverse, covering a wide spectrum of industries. The sector faces particular difficulties due to this cross-cutting, multi-level, and fragmented nature. Tourism services are often interdependent and a crisis in one sub-sector has its particularities. Ignoring these risks would be reckless. Because of the current situation, there have been calls for taking advantage of this period of stoppage to make far-reaching structural changes to the tourism sector, starting with a reflection on its sustainability.

Despite the uncertainty, one of the most likely consequences of this crisis is the bolstering of local tourism, understood as doing tourism and travelling near home. That's because, with greater social and environmental awareness (Lew, 2020), post-crisis tourists will probably choose to travel to destinations closer to their place of residence. In the context of growing insecurity and uncertainty, nearby destinations could be considered less risky by many potential tourists who, having been noticeably affected by the economic crisis resulting from the health crisis, have seen their purchasing power reduced. The managing director at the World Travels Tourism Council, Virgina Messina, believes that it could take up to 10 months for the tourism sector to recover once the virus is under control. There are still no accurate answers for the future of the pandemic and the evolution of the tourism industry. We don't know yet if tourist behaviour will permanently change. We cannot say what kind of restrictions could be adopted while ensuring safety at all times. We cannot determine if restrictions only are completely lifted once the vaccine would be implemented. (EUROSTAT, 2020) In the meantime, we can make use of the possibility to explore our own countries' touristic offerings and promote local tourism.

There may be restrictions on international (long-distance) travel, at least for a long while; in helping to reduce overall emissions, this would certainly be in line with both the promotion of more sustainable tourism and the concept of degrowth (Romagosa, 2020). Indeed, a study of Catalan government reflects that 45% of travellers expect to travel when the situation is safe (with vaccines), 59% of travellers thought of travelling during confinement, and 22% of travellers discard distant destinations (Agència Catalana de Turisme, 2020).

Re-thinking the tourism sector, rebuilding destinations and tourism system, and innovating in tourism are priorities and challenges for the future. This crisis is an opportunity to transform the relationship of tourism with nature, climate and the economy. It is time to rethink how the sector impacts our natural resources and ecosystems, building on existing work on sustainable tourism; to examine how it interacts with our societies and other economic sectors; to measure and manage it better; to ensure a fair distribution of its benefits and to advance the transition towards a carbon neutral and resilient tourism economy (Bhuiyan et al., 2020). Again, a post-pandemic situation will require sustainable tourism development based on tourists' behaviour, destination choice, and tourism planning and strategies in new consequences.

Spain prepared a Tourist Recovery Plan, based in 4 pillars (OECD, 2020a):

- Health – is essential to recover both domestic and international tourism demand, to develop socio-sanitary specifications to ensure that tourist destinations are safe and perceived to be safe.
- Support – several measures developed with other government ministerial departments (Finance, Economic Affairs, Labour, etc.), will provide financial, economic and social support for tourism companies and workers, notably affected by this crisis.
- Knowledge – the national government want to improve the tourism knowledge model, improving data processing, creating new indicators and designing new observation mechanisms.
- Promotion – Spanish government is developing national and international promotion campaigns to activate demand at the appropriate time.

METHODOLOGY

The main objective of this is study is to explore the effects and impact of the pandemic in the tourist sector in Barcelona. This destination has a larger number of touristic companies located at. According to the International Labour Office (2017), the main branches in tourism are the following: accommodation, food service, travel agencies and leisure and entertainment tourist attractions. Barcelona City Council offers some data related to tourist companies according to these branches:

Table 3. Companies in Barcelona (2019)

	No Employees/ Self-employers		Microbusiness		Small Companies		Medium Companies		Large Companies		Total
			1-9 Employees		10-49 Employees		50-249 Employees		> 250 Employees		
Accommodation	365	29%	613	49%	207	17%	52	4%	16	1%	1253
Food service	2831	29%	6404	64%	609	6%	76	1%	12	0%	9932
Travel agencies	664	57%	412	35%	63	5%	21	2%	1	0%	1161
Sport and leisure	612	43%	588	41%	158	11%	42	3%	19	1%	1419

Source: (Barcelona City Council, 2019)

These data compile not only the touristic sector but all sectors, that's why the authors of this study decided to focus on travel agencies, in which self-employers or MSMEs are the main kinds of companies. Many researchers also adopted a similar approach and found it useful as well as time and cost-saving (Duarte Alonso et al., 2018; Shafi et al. 2020). The sample population was conducted on 358 travel agencies located in Barcelona, which offer leisure experiences in the city such as tours, routes or visits to cultural and touristic attractions. Specifically, the sample consists of 108 tourist MSMEs, and 250 tour guides. To increase the response rate, it was decided to utilize the directory of Barcelona city Council and Tourism register of Catalonia provided by the Ministry of Business and Knowledge. We reached out to potential enterprises through a variety of social media platforms and email.

The questionnaire items were pretested among a small sample of business owners and academics to evaluate the clarity and relevance of the questionnaire items to identify and eliminate any potential issues, before formally launching the survey. The data were collected through a snowball sampling technique as it is widely used (Bo Liu et al., 2014; Shafi et al. 2020). We collected data during the period from September to November 2020 and obtained 194 responders. A descriptive analysis approach was employed for data analysis.

The survey included several questions regarding basic information about enterprises and their characteristics, the impact of the covid-19 outbreak on business, decline in sales and profit, survival period and appeal to the government to ease the burden. The questionnaire comprised a total number of 17 Likert scale questions, organized into 5 sections. The first section focuses on the consequences of covid-19 in the enterprise and employees; the second is about the consequences on tourist services, that enterprises provided; third part asks about the security measures adopted to avoid covid-19 by the organizations; the fourth section focuses on new virtual touristic products as new tourist product; the last part of the questionnaire focuses on the future viability of the business.

FINDINGS AND DISCUSSION

A vast majority of participants are 91% microbusiness and self-employers, followed by medium companies (6%) and small companies (3%). Microbusiness has fewer than 10 employees, small companies have 11-50 employees and mediums have between 51 and 250 employees. Around 90% of the responders are either owners of MSMEs or working in managerial positions.

Problems Faced by Micro, Small and Medium-Sized Enterprises (MSMEs) due to Covid-19

As, during the crisis, most businesses face a cash-flow shortage, we asked the participating enterprises to share their strategies to overcome the cash-flow shortage. The results show that 75% of the enterprises choose not to lay off employees, 19% choose to lay off more than 80% of employees, and 6% choose to lay off between 1 and 20%. Spanish government approved ERTE (Temporary collective redundancy Scheme), which could be a suspension of the contract or reduction in working hours by the appropriate percentage. Results show 41% of the participants impose ERTE. Considering the gravity of the outbreak, laying off employees was already expected. A recent study reveals that many small businesses are temporarily closed and have laid off their employees by 40% relative to January (Shafi et al. 2020).

Table 4. Degree of impact of Covid-19 by service

	Regular Tours	Private Tours	Educational Visits	Activities/Travel Incentive	Other Services
> 80%	66%	66%	66%	66%	66%
50% - 80%	9%	19%	12%	6%	9%
20% - 50%	19%	6%	19%	19%	19%
5- 20%	3%	3%	3%	3%	3%
Not cancelled	3%	6%	0%	6%	3%

As expected, covid-19 outbreak has a significant impact on MSMEs operating in Barcelona. Over 100% of the enterprises indicated that their businesses have been affected in some form by the ongoing coronavirus outbreak or due to lockdown. 91% of enterprises reveal a dramatic decrease (more than 80%) of sales compared to the previous year.

We asked specifically what services are more affected by cancellations or what kind of services have been locked down. As many of participants are a tour operator or tour guides, responders indicate all services were cancelled. Services provided by these enterprises are regular or private tours, visits, activities or travel incentive or other services. Table 4 exposes the results of cross-tabulation. Participants indicated 94% of services were cancelled. These findings indicate that most of the participating enterprises are struggling to survive.

Get Back to Work

In terms of security and safety, 66% of enterprises don't encourage employees to do a course in order to adopt security measures in the office. When we asked about measures adopted in the office to face covid-19, participating enterprises reported that they provide a mask for employees for free, hydroalcoholic gel, protective desk screens, reinforcement of the cleaning and disinfection protocols and teleworking.

Figure 4. Security measures

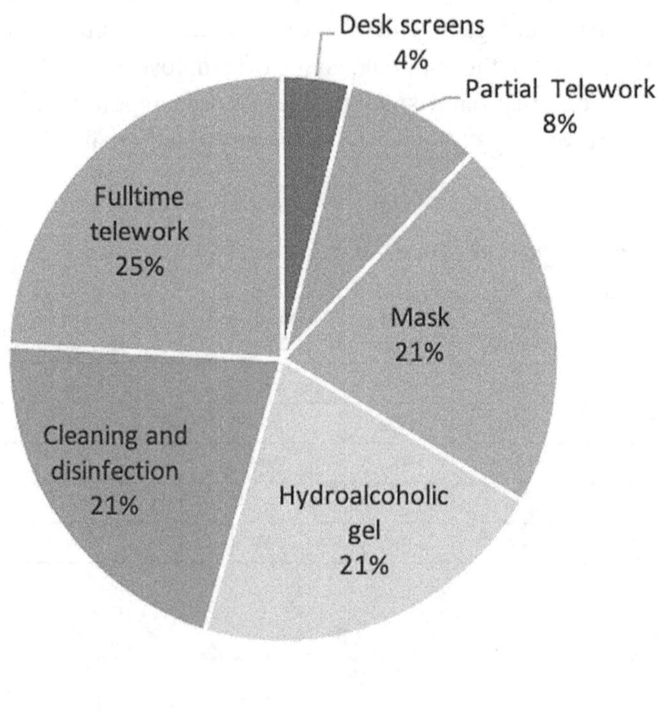

Fulltime telework began as a result of the pandemic and security reasons. Now 25% of employment is telework. Telework has grown very slowly in Spain in recent decades. Indeed, the implementation of teleworking practices in Spain is among the lowest in Europe. With the outbreak of the pandemic, telework has reached a tipping point as more companies, especially large enterprises and institutions, have introduced this work system to keep their employees safe while ensuring the continued delivery of critical services (European Commission, 2020). Five factors that influence telework: job factors, which include technological infrastructure; organizational factors; family or home factors; individual factors; and safe, environmental and legal factors Belzunegui-Eraso & Erro-Garcés, 2020). A positive impact of telework on worker efficiency has been tested. Although work efficiency may also decrease as telework reduce the number of in-person interactions. Also, telework can directly reduce costs by reducing office space and equipment (OECD, 2020b).

Spain made masks mandatory in public. In addition, the mask is mandatory as well in office when social safety distance is not possible. Some (21%) enterprises bring a mask and hydroalcoholic gel to employees to prevent the virus. Therefore, the other enterprises don't bring masks and hydroalcoholic gel to employees for free consequently, employees provide their own masks and hydroalcoholic to work. Only 4% of enterprises installed desks screens in offices. In order to prevent infection in the workplace, 21% of enterprises increase clean and disinfection. Cleaning helps to reduce the risk of spreading infection and the number of germs of surfaces, but cleaning alone does not kill germs. Disinfection means using chemicals to kill germs on surfaces, therefore disinfection reduces the risk of spreading infection. Nevertheless, this process does not necessarily clean dirty surfaces or remove germs, but by killing germs that remain on surfaces after cleaning. Therefore, cleaning before disinfection is very important as organic matter and dirt can reduce the ability of disinfectants to kill germs.

Strategies Adopted by MSMEs

The enterprises have chosen a different variety of strategies to curb the business crisis. Particularly, 66% of enterprises have shut down the business completely, while 34% have partially closed their business or offer virtual tourist experiences. In other words, 88% of enterprises don't offer services that they used to offer before covid-19.

Innovation is a crucial factor in enhancing the service quality of the tourism industry. The development of the Internet has brought never-before seen transformation and unparalleled opportunities to the tourism industry Current developments indicate that the Internet is a source of information and travel experience sharing. Many researchers in tourism literature argued that the virtual travel experience of exploring a destination can positively influence actual visitation (Chiao et al., 2018; An et al.,

2021). Tour guiding plays a significant role in determining whether or not tourists obtain meaningful and enjoyable visiting experiences.

46% of enterprises made 1-5 virtual tours, 45% made between 6 and 20, and finally, 9% of responders made more than 20 virtual tours. The 8% of virtual tours have been published on Instagram (80%) and Youtube (18%), and consequently, access was free. On the other hand, 97% of non-free tours don't refund the price of tickets. These tours were target to an international audience or both, national (15%) and international (85%). As social media is free and there are many profiles, virtual free tours had many attendances. According to the results, 34% of tours had between 50-150 attendants and 33% of virtual tours had between 31 and 50. 33% of tours had between 2 and 30 participants. Most of these virtual tours are presented in English (32%). There are also some virtual tours in Spanish (21%), Catalan (28%), and other languages (19%).

Appeal for Assistance

When asked about the assistance required from local, Catalan or Spanish government, over the 38% participating enterprises indicated the need for the provision of cash-flow and ask for grants and subsidies, 17% suggest an improvement of communication of new measures, 15% want an improvement of the application process for grants, while 23% asked for a practical, no theoretical, tourist recovery plan with economic strategy. Finally, 6% expressed their interest in doing courses and training.

CONCLUSION

This pandemic has several effects on the tourism sector. In Barcelona, the tourism sector represents around 12% of its GDP and creates roughly 9% of the city's jobs. This study has been conducted to examine the impact of covid-19 on touristic MSMEs operating in Barcelona. Findings reveal several issues faced by MSMEs and self-employers due to the current situation.

Tourism is a fast-evolving and competitive service sector requiring adaptability to market trends and demand. According to the survey, many MSMEs have been several affecting. Companies are facing a variety of issues such as a more than 70% decrease in demand, 80% reduction in sales or benefits.

Financial help from the government to small business proved partial and insufficient. The bailouts have highlighted the increasing reliance of the market on the state and the dependence of the companies on aids and grants to sustain. 41% of MSMEs waits for the recovery to continue and apply for ERTOS, others cannot stay in this situation and must close the business. Conventional policy measures may not be able to overcome this crisis, as companies are not adapting the business to the new situation.

This study shows that not all companies are working towards that holistic and innovative mindset. Only 34% of organizations offer new touristic products, hence results reveal a lack of innovation and resilience to adjust to a new way of life and businesses. Authors who have investigated risk issues in tourism and especially covid-19 indicate that an important part of the population will need security measures to restart their activities and tourist-cultural visits internationally. Nevertheless, according to this study, only 15% of companies offer products for national tourism. For this reason, companies must offer new tourism products aimed at local visitors and should look for inexpensive formulas that allow cancellations and refunds, since 97% do not have cancellation or ticket return policies. Otherwise, many people could be laid off and touristic companies could disappear in the coming months.

This crisis has revealed the need to rethink tourism. National and Catalan governments should plan a new strategy for the tourism sector, not only based on subsides but to face a new reality. This plan must include digitalization and innovation (UN, 2020). The recovery of tourism depends on the capacity to take advantage of technology to better understand travellers and trends; create innovative experiences and use a digital platform to enhance the competitiveness and agility of MSMEs to reach customers.

Re-opening of borders and tourism activities are risky as they led in some places to rising infections, leading to new localised restrictions. National and Catalan governments need to plan a new tourism strategy based on national or domestic tourism for the imminent future. In other destinations, like China, the tourism industry has been gradually and partially regaining its momentum in closer-to-home leisure destinations during weekends (Richards, et al., 2020).

Covid-19 is not only a risk but an opportunity to start a new strategy based on sustainability and its contribution to SDGs (UN, 2020; Romagosa, 2020). The Spanish government have a great responsibility to redirect tourism towards a sustainable and resilient profile. This support should be responsible to people, the planet and the local economy. To sum, the new strategy needs to fit tourism for a future, that is constantly changing and full of challenge.

Like many other studies, this presents some limitations such as geographical and temporal. These limitations do not detract from it, what we invite is to expand this study both in the typology of companies and also in the analysis of their forms of communication. This chapter contains a survey focuses on SMEs in Barcelona. This study could be completed with a larger sample and compare results. Moreover, we should wait to see how this pandemic and crisis affect in a long term.

REFERENCES

Agència Catalana de Turisme. (2020). *Informació de deguiment de la COVID-19 en la activitat turística de Catalunya.* http://act.gencat.cat/informacio-seguiment-activitat-turistica-covid-19/

An, S., Choi, Y., & Lee, Ch. (2021). Virtual travel experience and destination marketing: Effects of sense and information quality of flow and visit intention. *Journal of Destination Marketing & Management, 19*, 1–10. doi:10.1016/j.jdmm.2020.100492

Banerjee, S., & Chua, A. (2020). How alluring is the online profile of tour guides? *Annals of Tourism Research, 81*, 1–11. doi:10.1016/j.annals.2020.102887

Barcelona City Council. (2019). *Statistics of enterprises and establishments (DIRCE).* https://www.bcn.cat/estadistica/angles/dades/economia/dirce/a2019/t111.htm

Belzunegui-Eraso, A., & Erro-Garcés, A. (2020). Teleworking in the context of the Covid-19 crisis. *Sustainability, 12*(9), 2–18. doi:10.3390u12093662

Bhuiyan, A. M., Hassan, S., Darda, M. A., & Habib, M. W. (in press). *Aspects of Sustainable Tourism Development and COVID-19 Pandemic.* https://www.preprints.org/manuscript/202008.0418/v1

Bo Liu, H., McCarthy, B., Chen, T., Guo, S., & Song, X. (2014). The Chinese wine market: A market segmentation study. *Asia Pacific Journal of Marketing and Logistics, 26*(3), 450–471. doi:10.1108/APJML-07-2013-0089

Chiao, H., Chen, Y., & Huang, W. (2018, November). Examining the usability of an online virtual tour-guiding platform for cultural tourism education. *Journal of Hospitality, Leisure, Sport and Tourism Education, 23*, 29–38. doi:10.1016/j.jhlste.2018.05.002

Cohen, E. (1985). The tourist guide: The origins, structure and dynamics of a role. *Annals of Tourism Research, 12*(1), 5–29. doi:10.1016/0160-7383(85)90037-4

Duarte Alonso, A., Sakellarios, N., Alexander, N., & O'Brien, S. (2018). Strengths, innovation, and opportunities in a burgeoning industry: An exploratory study. *Asia Pacific Journal of Marketing and Logistics, 30*(2), 276–296. doi:10.1108/APJML-05-2017-0105

European Commission. (2020). *Science for Policy Briefs. Telework in the EU before and after the COVID-19: where we were, where we head to.* https://ec.europa.eu/jrc/sites/jrcsh/files/jrc120945_policy_brief_-_covid_and_telework_final.pdf

Eurostat. (2020). *Data on tourism.* https://ec.europa.eu/eurostat/web/tourism/data/database

Haleem, A., Javaid, M., & Vaishya, R. (2002). Effects of COVID-19 pandemic in daily life. *Current Medicine Research and Practice, 10*(2), 78–19. doi:10.1016/j.cmrp.2020.03.011 PMID:32292804

Houge, S., & Raymond, E. (2020). A conceptual model of adventure tour guide well-being. *Annals of Tourism Research, 84*, 1–4.

INE. (2020). *Resultados mensuales detallados.* https://www.ine.es/jaxiT3/Tabla.htm?t=13884

International Labour Office. (2017). *Pautas de la OIT sobre trabajo decente y turismo socialmente responsable.* ILO.

Larsen, S., Doran, R., & Wolff, K. (2017). How psychology can stimulate tourist experience studies. In N. Scott & J. Gao (Eds.), *Visitor experience design* (pp. 13–29). CABI. doi:10.1079/9781786391896.0013

Lew, A. (2020). *How to create a better post-COVID-19 World.* https://medium.com/new-earth-consciousness/creating-a-better-post-covid-19-world-36b2b3e8a7ae

Li, J., Feng, Y., Li, G., & Sun, X. (2020). Tourism companies' risk exposures on text disclosure. *Annals of Tourism Research, 84*, 1–15. doi:10.1016/j.annals.2020.102986 PMID:32834224

Li, J., Nguyen, T., & Coca-Stefaniak, J. A. (2021, January). Coronavirus impacts on post-pandemic planned travel behaviours. *Annals of Tourism Research, 86*, 102964. doi:10.1016/j.annals.2020.102964 PMID:32836567

Ministry of Industry, Trade and Tourism. (2020a). *Good practice guide for commercial sector establishments.* https://www.mincotur.gob.es/en-us/COVID-19/Paginas/COVID-19.aspx

Ministry of Industry, Trade and Tourism (2020b). *Guides for reducing spread of coronavirus SARS-COV-2 in the tourism sector.* https://www.mincotur.gob.es/en-us/COVID-19/Paginas/Guias-sector-turistico.aspx

Observatori de Turisme. (2019). *Barcelona, 2019. Tourism activity report.* https://www.observatoriturisme.barcelona/sites/default/files/IAT19.pdf

Observatori de Turisme. (2020). *Perspectives d'activitat turística a la Destinació Barcelona.* https://www.observatoriturisme.barcelona/sites/default/files/201218_Perspectives%20d%27activitat.pdf

OECD. (2020a). *Tourism Policy Responses to the coronavirus (COVID-19).* https://www.oecd.org/coronavirus/policy-responses/tourism-policy-responses-to-the-coronavirus-covid-19-6466aa20/

OECD. (2020b). *Productivity gains from teleworking in the post COVID-19 era: How can public policies make it happen?* https://www.oecd.org/coronavirus/policy-responses/productivity-gains-from-teleworking-in-the-post-covid-19-era-a5d52e99/#section-d1e675

Qiu, R., Park, J., Li, Sh., & Song, H. (2020). Social costs of tourism during the COVID-19 pandemic. *Annals of Tourism Research*, *84*, 1–14. doi:10.1016/j.annals.2020.102994 PMID:32834228

Reddy, M., Boyd, S., & Nica, M. (2020). Towards a post-conflict tourism recovery framework. *Annals of Tourism Research*, *84*, 15–30. doi:10.1016/j.annals.2020.102940 PMID:32367898

Richard, T. R., Park, J., Li, S., & Song, H. (2020). Social costs of tourism during the COVID-19 pandemic. *Annals of Tourism Research*, *84*, 1–14. PMID:32834228

Ritchie, B. W., & Jiang, Y. (2019). A review of research on tourism risk, crisis and disaster management: Launching the annals of tourism research curated collection on tourism risk, crisis and disaster management. *Annals of Tourism Research*, *79*, 1–15. doi:10.1016/j.annals.2019.102812

Romagosa, F. (2020). *The COVID-19 crisis: Opportunities for sustainable and proximity tourism.* Tourism Geographies An International Journal of Tourism Space, Place and Environment.

Romao, J. (2020). Tourism, smart specialisation, growth, and resilience. *Annals of Tourism Research*, *84*, 31–46. doi:10.1016/j.annals.2020.102995 PMID:32834229

Shafi, M., Liu, J., & Ren, W. (2002). Impact of COVID-19 Pandemic on Micro, Small, and Medium-Sized Enterprises operating in Pakistan. *Research in Globalization*, *2*, 1–4.

Tsaur, H., & Teng, H. (2017). Exploring tour guiding styles: The perspective of tour leader roles. *Tourism Management*, *59*, 438–448. doi:10.1016/j.tourman.2016.09.005

UN. (2020). *Policy Brief: Covid-19 and Transforming Tourism.* https://www.un.org/sites/un2.un.org/files/sg_policy_brief_covid-19_tourism_august_2020.pdf

UNCTAD. (2020). *World Investment Report.* https://unctad.org/webflyer/world-investment-report-2020

UNWTO. (2020a). *International tourist numbers could fall 60-80% in 2020, UNWTO reports.* https://www.unwto.org/news/covid-19-international-tourist-numbers-could-fall-60-80-in-2020

UNWTO. (2020b). *Tourism and COVID-19, Issue 1. How are countries supporting tourism recovery?* https://www.e-unwto.org/doi/book/10.18111/9789284421893

UNWTO. (2020c). *International tourism and COVID-19.* https://www.unwto.org/international-tourism-and-covid-19

UNWTO. (2020c). *Covid-19 measures to support the travel and tourism sector.* https://webunwto.s3-eu-west-1.amazonaws.com/s3fs-public/2020-06/europe.pdf

Weiler, B., & Black, R. S. (2015). The changing face of the tour guide: One-way communicator to choreographer to co-creator of the tourist experience. *Tourism Recreation Research, 40*(3), 364–378. doi:10.1080/02508281.2015.1083742

Weiler, B., & Kim, A. K. (2011). Tour guides as agents of sustainability: Rhetoric, reality and implications for research. *Tourism Recreation Research, 36*(2), 113–125. doi:10.1080/02508281.2011.11081313

Wolff, K., Larsen, S., & Øgaard, T. (2019). How to define and measure risk perceptions. *Annals of Tourism Research, 79*, 1–9. doi:10.1016/j.annals.2019.102759

ADDITIONAL READING

Belzunegui-Eraso, A., & Erro-Garcés, A. (2020). Teleworking in the context of the Covid-19 crisis. *Sustainability, 12*(9), 2–18. doi:10.3390u12093662

Lew, A. (2020). How to create a better post-COVID-19 World. https://medium.com/new-earth-consciousness/creating-a-better-post-covid-19-world-36b2b3e8a7ae

Qiu, R., Park, J., Li, Sh., & Song, H. (2020). Social costs of tourism during the COVID-19 pandemic. *Annals of Tourism Research, 84*, 1–14. doi:10.1016/j.annals.2020.102994 PMID:32834228

Richard, T. R., Park, J., Li, S., & Song, H. (2020). Social costs of tourism during the COVID-19 pandemic. *Annals of Tourism Research, 84*, 1–14. PMID:32834228

KEY TERMS AND DEFINITIONS

Economic Impact: Effects on the economy are either direct or indirect. Economic impacts associated with tourism arise as a result of the demand and supply relationships in the industry, the associated visitor and investment expenditure patterns that they stimulate, and the structure of the economy.

ERTE: Temporary lay-off plans. ERTE allows companies to temporarily send workers home or reduce their working hours and readmit them on their previous conditions after a certain amount of time.

MSME: Micro, small, and medium-sized enterprises.

OCDE: Organisation for Economic Co-operation and Development is an intergovernmental economic organisation with 37 member countries, founded in 1961 to stimulate economic progress and world trade.

Pandemic: An epidemic occurring worldwide, or over a very wide area, crossing international boundaries and usually affecting a large number of people.

Tourist Destination: Is a unique place where a visitor spends at least one night and exhibit tourism products such as attractions, support services, and tourism resources complete with defined management, physical and administrative boundaries, and a well-known image.

UNWTO: The World Tourism Organization (UNWTO) is the United Nations agency responsible for the promotion of responsible, sustainable and universally accessible tourism.

Chapter 4
Public Health Risk Environment for Bulgarian SMEs (Guest Houses and Family Hotels) in the COVID-19 Pandemic

Teodora Kiryakova-Dineva
https://orcid.org/0000-0002-1374-9999
South-West University Neofit Rilski, Blagoevgrad, Bulgaria

Ruska Bozhkova
South-West University Neofit Rilski, Blagoevgrad, Bulgaria

ABSTRACT

At a time of the global health pandemic, the most affected areas are economy and social life. Along with the practical limitations of travel, regarding personal security reasons and the objective risks for the environment, the world of tourism has changed. However, under the circumstances, some small accommodation units have managed to survive, like the Seamen between Scylla and Charybdis – the mythical situation. The purpose of this chapter is to delve into the public health risk environment for Bulgarian SMEs in tourism (guest houses and family hotels) during the COVID-19 pandemic. The extent of the analysis includes hotels and guest houses in the south-western part of Bulgaria that managed to keep operating despite the global pandemic situation.

DOI: 10.4018/978-1-7998-6996-2.ch004

INTRODUCTION

The world economy uses a rich terminology and a variety of applied apparatus for determining the economic growth of SMEs (Nesher, 1971; Zuvekas, 1979; McCartney, 2015; Thomas 2020). Some of these approaches analyze the development in view of specific touristic activity (Williams and Shaw, 1998; Mazzanti, 2002; Brau, Lanza and Usai, 2008) and others relate it to international economy and tourist flows (Candela and Figini, 2012; Cooper and Hall, 2019; Aliyev and Ahmadova, 2020). Prognostics and innovations in tourism are other well-known lines of research which are fully applicable for the scope of SMEs growth and sustainability (Frechtling, 2001; Li, 2006; Dimitrov, 2010; Elhadary, 2018).

The World Tourism Barometer (2019) monitors the number of visits by foreign tourists worldwide and identifies emerging and leading destinations. For example, in 2019, there were no significant changes in the preferences of tourists regarding travel choices. In the same way, Europe continues to be the most visited continent in the world, and data on the number of trips of different groups of nationalities and regions with the highest growth scores can be specified. The state of Bulgarian tourism is also most accurately represented through official statistics of the Bulgarian National Statistical Institute. The same database refers to the International Relations of the SMEs in tourism, and reflect the trends identified by both the World Tourism Organization and the European Tourism Commission. Several national strategies are being developed and constantly updated (National strategy for sustainable development of tourism in the Republic of Bulgaria 2014-2030 and its Action Plan; a certain number of Regional strategies for tourism development). The importance of the small and medium enterprises in tourism (SMET) is mainly visible in the sections that address the managerial approaches to different tourist activities. In this way, there are the prerequisites in order to comply with the main recommendations for inclusive growth of the Bulgarian tourism industry and for ensuring relevance to the new trends of the global tourism (Talwar, 2010). Roads to success can be found by strengthening the sub-regional specifics provided by the SMETs in the economic sector and following the dynamics of key factors (capacity building, management, sustainable growth). The Bulgarian SMEs in tourism are constantly adopting actions for acquisition of greater revenues to the gross domestic product and their economic role in the last few decades is significantly larger, since Bulgarian tourism developed more or less in the regime of the centrally-planned economy. The extent of the restructuring of the tourism industry gave rise to continued expansion and diversification in tourism, making it one of the fastest growing industries analogous to the rest of the world economy. In the overall process of the industrial development, Bulgarian SMETs are greatly dependent on external factors. Moreover, since only recently they gained economic relevance and they became more open to hazards, risks and disasters.

In fact, there are several studies that claim tourism is in a process of constant growth (Ivanov and Webster, 2011; Aliyev and Ahmadova, 2020), and the particular studies contribute to the competitiveness of the sector. As a result, there are intensified interrelations, among which the improvement of production processes (Pett & Wolff, 2009) as being the outstanding one, the good orientation of the family business and the inheritance of SMEs as a second (Getz, & Carlsen, 2005; Cates, & Sussman, 1993; Kyurova & Koyundzhiyska-Davidkova, 2017). Third, an additional impetus is given by the new technologies and the way information is speeded and handled (Denchev et al., 2020; Vasenska et al., 2021). The emergence of the global tourist, but also the fact that the tourist of modern times already has a lot of data available, and can characterize the trip individually, is also of great importance.

The tourism product becomes a very broad concept, in which many and essentially different activities are integrated (Benur & Bramwell 2015; Ritchie & Jiang, 2019). Combining more than one type of tourism, is already an imperative when choosing a trip. On the other hand, tourist motivation and tourist demand are highly sensitive to the intangible assets of the destination. In this sense, the list of factors traditionally contributing to the development of tourism, including the entry of new generating markets, visible and invisible competition between destinations, the transport facilities and the possibility of low-cost airlines, also have a positive effect on national and regional tourism. In its expansive development, tourism has only one Achilles' heel: the security factor. This factor is a weakness in some situations (Pizam & Mansfeld, 2006), and in others it is interpreted in the scope of threats (Paraskevas & Arendell, 2007; Baker, 2020).

So far, the main threats and risks that the tourism industry faces have been wars (Singh, 1998; Lisle, 2016), looting, riots (Fallon, 2004), and natural disasters (Abd-Elhamid, Fathy & Zeleňáková, 2018) such as earthquakes (Mazzocchi & Montini, 2001; Henderson, 2013), floods (Kellens et al., 2011) and devastating fires (Hystad & Keller, 2008; Sanders & Laing, 2009). There have also been threats to national and international tourism associated with social (Jackiewicz, 2005), political (Papatheodorou, & Pappas, 2017) and religious (Raj & Griffin, 2017) factors as well as risks related to migration flows and refugees (Krasteva & Pantelis, 2017). Not only are SMETs affected negatively by the different types of threats, but also the whole tourism industry.

In order to analyse the state of the SMEs (guest houses, family houses) in the Bulgarian tourism during the COVID-19 pandemic, the present chapter applies a twofold analysis. Firstly, through an exploratory research, main and secondary resources related to the existence and survival of Bulgarian SMEs in the regional tourism industry (two main legislatives acts in particular) are going to be presented.

The assessment of the public health risk environment in which these SMEs struggle to survive belongs to the scope of the secondary resources.

Acknowledging the obtained data from official documents (legal sources) and from the practical point of view (conducted interviews) the authors shall discuss the preconditions for the survival of the Bulgarian regional SMEs in terms of hospitality. The next part of the analysis is focused on the observations based on in-depth interviews with tourists visiting regional guesthouses and small family hotels. Furthermore,_some insights from additional interviews with stakeholders and practitioners in the regional tourism industry are going to be discussed. In such a deductive approach, the domain of public health risk environment is going to be defined and presented as a key area for future interventions. In this environment, some common principles are becoming obvious and interesting interrelations for achieving security and sustainability can be revealed. A possible way to get oriented in the unpredicted situation of public health risk has been identified.

BACKGROUND: MAIN FEAUTURES OF THE BULGARIAN SMES IN TOURISM AND CONDITIONS FOR SURVIVAL IN TIMES OF PUBLIC HEALTH THREATS

Main Legislative Regulations for the Function of the SMEs in the Bulgarian Tourism Sector

The Bulgarian legislation, as well as other international norms, apply a definite system for classification of accommodation. Bulgarian legislation through the basic Law on Tourism defines three main places for accommodation, which it classifies by classes. Class "A" includes hotels, motels, apartment tourist complexes, holiday villages, tourist villages and villas; Class "B" includes family hotels, hostels, holiday resorts, guest houses, bungalows and campsites; Guest rooms and guest apartments for class "C" (Tourism Act Article 2 and Article 123).

The number of stars that one type of accommodation can receive is also definitely determined. For example, holiday villages and villas can receive from three to five stars, motels from one to three, tourist villages from two to four stars, motels from one to three stars. Depending on their category, hotels can receive from one to five stars. Each star corresponds to a certain category. In Bulgaria, this classification is certified by the State Tourism Agency.

The categorization in class "B" is interesting. In this group, the accommodation units are classified in nine categories. Eight of the categories can receive from one to three stars. These are family hotels, hostels, holiday resorts, guest houses, bungalows and campsites. Boarding houses are categorized with only one or two stars.

Additionally, the types of hotels are determined according to the period of operation (year-round hotel or seasonal hotel); territorial location (sea hotel; mountain-hotel; city hotel; suburban hotel), and operational purposes (apartment (studio) hotel, hotel residence, business hotel or "balneo hotel", namely hotel with certified medical SPA area; SPA hotel - hotel with certified SPA area; wellness hotel - hotel with certified wellness area or thalassotherapy hotel - hotel with certified thalassotherapy area).

A closer look at both the Bulgarian and international certification bodies can hardly indicate the existence of norms that classify the accommodation facilities according to their size or number of rooms. In Europe, the size is more modest than in the United States where, not surprisingly, hotels with several thousands of rooms crowded on the shores of Miami or Las Vegas can have less stars. As a rule, the European small hotel has no more than 50 rooms, while its American counterpart with 150 rooms is also considered a very small accommodation unit. On the other hand, discussing the role of SMEs in tourism, it is necessary to take a closer look at another feature, namely the criteria for managing and operation of hotel business. In the Bulgarian Tourism Act, it is said that the hotel or restaurant business on the territory of the Republic of Bulgaria is strictly categorized or registered under this Law, regarding tourist sites (Art. 111).

It is also determined that a hotel or restaurant business may be run by a person who is a trader within the meaning of the Commercial Law or is a legal entity who has the right under another law to carry out economic activity. Another condition is also that this person hires staff with the required education and language qualifications, and management experience is also required for the managing staff. An interesting feature is that hotel management in guest rooms, guest apartments and guest houses may be carried out by a person who is not a trader within the meaning of the Commercial Law, including the owner. In addition, Article 123 stipulates that "the specifics of the requirements for categorized accommodation and catering establishments, the procedure for determining the category, as well as the conditions and procedure for registering guest rooms and guest apartments shall be determined by an ordinance adopted by The Council of Ministers".

In order to give examples of the Bulgarian Tourism Act and its interpretation we should also add the interpretation of the term "small and medium enterprises". The definition of the term "small and medium-sized enterprises" (SMEs) varies from country to country. It also varies according to the types of industries, as in the present case, with tourism. Key indicators are the number of employees, the annual sales (turnover), the value of assets and net profit. The classification can be done by a single indicator or by a combination of indicators. As defined in the Bulgarian Small and Medium-Sized Enterprises Act, small and medium-sized enterprises include enterprises with a staff of less than 250 people, and a medium-sized enterprise is a company with fewer than 250 employees and with a turnover of less than 49, 851

172. 00 euros (97, 500 000.00 BGN) or with a balance sheet total not exceeding 42, 948 702. 00 euros (84, 000 000.00 BGN). A small business is an enterprise with no more than 50 employees, and the total balance sheet is less than 10 million euros, and the micro-business must have less than 10 employees and less than 2 million euros in the final balance sheet.

With this definition, the evaluation of SMEs in tourism is reasonably difficult, because apart from the main factor for annual turnover, value of assets and number of staff, it is difficult to find any precise limits. Again, according to the legal criteria, it is clear that SMEs in Bulgarian tourism will be the accommodation for group B. In particular, the research interest here is focused on accommodation types for class 'B'- family hotels, hostels, holiday resorts, guest houses, bungalows and camping.

There are several reasons to focus on them; at first, unlike high-end hotels, they cannot impress with service capacity, amenities and design, novelty and technology. SMEs in tourism, and specifically identified by us in group "B" can have from one to three stars. In practice, they cannot compete with Group A hotels with one to five stars, unless they rely on their uniqueness and originality. But not only this type of hotels manage to stand out in the struggle for survival and justifiably be considered prototypes of hotels of the XXI century: they are comfortable and offer services at a reasonable price, have everything you need for recreation, there is no restaurant, and usually the tourist meets a small part of the other guests within the premises. It is assumed that guests visit a restaurant that is located nearby or that they will combine their meals elsewhere, where they will combine their trip with tourist activity and service that will be tailored to their individual needs. The uniqueness of small hotel accommodation is the main tool for effective market policy in modern conditions. The capacity, volume, number of staff and category and stars discussed here can hardly be correlated with the realized quality, commitment and reaction of the staff. Ultimately, there is no guarantee that the satisfaction of tourists in a class "A" accommodation will be higher than the satisfaction of a stay in a class "B" accommodation.

This feature turns out to be verifiable (even especially profitable) also due to another coincidence, which this time is not due to the peculiarities of the tourist flow, nor to the marketing strategy or to the instruments for market economy. This is the unfavorable situation with the coronavirus, which forced people to obey to restrictive measures, forget about their vacation plans and cancel their reservations. As citizens, not tourists, people were exposed to months of social isolation. They changed their usual course of life, social contacts, and way of working. This situation of global threat to public health has changed the behavior of tourists. The current study aims to explore the characteristics that allowed the particular establishments to survive the crisis of Covid-19. These establishments are categorized as group "B", in the range of close locations for regional and national tourists. Accommodation in class "B" - guest houses, small family hotels, hostels, bungalows and campsites, presumably meets the requirements of social distance. The tourist in the era of Covid-19, in many of the options for choosing a vacation in the summer of 2020, followed the requirement to observe social distance, not because of the availability of space, but because of the possibility of overcrowding. Thus, the limited capacity in the accommodation was used as a powerful tool by small and medium enterprises in Bulgarian tourism. Small guest houses and family hotels have another advantage over large hotels; they bring tourists as close as possible to the nature and uniqueness of the place, as they reflect the spirit of the place to a greater extent. The hotel staff are local people, there is no staff who do not feel connected to the traditions and culture of the hosts, which is noticed and appreciated by guests.

The tourist in the era of Covid-19 was reoriented to nearby places and small hotels. It cannot be claimed that this was the middle-class tourist, whose stay does not require the creation of certain criteria in the arrangement of the morning buffet, the gourmet assortment in the hotel or the arrangement of the green space in the hotel. According to our observations, the tourists in 2020 were guests who needed moderate comfort, good food, cleanliness and safety in relation to public health. They managed to find all this in small family hotels, guest houses and friendly apartments for rent, which usually remain empty due to the conjuncture and the unique offers of the larger hotels with more stars. The conditions for secure and safe tourism were created on the basis of internal rules and orientation, which are not influenced by advertising or recommendation from relatives.

SMETs in Bulgarian hospitality had a hard time surviving in 2020. But they did not close their doors. Instead, they welcomed tourists who needed to go on vacation and immerse in nature. These Bulgarian SMETs have realized their advantage and built their network with future clients. The visions and opportunities for a safe and consistent with public health tourism played an important role in the pre-orientation period preceding the next tourism season.

The particular SMETs tend to offer more reasonable prices and provide a special atmosphere that is difficult to create in bigger hotels. In addition, the share of variable costs in the turnover structure of such a hotel increases significantly, as with the existing volume of the expenses it is difficult to achieve optimal prices for detergents, laundry services etc. Small hotels face great difficulties in the human resources sector, and the staff must have a number of skills to allow the hotel to remain profitable. With a small capacity, a guest house or class "B" hotel can ensure compliance with all anti-epidemiological measures more easily.

Main Features of the Bulgarian Tourism Guesthouses and Family Hotels

As mentioned above, a guest house or class "B" hotel with a small capacity, can ensure compliance with all anti-epidemiological measures. When considering the destabilization of a tourist destination, experts analyze different types of hazards and define the main types of risks. Some of the problems are justified in view of shortcomings and lack of risk management and marketing. Another direction in the study of risk determinants for Bulgarian SMETs, is linked to the geopolitical environment. A third area of identified risks emerges from several distinctions in the social and cultural environment. In brief, the variety of hazards and types of risk for the small and medium sized hospitality establishments are complex. In the present study, an attempt is made to interpret the main parameters for their activity in times of a global pandemic.

Economics literature, and in particular the studies addressing SMEs in the context of political crisis, point out that many risks are caused by legislation, regulation and deregulation, as well as by the behavior of tourists during the global pandemic. Wherever the elements of public health risk within the tourist activity are complex, the authors try to at least determine the main parameters, which most often consist of a set of unpredictable factors. Secondly, an attempt is made to outline the theoretical and conceptual frameworks of the existing and developed definitions, analyses and means for measuring the risk of threats related to Covid-19. These two primary tasks are in themselves full of ambiguity. On the one hand, there are main risk factors for the economic activity of SMETs in Bulgarian tourism, and on the other, there are not specific parameters of public health risk environment for the global tourism industry, an area which is multi-faceted but includes the transition from local to global.

The given definitions of risk to public health through the implementation of the economic activity of SMEs are determined by a number of methodological approaches, which are used to examine the activity of market players in a pandemic, including at national level, namely ministries and national security authorities. As a result of these features, it is difficult to determine the exact rules based on the decisions made upon the activities of SMEs in the field of tourism operating in conditions of risk. In this case, namely in a situation of uncertain environment for public health, the approach to setting scenarios turns out to be the most appropriate for SMEs in tourism. There are two scenarios; the first scenario is that if the SME conducts foreign economic activity, as it is usually the case with tourism, then the business is accompanied by an increased level of risk and uncertainty. Therefore, SMEs in international tourism will be forced to make a comprehensive assessment of the external environment in which they are forced to operate on a daily basis. Here the models generally have an application in the form of PEST models.

The second scenario is the scenario for pro-active solutions to avoid the risk of geopolitical proximity. Threats and uncertainty may not arise from specific events, but rather from events that are usually temporary. Overcoming them is a complex task - they can be presented quantitatively and be measured against the price basis. In our further analysis of the reasons for these risks, in our view, factors appear that counteract the current situation with a global risk of the spread of COVID-19. Scenario 2 identified by us, is only part of the possibilities for overcoming these risks. Here we define it as a pro-activity in attracting tourists from nearby regions.

In 2011 a study on the characteristics of Bulgarian SMEs identified key factors for sustainable development and competitiveness, which are defined as Access to finance; Innovations; Intellectual Property; Internationalization and Good Practices. Certain guides have been constructed and estimated for each of these factors. The aim of the econometric research is to trace the dynamics of individual groups of SMEs and to explore the factors of development and competitiveness and economic performance of enterprises in different sectors. The study found that Bulgarian SMEs mainly use financing through their own funds. Another feature is the difficult access to funding. SMEs also fear bankruptcy and do not have sufficient resources to finance the necessary investments. Again, according to the same study, the most common innovation activity in Bulgarian SMEs affects business processes in order to optimize and reduce costs. The innovation activity is identified as key risk area where urgent action is needed. At the same time, this area stands open for support by joint initiatives in the field of education between the individual enterprise and the academic institutions (Study: Analysis of the state and factors for development of SMEs in Bulgaria: 2011-2012, conducted for the Executive Agency for Promotion of Small and Medium Enterprises).

These characteristics can also be defined as areas in need of improvement and active intervention at present. The reasons for this are the financial instability, labor intensity in economic processes and return on funds, which are a direct investment of the owner of the SMEs. Bulgarian SMEs in the context of the crisis provide numerous summaries and recommendations, balance of payments analyses, inflation and the labour market. There is a proposal of business demography, manufacturing in the SMEs sector and sustainability. What this and many other studies fail to discuss, is the role of risk and unpredictability in the context of public health.

The Reconceptualization of Risk in the Scope of Public Health Risk Environment

In many studies, the term *risk* is associated with the risk of potential losses in various areas of activity (loss of material and financial resources, lack of income, unforeseen additional costs, etc.) [Bittle, 1985; Moore, 1983]. In other words, the main prerequisite for risk analysis in any economic system is its identification with a quantitative or cost hazard measure. Accordingly, the content of tourism risks covers all types of threats against participants in the tourism market. Therefore, it seems appropriate to divide the whole set of tourist risks into two groups: the risks to which tourists are exposed when planning and conducting a tour (particular tourist risks), and the risks from the activities of tourist enterprises in designing, promoting, touring and providing entertainment services (economic or business risks).The first group encompasses unfavourable events related either to material and financial losses (losses and damages to property during tourist trips, financial losses, thefts, fines), or to threats regarding the life and health of tourists. The second group includes the risks caused by the nature of tourism and recreation activities within the enterprises. These risks within the enterprises, seem more interesting for analysis due to the complexity of economic relations.

Specific characteristics of economic risks in the economic sector of tourism in general and in view of the Covid-19 crisis indicate that:

- Tourist risks are associated with damage not only to tour operators and travel agents, but also to service activity in the whole business. Also related to the risks in the tourism industry are the tourists themselves, as well as the entire public health environment;
- As in some cases tourist risks include those regarding the life and health of tourists, particularly strict requirements must be imposed on the activities during the accommodation, especially where tourism is widespread;
- Tourism risk derives from both financial and economic-related risks for the economic activities of SMEs in tourism, but still to a greater extent the latter are the natural sources of danger that do not depend on the people and operators of tourism services.
- The activities of SMEs in tourism are related to actors from the macro environment, who are also subject to mutual economic and risk of loss in compliance with public health measures in order to limit the spread of Covid-19.
- Resources for economic activity of SMEs in tourism in times of crisis are limited.

The main specific characteristics of the risks in carrying out tourist activities in a period of threat to public health, and above all, in view of the functional characteristics of the tourist complex, contain multidimensionality. SMEs in tourism are related to the conjugation of industries included in the tourism industry, which enhance the effect of tourism due to the action of the multiplier mechanism, but in times of crisis this effect is negative. The main disadvantage of the public health risk multiplier in the tourism industry is the specificity of the collapse due to the absence of any tourist flow. The lack of tourists also affects transport and ancillary activities in relation to SME. The financial loss from this type of activity is further intensified. Collective accommodation is subject to risks in recruiting and providing services due to restrictive measures and morbidity. SMEs that carry out attractions and leisure activities presumably contain a danger - the likelihood of harm to life and health in the operation of technical devices. At the same time, all enterprises in the tourism industry are affected by the risk of covid-19, which can largely be defined as universal, i.e. In case when the world has to face global financial, geopolitical, socio-cultural, behavioural and other risks.

Overview of the Practical Implication Regarding the Risks for Bulgarian SMETs (Regional Guesthouses and Family Hotels) Due to Threats in the Public Health Environment

The activity of SMEs in tourism in the conditions of a global pandemic is difficult, and the risks can be neither avoided, nor minimized. However, in the process of studying the risks to public health, the development of scenarios can be successfully applied, which would lead to a corresponding better knowledge of the risk factors The model for assessing political risks proposed by McKellar (2010: 87) is applicable in this case. The model covers three conditional stages; making a hypothesis, determining the relevant risk factors and accessing the risk. This model can also be adopted when assessing the risks to public health in the environment of a global pandemic and uncertainty of tourist behavior.

METHODOLOGY

Methodology Approaches in the Research of Risk for SMEs in Tourism Sector

This chapter is focused on examining public health risks environment for Bulgarian SMEs in tourism during COVID-19 pandemic. Thus, the present study falls in the scope of social and economic science research. Social studies, as well as business and management studies have many research directions in their scientific disputes on methodologies for business environment analysis. As regards their examination techniques, many possible approaches are conceived as being applicable, having one thing in common, namely the fact that research has to be done more or less in relation to the behaviour of tourist. Since tourism is purely social phenomenon, three main methodological techniques can be regarded as mostly commonly used. According to Veal (2018), these three types of research can be identified corresponding to the functions (i) describing what; (iii) explaining how and (iii) evaluating. As a result, descriptive, explanatory and evaluative research can be respectively found in the mainstream of the studies related to SMEs in tourism as well.

Figure 1. Conceptual framework of health risks in tourism
Source: authored model of the stages in the public health risk environment in tourism

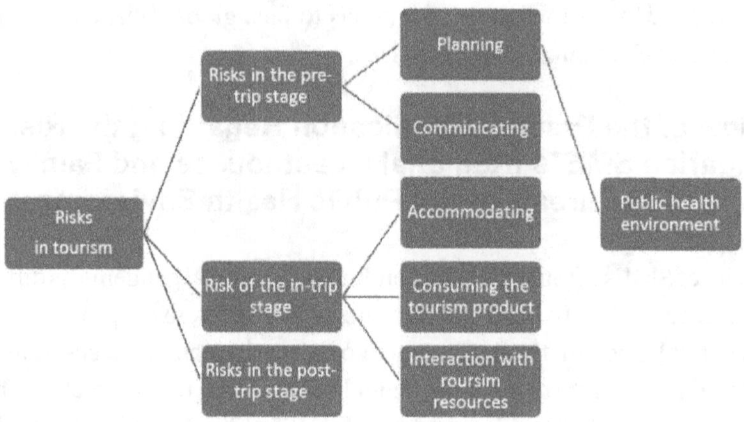

It is an old practice to use the descriptive analyses for the needs of national and regional SMEs in tourism (Akis and Warner, 1994). In such cases, researches make suggestions for the diversification and further tourist product development offered by the SMEs in tourism sector. In a similar way, the inclusion of risk for the TSMEs can be viewed in scope of the competitiveness of the organizations (An and Thi, 2012). According to Komppula (2014) and via the means of descriptive analysis, the importance of the entrepreneurial environment can be successfully related to the risk-taking of SMEs in rural tourism. When applying the explorative research method other researchers key determinants of systematic risk (Kim et al., 2012), others prove that risk can arise from fuzzy perspectives (Sheng-Hshiung et al., 1997) or even specific text disclosures (Li et. al., 2020).

Much in the same vein are studies showing results from explorative research. Exploratory studies in tourism are acknowledged to contribute most to the design of mixed methods research and to be receive more and more attention within scientific field of the leisure and tourism (Mason, P., Augustyn, M., & Seakhoa-King, 2009; Shivakumar & Agarwal, 2014; Egger, Lei & Wassler, 2020). Exploratory study approaches can be also found for the investigation of the business processes of the SMEs, including risk assessment. Interesting aspects in view of future tourism development and risk minimization can be derived in view of the brand building (Spence & Hamzaoui and Essoussi, 2010; Chung et al., 2019; Yarkova et al. 2019) and innovations (Lage & Alturas, 2013; Freitas et al., 2018; Caloffi et al., 2019). In their research Setyawan et al. (2017) reveal the disaster as business risk in SME and in the same time apply an exploratory study.

For example, in their study Braun and Hollick (2008) estimate factors as capacity building, others relate tourism growth to sustainable economic development of the SMEs (Mazzanti, 2002). The contexts of estimate research differ from application of the information and communication technologies (ICT), innovativeness and reach to the increasingly networked economy and E-Commerce Adoption. Some other research is devoted to the field of risk management during crisis. Such suggestions are also delivered in the case of technology risks and lack of innovation (Verreynne, 2019).

When dealing with the risk's evaluation, many studies discuss the financial risk as being the main type of risk, and others add to the main line of research such other risks as destination risks, geopolitical risks and cultural-religious risks. In this research the authors concern the public health risk environment and define it as a main factor for economic survival of the SME in the tourism during the coronavirus pandemic. Although there is much research devoted to COVID-19 and the tourism industries, there is still limited research addressing how health risks, i.e. COVID-19 influences the tourists' intentions and decision-making process in relation to the tourism small and medium enterprises.

Methodology of the Present Study of the Public Health Risk Environment for SMEs in the Bulgarian Hospitality Sector

Acknowledging that a variety of scientific approaches applies large-scaled samples and questionnaires, and makes use of mixed methods to study risks for SMEs in tourism so to achieve statistically grounded results and practical implications, here we opted for a descriptive analysis. Two methodological approaches were adopted with a view to reveal and discuss the most pertinent manifestations of public health risk environment, as well as its influence on the SME in the hospitality and travel industry in south-west Bulgaria during the Covid-19 crisis. Firstly, an exploratory research was carried out. This was achieved by using the means of the stationary observations of the regional SMEs (gust houses and family hotels). The stationary analysis allowed to investigate two main sub-research questions raised, namely (i) what main directions may be relevant to the activity of SMEs in tourism at regional level and (ii) what specific risks can be identified in the created public health risk environment. To this end, seven in-depth interviews were carried out with representatives of the hospitality and tourism agency practitioners' we conducted an exploratory research. The next step of the analysis estimated some possible business principles in providing tourists services in times of pandemics in order to minimize the risks in the public health environment related to the small and medium enterprises in the tourism industry in the south-west Bulgaria.

RESULTS

Main Results of the Stationary Observations of the Regional SMEs in Tourism Destination South-West Bulgaria

In Bulgaria, there are some predominantly developed forms of tourism, including seaside and recreational tourism, mountain tourism and spa tourism. The main touristic regions are the Black Sea coast, and the mountains Rila, Pirin, the Rhodope Mountains and the Balkan Mountains.

When the Covid-19 pandemic hit the world, most hotels in the native resorts were empty of tourists. The state of emergency in Bulgaria lasted from 13th March to 19th June 2020. During the quarantine, the bigger hotels in the main tourist' destinations were unprepared to reopen and serve the tourists. The lack of tourists threatened the existence of big hotels. In contrast, even during quarantine, small towns developed active rural tourism and forms of special interest tourism, such as nature-based tourism, pro-active tourism, off-road tourism, etc., therefore guest houses had high occupation rates throughout the year.

Due to the proximity of the mountain resorts and tourist centres like Bansko and Razlog, many Bulgarian and foreign tourists visited smaller villages like Banya and Eleshnitsa. To attract more tourists, locals set up guest houses with public health friendly environment and different alternative tourist activities, i.e. tasting and cooking traditional dishes, horse riding and tea picking.

Small hotels and destinations used celebrities' testimonies in their promotional campaigns. In the public there were often mentioned the opinions of social and political celebrities. This is the standpoint of a representative of the mayor administration of the village of Eleshnitsa:

Tourism both in the whole region and in Eleshnitsa shows an increase in the tourist flow, especially in the summer season, which is favorable for the village. There are currently 5 guest houses and one hotel. All of them offer up-to date tourism services and to the greatest extent comply with security measures. Social distancing is guaranteed, there are limited guests, because of the capacity of the guest houses themselves",

"The tourist risk of Covid-19 is of a high level of danger and does not depend on the organizers of the tourist activities, nor on the consumers, nor on us, the policy makers".

According to the opinions of two other tourists, the potential of smaller tourist regions Eleshnitsa, Dobrinishte and Banya, and Class "B" accommodations located in the smaller touristic accommodating areas to prosper during the health crisis:

"The price is good. Here it is relatively less risk compared to the bigger resorts Sandanski and Velingrad.",

"I prefer the place because of its friendly style. All of the measures are secured. It is close to Bansko and more accessible. Water is everywhere, there are no big hotels. We walk and we are close to nature. Two lifeguards take care of the safety of children and the elderly while they are in the water. The mineral pool will be open until September, and entrance prices will not be changed".

Much in this vein, the implementation of tourism in the world is associated with image making, marketing vision of destinations and tourist sites. As a rule, the better the hotel's image, the more tourists are inclined to look for the number of stars of the hotel, and willing to pay a premium price for their stay.

Very often, reputable high-end hotels have loyal customers who visit the same chain of hotels around the world. Too often, the hotel brand creates the misconception of reducing the variety of middle-class hotels.

The creation of hotel chains and higher-class accommodation due to its standardization in service cannot meet the different demands of tourists. This is the main reason why there are small and medium enterprises in the accommodation sector. In other words, due to the difference in the profile of today's tourist, the peculiarities of the destination and the characteristics of the tourist flows, small and independent hotels, guest houses and other accommodation from class "B" find their niche.

Identified Main Risks Attributed to the Public Health Environment of the SMEs in the Southwest Bulgarian Tourism

The analysis and assessment of the risk related to public health can be identified as major factor both for the tourism activity of the SMEs and as a component of the overall situation in the COVID-19 pandemics. After summarizing the opinions given by our respondents, six main groups of risks were identified:

1. Risk of being unable to travel in the period for which the reservation was made.
2. Risk of inapplicable customers programs, which do not take into account the components of the public health risk environment.
3. Risk in the implementation of adaptability to the environment and the situation of high instabilities in the world market due to rapid risks, including geopolitical, socio-economic and threats.
4. Risk in the work environment regarding the health of both tourists and employees.
5. Risks related to gaps in the information about predicting both national and international tourist flows.
6. Risks of increasing losses and declining profits due to reduced tourist flow.

Identified Principles on how to Deal with Risks Attributed to the Public Health Environment of the SMEs in the Southwest Bulgarian Tourism

The principles of analysis and assessment of the risk component to public health as a major factor in tourism are summarized by deduction from the in-depth interviews. Summarizing the viewpoints of the interviewees, three main principles for risk analysis attributed to the public health risk environment, are regarded as significant. When reflecting on scientific knowledge according to the risk analysis Aven and Renn (2018) comment on eight principles applicable to the wide scope of risk. They consider the general principles (openness and transparency, involvement, proportionality and consistency) and discuss the scope of the "good governance". Adopting the view that acting under conditions of risk uncertainty (Frewer et al. 2002) and the nature of policy guidance both on supranational (specific restrictive measures during COVID-19) and organizational level (measures undertaken by SMET), the authors consider the proposed principles as applicable for effective treatment of health risk environment in society. Although relationships between causes and effects in the whole tourism industry can hardly be defined, the authors' point of view reflects on discussions based on the in-depth interviews. These principles follow the proposed by Aven and Renn (2018) principles:

1. The principle "protection of values like human life and health" (Aven and Renns' principle 3),
2. The principle "risk-informed, cautionary/precautionary and discursive strategies" (Aven and Renns' principle 4) and
3. The principle "openness and transparency about the understanding of the nature of risks to the public" (Aven and Renns' principle 6).

Since the goal of the specific principles is not to give an exhaustive list and a justification for the understanding of the issue, the idea applied here is rather to consider them in order to ensure health risk environment for SMET in the Bulgarian national and regional tourism industry.

Bulgarian regional SMEs in tourism could profit, if they examined the proposed risk principles and if they applied them in future measures for risk management. A minimisation of the volumes of risks and their economic losses can be thus expected. SMEs in the regional tourism industry can gain competitive advantages if they also take into consideration a comprehensive variety of risks attributed to the public health environment. In this way, investment activities for the SMEs in tourism can be better guaranteed. On the other hand, using the principles for risk minimisation, can influence the tourist flow and shall be a great help in securing clients. Considering other aspects of the risk, new models related to seasonality and tourist behaviour can be developed. According to our interviewees, none of the SMEs in the regional Bulgarian tourism industry works with the correct application of risk management. As a result, they are not able to adequately forecast problems connected with the labor market and the development of new tourism products.

Identified Areas for Future Research and Intervention Regarding Public Health Environment of the SMETs

At the current stage of development of an economy based on public health and striving for the principles of sustainable development, it should be expected that the characteristics of SMEs will acquire new forms of governance and management. It is certain that the small hotel business, like all small businesses, will continue to experience serious difficulties in its micro and macro environment. So far, the position of the small hotel business in the overall structure of the hotel industry in the country is somewhat controversial. On the one hand, the role of the small hotel business in the development of the economy in areas with strong tourist specialization remains uncertain. In the absence of an effective system of meters, one can speak of the role of the small hotel business only as an important consumer who makes significant purchases for their own production needs, a flexible producer who responds quickly to changes in market conditions and gives the economy the necessary resilience, mass employer. These parameters characterize the farming aspect of the operation of small hotels.

Foreign experience shows that small hotels, complementing large and medium-sized hotels, are less economically dependent on changes in the hotel services market and have the ability to quickly reorient prices. Combining small hotels in a chain with the establishment of uniform service standards and centralized (through the hotel management company) provision of a number of hotel services, including reservations, logistics, security, etc., significantly reduces management costs and increases the profitability of small hotel enterprises, especially taking into account the established administrative and economic mechanisms to support small business.

No less important is the problem of the instability of the small hotel business caused by the uncertainty of the external environment. The internal problems of the operation of small hotels can very often be solved by the enterprises themselves at the local level, while overcoming most of the external problems and risks.

- Features of the regional factor in tourism - a chance for communal policy at the micro-destination level. Entrepreneurship in the context of small and medium-sized enterprises to overcome risks and threats of a new kind. Formation of new tourism, responsible for avoiding the threats of a new kind.
- Small businesses in the context of safe tourism and ways of overcoming risk, threats and disaster.
- Safe tourism through small and medium enterprises - a profitable concept or an emerging trend in micro destinations.
- Micro-destinations in Bulgaria and collaboration with regional SMEs in tourism industry are able to compete risks of different nature, including those related to public risk health environment.

CONCLUSION

In the context of an unprecedented global pandemic, for all political and economic experts, stakeholders and practitioners, it is clear that the scope of *risk* is more than a matter of definition. The scope of *risk*, together with goals like risk elimination, restrictions, soft and hard measures, have a negative effect, especially for the service industry, including the tourism industry. The reduction of the tourist flow is the beginning of the domino effect, which will negatively affect all spheres of the economy. The current epidemiological crisis is a trial for the resilience not only of the health systems, but also of the political and socio-economic systems. SMEs in tourism worldwide are experiencing tremendous hardship. For this reason, the present study aimed to examine the public health risk friendly environment created by SMETs in Bulgarian regional hospitality during the COVID-19 pandemic. The relevant framework was achieved through stationary observations (legal documents) and exploratory research (deductive analyses) of the area of public health risk environment. In terms of risk, a broader concept of *public health risk environment* was presented, and the main research question was asked about the assessment of this environment. By adopting the *public health risk environment* as research settings, it was possible to assess the state of development of SMEs in times of a health crisis. Both from a theoretical and from a practical point of view, the existence of some important prerequisites facilitating the survival of SMEs in the regional Bulgarian tourism industry were corroborated. As a first a main prerequisite, the characteristics of the SMEs in tourism as smaller economic entities gained importance (due to the comparative analysis of two main official documents, the Bulgarian Tourism Act and the Bulgarian Act for the SMEs). To this end, an explorative analysis was carried out. The research object were regional guesthouses and family hotels (SMETs) in south western Bulgaria. In order to conduct this research, the methodology combined descriptive analysis of the normative basis - from the point of view of two act of the Bulgarian legislation, on the one hand, and from the viewpoint of tourism practitioners, business representatives and consumers (tourists), on the other hand. Another prerequisite was discovered against the backdrop of the function and characteristics of regional tourism SMEs in the category of class "B": family businesses, family hotels, hostels, guest houses and bungalows. These accommodation establishments seemed to be managed more easily in times of uncertainties and to act more flexible towards labor and seasonality factors, and consequently, can gain a competitive advantage. The management of these accommodations is easier to be adapted to the conditions of the identified by the authors, public health risk environment. The security requirements in the so postulated area and by following the principles proposed, seem to open a scope for possible interventions in the future. Additionally, the short distance for the domestic tourist meets the requirements of avoiding long-haul travel (especially international travel). Moreover, social distance is easier to be achieved in such small accommodations. To the main advantages of the regional SMEs in tourism belongs the possibility to offer new forms of tourism (pro-active tourism, nature-based tourism, wellbeing tourism, cycling and hiking tourism, creative tourism, etc.).

The provided explanatory research confirmed that the role of the public health risk environment for the SMEs in tourism will have an important role for the practice of the whole tourism industry. The results of the stationary monitoring and qualitative research showed that the so defined public health risk environment reduces the risks to micro-destinations, and to the whole tourism sector. Another main conclusion from the research is that regional tourism SMEs can survive in times of uncertainties because they are generally less exposed to risks caused by negative changes at the macro and micro level. Although the analysis of the public health risk environment was a difficult task for the authors, the identified key area is a part of the important risk management at a regional level. In the context of such complicated phenomenon, it is advisable for future research in the field of tourism economics, to focus on the evidence for successful SMEs' operations in the so proposed public health risk environment. Suggestions of how to minimize the risk to the tourist shall also be given when considering better ways for risks, crisis and disaster management of small and medium enterprises.

ACKNOWLEDGMENT

The authors would like to thank everyone who took their time to participate in the provided in-depth interviews and thus contributed to this research and producing this chapter. The authors express their gratitude to the editors of the book and the peer-reviewers of this study for their support and constructive advice when working on the chapter proposal.

REFERENCES

Abd-Elhamid, H. F., Fathy, I., & Zeleňáková, M. (2018). Flood prediction and mitigation in coastal tourism areas, a case study: Hurghada, Egypt. *Natural Hazards*, *93*(2), 559–576. doi:10.100711069-018-3316-x

Akis, S., & Warner, J. (1994). A descriptive analysis of North Cyprus tourism. *Tourism Management*, *15*(5), 379–388. doi:10.1016/0261-5177(94)90092-2

Aliyev, K., & Ahmadova, N. (2020). Testing tourism-led economic growth and economic-driven tourism growth hypotheses. *Toкrism*, *68*(1), 43–57. doi:10.37741/t.68.1.4

An, T. B., & Thi, D. B. V. (2012). Research on factors affecting competitiveness of four-star hotels in Thua Thien Hue. *Journal of Economic Development*, (261), 21–29.

Aven, T., & Renn, O. (2018). Improving government policy on Risk: Eight key principles. *Reliability Engineering & System Safety*, *176*, 230–241. doi:10.1016/j.ress.2018.04.018

Baker, D. M. (2020). Tourism and terrorism: Terrorists' threats to commercial aviation safety and security. *Tourism, Terrorism and Security*, 163-181. doi:10.1108/978-1-83867-905-720201012

Becken, S., & Hay, J. E. (2007). *Tourism and climate change: Risks and opportunities*. Channel View Publications.

Benur, A. M., & Bramwell, B. (2015). Tourism product development and product diversification in destinations. *Tourism Management, 50*, 213–224. doi:10.1016/j.tourman.2015.02.005

Brau, R., Lanza, A., & Usai, S. (2008). *Tourism and sustainable economic development: Macroeconomic models and empirical methods*. Edward Elgar.

Braun, P., & Hollick, M. (2008). Capacity building in SME Tourism networks. Encyclopedia of Networked and Virtual Organizations, 136-141. doi:10.4018/978-1-59904-885-7.ch018

Caloffi, A., Freo, M., Ghinoi, S., Rossi, F., & Russo, M. (2019). Is a policy mix more effective than individual policies for SME innovation? An exploratory analysis. *Fteval Journal for Research and Technology Policy Evaluation,* (47), 72-77.

Candela, G., & Figini, P. (2012). *The economics of tourism destinations*. Springer. doi:10.1007/978-3-642-20874-4

Cates, J. N., & Sussman, M. B. (1992). Family systems and inheritance. *Family Business Review, 5*(2), 205–226. doi:10.1111/j.1741-6248.1992.00205.x

Chung, J., Jin, B., Jeong, S. W., & Yang, H. (2019). Nie-based SME brand building in foreign markets: An exploratory study. *Journal of Product and Brand Management, 28*(1), 63–79. doi:10.1108/JPBM-10-2016-1331

Cooper, C., & Hall, C. M. (2019). *Contemporary tourism: An international approach*. Goodfellow Limited.

Denchev, S., Trencheva, T., & Trenchev, I. (2020). Application of the Game Engine in the Promotion and Training of the Cultural Heritage. *Conference Proceedings: 14th International Technology, Education and Development Conference*, 2700-2706. 10.21125/inted.2020.0806

Dimitrov, P. (2009). Measuring Occupational Health and Safety Risks in Tourism Companies. *Ege Academic Review, 9*(3), 889-902.

Egger, I., Lei, S. I., & Wassler, P. (2020). Digital free tourism – an exploratory study of tourist motivations. *Tourism Management, 79*, 104098. doi:10.1016/j.tourman.2020.104098

Elhadary, T. (2018). Sufi tourism: The impact of Sufi heritage on Islamic religious tourism. *Islamic Tourism: Management of Travel Destinations*, 110-123. doi:10.1079/9781786394132.0110

Fallon, F. (2004). After the LOMBOK riots, is sustainable Tourism Achievable? *Journal of Travel & Tourism Marketing*, *15*(2-3), 139–158. doi:10.1300/J073v15n02_08

Frechtling, D. (2001). The tourism forecasting process. *Forecasting Tourism Demand*, 45-57. doi:10.1016/B978-0-7506-5170-7.50007-6

Freitas, A., Muritiba, P., Muritiba, S. N., Lima, Y. O., & Riascos, L. A. (2018). Measurement of innovations. In *SME: Exploratory study with emphasis on manufacturing processes in the south of São Paulo*. doi:10.14488/ENEGEP2018_TI_ST_265_520_36054

Getz, D., & Carlsen, J. (2005). Family business in tourism. *Annals of Tourism Research*, *32*(1), 237–258. doi:10.1016/j.annals.2004.07.006

Harrison, D. (1993). Bulgarian tourism. *Annals of Tourism Research*, *20*(3), 519–534. doi:10.1016/0160-7383(93)90007-P

Henderson, J. C. (2013). The great east japan earthquake and tourism: A preliminary case study. *Tourism Recreation Research*, *38*(1), 93–98. doi:10.1080/02508281.2013.11081731

Hystad, P. W., & Keller, P. C. (2008). Towards a destination tourism disaster management framework: Long-term lessons from a forest fire disaster. *Tourism Management*, *29*(1), 151–162. doi:10.1016/j.tourman.2007.02.017

Ivanov, S., & Webster, C. (2011). Tourism's contribution to economic growth: A global analysis for the first decade of the Millenium. SSRN *Electronic Journal*. doi:10.2139srn.1962510

Jackiewicz, E. L. (2005). Tourism without threat? Excerpts from rural Costa Rica. *Annals of Tourism Research*, *32*(1), 266–268. doi:10.1016/j.annals.2004.05.004

Kellens, W., Neutens, T., Deckers, P., Reyns, J., & De Maeyer, P. (2011). Coastal flood risks and seasonal tourism: Analysing the effects of tourism dynamics on casualty calculations. *Natural Hazards*, *60*(3), 1211–1229. doi:10.100711069-011-9905-6

Kim, H., Kim, J., & Gu, Z. (2012). An examination of US hotel companies' risk features and their determinants of systematic risk. *International Journal of Tourism Research*, *14*(1), 28–39. doi:10.1002/jtr.828

Komppula, R. (2014). The role of individual entrepreneurs in the development of competitiveness for a rural tourism destination – a case study. *Tourism Management, 40*, 361–371. doi:10.1016/j.tourman.2013.07.007

Krasteva, R., & Pantelis, E. (2017). Bridging relationships between education, refugee crisis and tourism. *EDULEARN17 Proceedings*. 10.21125/edulearn.2017.1063

Kyurova, V., & Koyundzhiyska-Davidkova, B. (2017). Current issues in family business management. *European Journal of Service Management, 22*, 47–52. doi:10.18276/ejsm.2017.22-06

Lage, E., & Alturas, B. (2013). Information sharing among innovative SME: An exploratory study within the Portuguese SME innovation network. *Innovation through Knowledge Transfer*, 159-168. doi:10.1007/978-3-642-34219-6_18

Li, G. (2006). Forecasting tourism demand using econometric models. *Tourism Management Dynamics*, 219-228. doi:10.1016/B978-0-7506-6378-6.50033-0

Li, J., Feng, Y., Li, G., & Sun, X. (2020). Tourism companies' risk exposures on text disclosure. *Annals of Tourism Research, 84*, 102986. doi:10.1016/j.annals.2020.102986 PMID:32834224

Lisle, D. (2016). *Holidays in the danger zone: Entanglements of war and tourism*. University of Minnesota Press. doi:10.5749/minnesota/9780816698554.001.0001

Mason, P., Augustyn, M., & Seakhoa-King, A. (2009). Exploratory study in tourism: Designing an initial, qualitative phase of sequenced, mixed methods research. *International Journal of Tourism Research, 12*(5), 432–448. doi:10.1002/jtr.763

Maximiliano, E. K. (2018). Terrorism, tourism and hospitality. *The Challenges of Democracy in the War on Terror,* 105-122. doi:10.4324/9780429466182-7

Mazzanti, M. (2002). Tourism growth and sustainable economic development: A note on economic issues. *Tourism Economics, 8*(4), 457–462. doi:10.5367/000000002101298223

Mazzocchi, M., & Montini, A. (2001). Earthquake effects on tourism in central Italy. *Annals of Tourism Research, 28*(4), 1031–1046. doi:10.1016/S0160-7383(01)00008-1

McCartney, M. (2015). Economic growth and economic structure since 1750. *Economic Growth and Development*, 166-182. doi:10.1007/978-1-137-29031-1_9

Neher, P. A. (1971). *Economic growth development: A mathematical introduction*. J. Wiley.

Papatheodorou, A., & Pappas, N. (2017). Greek tourism, economic crisis and political turmoil, between Scylla and Charybdis. *Tourism (Zagreb)*. Advance online publication. doi:10.23912/9781910158814-3433

Paraskevas, A., & Arendell, B. (2007). A strategic framework for terrorism prevention and mitigation in tourism destinations. *Tourism Management*, *28*(6), 1560–1573. doi:10.1016/j.tourman.2007.02.012

Pett, T. L., & Wolff, J. A. (2009). SME opportunity for growth or profit: What is the role of product and process improvement? *International Journal of Entrepreneurial Venturing*, *1*(1), 5. doi:10.1504/IJEV.2009.023817

Pizam, A. & Mansfeld, Y. (2006). Toward a theory of tourism security. *Tourism, Security and Safety*, 1-27. doi:10.1016/B978-0-7506-7898-8.50004-7

Raj, R., & Griffin, K. (2017). Introduction to Conflicts, religion and culture in tourism. *Conflicts, Religion and Culture in Tourism*, 1-9. doi:10.1079/9781786390646.0001

Ritchie, B. W., & Jiang, Y. (2019). A review of research on tourism risk, crisis and disaster management: Launching the annals of tourism research curated collection on tourism risk, crisis and disaster management. *Annals of Tourism Research*, *79*, 102812. doi:10.1016/j.annals.2019.102812

Sanders, D., & Laing, J. (2009). In the line of fire: The challenges of Managing tourism operations in the Victorian Alps. *Tourism Review International*, *13*(4), 223–236. doi:10.3727/154427210X12741079930470

Setyawan, A. A., Isa, M., Wajdi, F., & Syamsudin, S. (2017). Disaster as business risk IN SME: An exploratory study. *International Journal of Research in Business and Social Science*, *6*(6), 52-63. doi:10.20525/ijrbs.v6i6.831

Sheng-Hshiung, Gwo-Hshiung, T., & Kuo-Ching, W. (1997). T. Sheng-Hshiung, T. Gwo-Hshiung, W. Kuo-Ching (2019). Evaluating tourist risks from fuzzy perspectives. *Annals of Tourism Research*, *24*(4), 796–812. doi:10.1016/S0160-7383(97)00059-5

Shivakumar, K., & Agarwal, R. (2014). Significance of user-generated content in travel and tourism: An exploratory study. *Journal of Tourism & Hospitality (Los Angeles, Calif.)*, *03*(03). Advance online publication. doi:10.4172/2167-0269.1000137

Singh, S. (1998). War, terrorism & tourism. *Tourism Recreation Research*, *23*(2), 83–83. doi:10.1080/02508281.1998.11014846

Spence, M., & Hamzaoui Essoussi, L. (2010). SME brand building and management: An exploratory study. *European Journal of Marketing*, *44*(7/8), 1037–1054. doi:10.1108/03090561011047517

Talwar, R. (2010). Scanning the horizon. *Trends and Issues in Global Tourism, 2010*, 237–268. doi:10.1007/978-3-642-10829-7_26

Thomas, K. C. (2020). *2 economic growth and economic development*. Competition Law in Developing Countries. doi:10.1093/law-ocl/9780198862697.003.0003

UNWTO world Tourism barometer and Statistical Annex, November 2019. (2019). *UNWTO World Tourism Barometer, 17*(4), 1-44.

Vasenska, I., Dimitrov, P., Koyundzhiyska-Davidkova, B., Krastev, V., Durana, P., & Poulaki, I. (2021). Financial transactions USING Fintech during the Covid-19 crisis in Bulgaria. *Risks, 9*(3), 48. doi:10.3390/risks9030048

Veal, A. J. (2018). *Research methods for leisure and tourism*. Pearson.

Verreynne, M., Williams, A. M., Ritchie, B. W., Gronum, S., & Betts, K. S. (2019). Innovation diversity and uncertainty in small and medium sized tourism firms. *Tourism Management, 72*, 257–269. doi:10.1016/j.tourman.2018.11.019

Williams, A. M., & Shaw, G. (1998). *Tourism and economic development: European experiences*. J. Wiley.

Yarkova, Y., Stoykova, B., Markov, N., & Bankova, M. (2019). Regional economy within the context of bioeconomy. *Trakia Journal of Sciences, 17*(Suppl.1), 5–13. doi:10.15547/tjs.2019.s.01.002

Zuvekas, C. (1979). Economic growth and economic development. *Economic Development*, 7-29. doi:10.1007/978-1-349-16275-8_2

ADDITIONAL READING

Asean. (2014). *Asean Member States*. Available in: https://www.asean.org/asean/asean-member-states

Botet, J. (2021). Dealing with unforeseen Circumstances. Implication of risk management in the COVID-19 public health emergency. *Risk Management*. Advance online publication. doi:10.5772/intechopen.97408

Edison, E., & Kartika, T. (2021). Tourism and tourism crisis management in the COVID-19 pandemic time. *Promoting Creative Tourism: Current Issues in Tourism Research, 641-644*, 641–644. Advance online publication. doi:10.1201/9781003095484-92

Godovykh, M., Pizam, A., & Bahja, F. (2021). Antecedents and outcomes of health RISK perceptions in TOURISM, following the COVID-19 pandemic. *Tourism Review*. doi:10.1108/TR-06-2020-0257

Gosheva, Z. (2016). Diversification of the tourist product - Hisar region. *New Knowledge Journal*, *5*(1), 11-16.

Kiryakova-Dineva, T., Kyurova, V., & Chankova, Y. (2019). Soft skills for sustainable development in tourism: The Bulgarian Experience. *European Journal of Sustainable Development*, *8*(2), 57. doi:10.14207/ejsd.2019.v8n2p57

Marinov, V., Vodenska, M., Popova, N., Assenova, M., & Dogramadjieva, E. (2016). Bulgaria. In: Vodenska (ed) Hospitality and Tourism in Transition in Central and Eastern Europe: A Comparative Analysis. Cambridge Scholars Publishing.

Maximiliano, E. K. (2018). Terrorism, tourism and hospitality. *The Challenges of Democracy in the War on Terror,* 105-122. doi:10.4324/9780429466182-7

Merad, M. (2021). Some elements of analysis of the bibliography on risk and resilience on covid-19. COVID-19: *Systemic Risk and Resilience*, 17-34. doi:10.1007/978-3-030-71587-8_3

Nutbeam, D. (2020). COVID-19: Lessons in risk communication and public trust. *Public Health Research & Practice*, *30*(2). Advance online publication. doi:10.17061/phrp3022006 PMID:32601649

KEY TERMS AND DEFINITIONS

Health Risk: Action taken in advance to make decrease one's chances of developing a disease. Often based on common disease.

Public Health Risk: General state of health risk related to the common population especially in times of pandemic.

SMET: Small and medium enterprise in tourism, mainly operating family business.

Chapter 5
Small and Medium Tourism Enterprise Survival in Times of Crisis:
"El Capricho de Gaudí"

Maria-Concepcion Lopez-Fernandez
University of Cantabria, Spain

Marta Perez-Perez
https://orcid.org/0000-0003-0342-504X
University of Cantabria, Spain

Ana-Maria Serrano-Bedia
University of Cantabria, Spain

Andrea Cobo-Gonzalez
Comillas Town Hall, Spain

ABSTRACT

This case study is intended to analyze, through a longitudinal analysis, the radical business model innovation experienced a Cantabrian tourism company that exploits a protected cultural property called "El Capricho de Gaudí." More specifically, this chapter focuses on the manager attitudes toward risk and their influence on response strategies implemented to redefine and renew this business model for competitiveness and survival. The results of the analysis contribute to generate knowledge about how a solid and well-established small experienced cultural-based tourism enterprise reformulates its business model to survive in time of crisis.

DOI: 10.4018/978-1-7998-6996-2.ch005

INTRODUCTION

The environment volatility and the current economic situation that is affecting the world with a high prevalence on tourism sector companies (Arcese, Valeri, Poponi & Elmo, 2020; Ritchie & Jiang, 2019; Cró & Martins, 2017) has forced business to look for new ways for creating value for customers and value capture in companies. Currently, we are witnessing the difficulties for the survival of many companies, especially small and medium enterprises (SMEs) in the service sector in general, and tourism in particular. Tourism is impacted by many external factors, including political instability and economic conditions (Okumus, Altinay, & Arasli, 2005), natural disasters, wars and, more recently, biological risks such as the Covid-19 health crisis (Arcese, Valeri, Poponi & Elmo, 2020; Gössling, Scott, & Hall, 2020; Jones, & Comfort, 2020). Within this context, small and medium tourism enterprises (SMTEs) are being more vulnerable to crisis impacts due to their limited capacities to reduce risk (Zeng, Carter, & De Lacy, 2005). Consequently, solid and well-established SMTEs have seen their profitability reduced, forcing them to rethink about how they must adapt to the new competitive conditions derived from the crisis that affects us (Morgan, Anokhin, Ofstein, & Friske, 2020; Osiyevskyy & Dewald, 2018). This fact has caused SMTEs to pay much more attention to analyzing their business models, which describe the bases on which a firm delivers value to customers and converts payment into profits (Teece, 2010). In the search for new ways to generate value and survive in time of crisis (Morgan, Anokhin, Ofstein, & Friske, 2020; Osiyevskyy & Dewald, 2018), innovation in business models as part of the firm´s strategic renewal process (Musa & Mamun, 2020), emerges as an option fully applicable to the tourism sector (Breier, Kallmuenzer, Clauss, Gast, Kraus & Tiberius, 2021; Bogers & Jensen, 2017; Alegre & Berbegal-Mirabent, 2016; Souto, 2015; Spieth, Schneckenberg, & Ricart, 2014; Zott & Amit 2010). In fact, business models are becoming a new conceptual tool complementary to the company's strategy that is receiving increasing attention from the scientific and professional community (Prescott & Filatotchev, 2021; Hjalager & Madsen, 2018; Massa, Tucci, & Afuah, 2017; Bocken, Short, Rana, & Evans 2014; Zott, Amit, & Massa 2011). That is because of its potential to analyze, compare and evaluate business performance and results (Massa, Tucci, & Afuah, 2017). Despite its relevance, the analysis of the strategic renewal process of business models has been less present both in the economic literature and in the business management literature (Teece, 2010). However, there is a growing streem of literature (Foss & Saebi, 2017; Foss & Saebi, 2018) analyzing the business model innovation, its relationship with the process of organizational change, the type of innovations generated as well as the outcomes derived from a company's business model innovation.

From this starting point, the objective of the present chapter is to analyze, through a longitudinal and in-depth analysis, the radical and profound renovation experienced by the business model of a Cantabrian tourism company that exploits a protected cultural property called "El Capricho de Gaudí". More specifically, this chapter focuses on the managers attitudes toward risk and their influence on response strategies implemented to redefine and renew this business model for being competitive and survive. The results of the analysis reveal at least three contributions to the literature: In the first place, this chapter contributes to generate knowledge about how a solid and well stablished small experience cultural based-tourism enterprise (Jensen, & Prebensen, 2015) reformulates its business model to survive in time of crisis (Morgan, Anokhin, Ofstein, & Friske, 2020). Second, it will make possible to identify the main obstacles faced by this company and how to overcome them. Finally, it will allow the identification of possible external and internal factors that may affect the business model innovation process and its interaction with the company's strategic decisions.

BUSINESS MODEL AND BUSINESS MODEL INNOVATION IN TOURISM ENTERPRISES

The concept of business model began to be used repeatedly in the literature on business administration with the arrival of the Internet in the mid-90s. Since its origins in the 1950s, when the idea of the business model was used initially in a wider sense (Drucker, 1954), both scholars and practitioners (Bigelow & Barney, 2021; Lanzolla & Markides, 2021; Prescott & Filatotchev, 2021) have devoted increasing attention to this concept, and a variety of definitions have appeared in the literature. However, although the term is fuzzy and its definitions, components, and theoretical underpinnings vary greatly, Massa, Tucci, & Afuah (2017: 76) identify three different approaches to the definition and study of business model: " (1) business models as attributes of real firms having a direct real impact on business operations, (2) business models as cognitive/linguistic schema, and (3) business models as formal conceptual representations/descriptions of how an organization functions.". We follow the first approach as long as we understand a business model as the way by which a company creates value for its consumers, attracts them to pay for the value, and obtains profits from the value deliveries (Zhang, Daim & Zhang, 2019; Teece, 2010). In other words, it is a 'story' that essentially explains how firms work (Arbussà Reixach, Bikfalvi & Marquès Gou, 2017). Thus, a business model is based on a series of hypotheses (more or less implicit) about consumers, their behavior, the price they are willing to pay, the analysis of how the competition is (or is not) meeting the needs of consumers and a reflection on the best way for the company to create value for customers by adapting the different areas of the company and assuming a certain cost structure. Actually, some scholars argue that the academic and practical success of the business model concept is rooted in its ability to challenge asumptions of traditional strategic theories that assumed that "value is created exclusively by producers, not by customers and competitive advantage is single sourced resource based only or activities based only" (Massa, Tucci, & Afua, 2017: 75). Instead, business model open the door to consider that value can be created both in the supply (producers) and demand (customer and other members of the value chain) sides and that competitive advantage can be also resource based and activities based, in both sides (supply and demand) (Massa, Tucci and Afuah, 2017).

Value creation is linked to the innovation of a business model that could be analyzed as a different configuration of what is done in the company and how it is done, to provide a new value proposition to customers (Fissi, Romolini & Gori, 2020; Hjalager, Kwiatkowski & Østervig Larsen, 2017; Souto, 2015). In other words, innovation is relevant for a new business model, as this activity represents a reworking of a company's general value chain (Runfola, Rosati & Guercini, 2013).

A recent review of previous research in business model innovation realized that "research streams have largely evolved in relatively isolated silos (as indicated by little cross-citation among the streams) and do not seem to build off one another." (Foss & Saebi, 2017: 208). The first research line focused on conceptualization. The second research line focused on the relationships between business model innovation and organizational change process highlighting the uncertain, dynamic and exploratory nature of business model innovation. The third research line identified by Foss and Saebi (2017) included works focusing in describing the final result of the business model innovation process, often contextualized in a specific industry such as tourism (Souto, 2015). Finally, the fourth research line focused on the effects of the business model innovation process either on the company's performance or on the innovation performance (Desyllas, Salter, & Alexy, 2020). A fruitful research on business model innovation require an integrated exploration of antecedents and outcomes besides the identification and analysis of the contingencies that can affect the process and the boundary conditions in which the business model innovation occurs (Foss & Saebi, 2017; Saebi, Lien, & Foss, 2017; Foss & Saebi, 2018).

Innovations, in tourism sector, are defined as "everything that differs from business as usual or which represents a discontinuance of previous practice in some sense for the innovating firm" (Hjalager, 2010, p. 2), and occur in the form of product/service, process, managerial, marketing, or institutional innovations. The establishment, implementation and renewal process of a business model over time is a critical task to achieve the success of any organization in general, and tourism sector in particular (Fissi, Romolini & Gori, 2020; Cheah, Ho & Li, 2018; Sundbo, Orfila-sintes, & Sorensen, 2007; Mihalic, Zabkar & Knezevic Cvelbar, 2012), both at the time of its creation and throughout its life cycle. The current competitive environment, the globalization, the highly competitive pressures and the dynamism of new technological opportunities, have evoked that product innovation was not enough to being competitive (competitors can quickly copy innovations). Thus, business model innovation emerges as a great opportunity to build sustainable competitive advantages (Mendoza, Gallego-Schmid & Azapagic, 2019; Ambrož & Omerzel, 2017; Foss & Saebi, 2017; Foss & Saebi, 2018; Massa, Tucci, & Afua, 2017; Amit & Zott, 2012). However, the high environmental dynamism has also decreased the average life span of business models from 15 to less than 5 years (Wiprächtige, Narayanamurthy, Moser, & Sengupta, 2019). Against this backdrop, it seems logical that companies adapt themselves through faster moving cycles, thus redefining their business models in order to identify products and processes obsolete faster than ever. In this sense, business model innovation, defined as a new configuration of what is done and how it is done in the company, in order to provide a new value proposition to customers, emerges as "a new holistic, integrated and systematic way for organizations to provide the operation of innovations in order to create value in a dynamical environment through collaboration with internal and external stakeholders" (Mihardjo & Alamsjah, 2019, p-69).

In the specific tourism sector business model innovation occurs when a firm adopts a novel approach for commercializing all of its underlying assets and challenging its hypotheses about creating and capturing value (Hjalager & Madsen, 2018) being considered an essential requirement to maintain competitiveness (Souto, 2015; Thomas and Wood, 2014), thus stimulating incremental and radical innovations (Breier, Kallmuenzer, Clauss, Gast, Kraus, & Tiberius, 2021; Han, Jung, & Gibson, 2013). In this vein, recent studies have identified that business model innovation can emerge as as a potential solution to recover from the COVID-19 crisis (Kraus, Clauss, Breier, Gast & Tiberious, 2020), due to the fact if a business model is innovated through substantial changes in the elements and/or their configuration (Foss & Saebi, 2017), new opportunities can be addressed that increase firm performance (Desyllas, Salter, & Alexy, 2020) and may help tourism firms (Breier, Kallmuenzer, Clauss, Gast, Kraus, & Tiberius, 2021).

Along the history some examples of business model innovation exists within the tourism sector. However, in most cases and due to often limited financial opportunities and capacities, these are mostly incremental innovations associated with technical advancements (Pikkemaat & Peters, 2006). For example the advancement of information and communication technologies (ICTs) in general, and cloud computing technologies in particular, motivates tourism sector to leverage smart technologies to enhance the tourist experience (Buhalis, 2019). As a consequence it has been a relevant growing of "smart tourism" that enable colossal data transformation into value proposition (Mehraliyev, Chan, Choi, Koseoglu & Law, 2020) through the use of augmented reality (Cranmer, tom Dieck, & Fountoulaki, 2020; Loureiro, Guerreiro & Ali, 2020; Yung & Khoo-Lattimore, 2019; Han, Jung & Gibson, 2013; Kounavis, Kasimati, & Zamani, 2012), mobile technologies (Dorcic, Komsic, & Markovic, 2019) or audio guides (Lee, 2017) among others. Moreover, the use of multimedia applications to present tourist attractions appears recently as a common practice among tourism enteprises. In this vein, innovative museums have incorporated the use of a multimedia applications to present the collections to which access is difficult (for example, delicate museum pieces) (Serravalle, Ferraris, Vrontis, Thrassou & Christofi, 2019). The use of this type of multimedia application by a number of museums belong to a common network opens the door to realize virtual exhibitions, expanding the displayed heritage by adding pieces or collections belonging to other facilities (Szromek & Naramski, 2019), thus incementing customer value experience.

In addition to incremental innovations, the literature also offers examples of radical innovations in the tourism sector. For example, in the hospitality industry The Italian Albergo Diffuso (AD) experience (Fissi, Romolini & Gori, 2020), which are based in historical villages and allow tourists to visit a historical residence and a stay at a rural farmhouse (Paniccia & Leoni, 2017), can be seen as a radical innovation that is able to introduce a new system in the hotel business model to meet tourists' new requirements and to change the living experience, creating a paradigm that involves new services and new experiences (Souto, 2015). The innovation of the AD can be identified within the types of services offered to customers, which are tailor-made to their needs and aim of bringing the guest closer to the territory and retaining them long term. In particular, the services are offered in collaboration with the businesses and inhabitants of the village; without them, the AD could not fully carry out its facilitating role. From this perspective, AD provides the opportunity to realize the conversion of rural villages, which may have few inhabitants, into tourist locations, even if the territory is outside of touristic flows (RandellI, Romei & Tortora, 2014). The market demand for sustainable tourism experiences, along with the rapid growth of the digital technology in tourism, for instance in the form of global distribution and central reservation systems, and the intensive presence on social media platforms (Gretzel, Sigala, Xiang & Koo, 2015; Krajcik, Kljucnikov & Rihova, 2019) has also encouraged the development of new firms inspired by organizational paradigms such as the circular economy and sharing economy (Rosato, Caputo, Valente & Pizzi, 2021; Gössling & Hall, 2019). Thus, the report "Green business model innovation in the tourism and experience economy" (Andersen & Andersen, 2012) describes in detail ten company case studies from Austria (the Boutique Hotel Stadthalle), Portugal (Zmar Eco Campo), Denmark (Tivoli Gardens), Finland (Moomin World park), Mexico (Experiencias Xcaret recreational activities), Norway (XXLofoten), Sweden (Vildmark i Värmland), Iceland (Íshestar Travel), Russia (PAC tour operator) and South Korea (COEX, an international hosting) that applicate a variety of business model innovation as a tool for green growth the sustainability of different types of enterprises of tourism sector. Regarding the sharing economy, the emergence of Airbnb is an example of a disruptive business model innovation (Christensen & Raynor, 2003) in hospitality industry (Guttentag, 2015). Airbnb has used an innovative internet-based business model whose permits ordinary people to rent out their residences as tourist accommodation, thus cannibalizing guests from the traditional accommodation sector. The fast growing of this company and its unique appeal to tourists has a resonance to all major tourism stakeholders, including tourists, residents, tourism bureaus, hotels, and governments (Guttentag, 2015).

These previous examples shows that business model innovation, although it is not without risks (Crick, Chaudhry, & Crick, 2018; Laukkanen, & Patala, 2014), can help tourism enterprises to increase profitability and ensure attractiveness vis a vis customers, staff and the surrounding community. In other words, business model innovations allow tourism enterprises to deal with a high variety of changes such as instability and economic conditions derived from the impact of new technologies, new tourist habits (Okumus, Altinay, & Arasli, 2005; Stamboulis & Skayannis, 2003) or the Covid-19 health crisis (Arcese, Valeri, Poponi & Elmo, 2020; Gössling, Scott, & Hall, 2020; Jones, & Comfort, 2020; Kraus, Clauss, Breier, Gast & Tiberious 2020).

CASE STUDY

This chapter analyzes, through an exploratory descriptive method, the radical and particular transformation of the business model of a Cantabrian cultural experience-based tourism enterprise called "El Capricho de Gaudí" which is located in the town of Comillas, Cantabria, Spain. This experience-based tourism enterprise has experimented a fastest growing derived form the implementation of differente business model innovations.

We used the case study method, which is commonly used in a wide range of disciplines in the social sciences including the management field (Yin, 2003; 2018) and tourism in particular (Reinhold, Zach & Krizaj, 2018). More precisely, the case study research design has been considered suitable for investigating current topics of new business models in the tourism sector, including hotels (Runfola, Rosati & Guercini, 2013), AD experience (De Montis, Ledda, Ganciu, Serra & De Montis, 2015), hospitality (Breier, Kallmuenzer, Clauss, Gast, Kraus & Tiberius, 2021), sport tourism companies (Peric & Wise, 2015) or health resorts (Szromek & Naramski, 2019), among others.

Case study is identified as a qualitative form of research design (Çakar & Aykol, 2020; Hyett, Kenny & Dickson-Swift, 2014; Riege, 2003) that usually includes a specific feature and well-structured narratives that naturally reflect the complexities and contradictions of real-life (Flyvbjerg, 2006). Case studies can include a single case or multiple cases (Gustafsson, 2017). Yin (2014) further proposed four types of case study designs: single-case holistic designs, single-case embedded designs with subcases which signal embedded units in the case of interest, multiple-case holistic designs and multiple-case embedded designs, with more than one case to be investigated (Hoorani, Nair, & Gibbert, 2019). In this chapter we chose a single case study design, which is considered appropriate for critical, unusual, common, revelatory, or longitudinal and in-depth analysis (Yin, 2018), as is this the case, and the most common approach adopted in hospitality and tourism research (Çakar & Aykol, 2020). The case under analysis was managed by collecting data from different sources (official web sites, corporate reports and documents, interviews) in order to favor the triangulation of data collection, as in previous studies (e.g. Crick, Chaudhry & Crick, 2018; Fissi, Romolini & Gori, 2020; Breier, Kallmuenzer, Clauss, Gast, Kraus & Tiberius, 2021). Employing data triangulation strengthens the robustness of the results and makes research more rigorous and credible to avoid measurement bias (Creswell, 1998; Stake, 2005; Creswell & Poth, 2018. In particular, regarding the interview, an in-person interview of approximately one hour of duration with the CEO was conducted around the headquarters of the firm. The interview captured issues associated with the objectives and the business model employed. In a final step, the information was submitted to a content analysis to study the key points of the firm business model: the business idea, innovation and value creation.

As utilisation of manual coding to split the data around emerging issues appeared to be appropriate in the context of a single case, no computer-aided qualitative data analysis software such as NVivo (Sinkovics & Alfoldi, 2012) was necessary. Moreover, since the interpretation of qualitative data can differ depending on various factors (including who undertakes the analysis (Leitch & Palmer, 2010)), the data were compared independently by the researchers. No noticeable differences emerged.

"El Capricho de Gaudí" has experienced a particular evolution with an interesting background to learn. That is because along its history it has implemented a radical business model innovation, which is less frequent than continuous incremental innovation (Hjalager & Madsen, 2018), going from being a restaurant to operating the building as a visitor center. Its origins, factors as well as the analysis of the two different business models are presented below.

To overcome limitations of previous literature, we try to offer insights in the three main streams identificated by Foss and Saebi (2017). Thus, we first offer a detailed and contextualized description of the business model process of change and innovation. Then we analyse how business model innovation relates to the organizational change process and specifically we identify the barriers to change faced by the company, the organizational and managerial capabilities required to change and the relevance of experimentation. Finally, we explore and present the organizational and performance implications of the process of business model innovation.

El Capricho de Gaudí: Its Origins

"El Capricho de Gaudí", is a modernist and singular building designed by Antoni Gaudí that has been exploited as a cultural visitor center since 2010. Its construction took place between 1883 and 1885 being designed as a summer house of Máximo Díaz de Quijano. However, the death of Máximo Díaz de Quijano evoked that the building was inherited by his sister which used it as a holiday accommodation until its closure in 1969. In this year the building was declared such as a UNESCO World Heritage Site. Then, in 1975 the City Council of Reus (Tarragona), the birthplace of Antonio Gaudí, tried to negotiate, without success, the transfer of the building stone by stone. In 1977, the businessman Antonio Díaz acquired the ownership of the building that presented a high degree of deterioration due to its abandonment from 1969. In 1989, the descendants of Antonio Díaz turned the building into a first class restaurant (1989-2010), thus beginning the exploitation of this cultural asset and designing the first business model of this SMTE. In 1990, the building was acquired by the Japanese Kurosawa family (Mido Development group) who continued operating the restaurant until 2010. However, due to profitability problems, the owners closed several months of 2010 to reformulate the business model as a visitor center. Along its history the tourist exploitation of this cultural asset has been affected by a series of factors, which are described below:

- The architectural building's value due to its uniqueness, level of protection and architect call effect. The relevance of "El Capricho" lies in the fact that it is, together with the Casa Vicens in Barcelona, the first building that Gaudí, who was the highest representative of modernism and one of the main pioneers of the artistic avant-garde of the 20th century, built. In this building the sun appears as the main theme, leaving evidence in all the decoration in which the sunflower theme is repeated in the exterior ceramics, highlighting the bright colors of the entire building in green, red and yellow. The sun also refers to its location, since each room is oriented according to the path of the sun and the activity carried out in it. Its U-shape is Gaudí's response to the contractor's interest in protecting a large south-facing greenhouse from the north wind, where he treasured the plants he brought from overseas. The contractor's taste for music is also reflected in numerous interior details.

- Geographic location: a monumental and seafaring town (Comillas) with a long tourist tradition. It is one of the most beautiful and well-known towns in Cantabria, with a magnificent architectural complex framed by an exceptional natural and scenic environment. It is also considered one of the most prestigious summer places of the Cantabrian coast.

- Seasonality: Tourist seasonality is one of the main characteristics and weaknesses of tourist activity. It is due to the concentration of tourists in certain months of the year, thus affecting the economic sustainability of businesses. In Cantabria and generally in the north of Spain, the high season is limited to summer months -July and, especially, August. In Comillas, seasonality affects in a special way, since the vast majority of tourist services such as bars, restaurants, hotels and monuments remain closed during the winter period.

El Capricho de Gaudí's Business Model as a Restaurant (1989 - 2010)

The exploitation of this cultural asset begins with its transformation into a high-class restaurant by the descendants of Antonio Díaz, who run it for two years. In 1990, a Japanese family Group, acquired the building. This Japanese family does not change the restaurant business model at the beginning but transforms it into an independent and autonomous company.

During the first two years, El Capricho de Gaudí basically changed owner but neither the structure nor the business model changed. The traditional cuisine was maintained, although progressively French and Japanese cuisine were introduced, with some renowned chefs that gave the restaurant more exclusivity, since it offered an image of great quality. Thus, El Capricho de Gaudí was turned into a high-quality and exclusive restaurant within an emblematic and different place, a singular modernist building, unique in Comillas and the northern of Spain, with a great tourist interest. The bases of the Business Model as a restaurant are summarized in Table 1.

Table 1.

How the Company creates value?	Offering an exclusive quality cuisine within an iconic modernism building of the architect Antonio Gaudí.
Advantages	Offers a high quality menu Restaurant inside of a Modernism building that was the unique Gaudí 's building in Cantabria.
Position regarding the competition	Differentiation: offering high quality cuisine in an exclusive place.

Source: Authors

As the years go by, the owner group observed that customers were not willing, in general, to pay a high price to eat in the restaurant and that the organization lacked the ability to respond and adapt to meet the new needs of its customers. Against this backdrop they decided to improve the building's facilities, internally and externally, to offer a better image of the complex and better services to the customer including a gift shop within the restaurant in order to find exclusive gifts from El Capricho de Gaudí.

Despite all these characteristics and the exclusivity of the building, the restaurant had difficulties to consolidate its profitability, especially with the arrival of the economic crisis since 2007 (Morgan, Anokhin, Ofstein, & Friske, 2020; Osiyevskyy & Dewald, 2018). The family group is finally forced to close the doors of the Restaurant in 2009. During the period of time that it was closed, the owners of El Capricho de Gaudí suffered pressure from the tabloid press, coupled with a deal by the press with racist and xenophobic connotation.

El Capricho de Gaudí: Metamorphosis to as Visitable Cultural Center 2010-2020

After the activity cessation in 2009, the building was closed for a few months. During these months the family considered various alternatives for implementing a radical innovation into the business model operated in previous years. The building's owners were aware that the value created by "El Capricho" company was based, firstly, on the architectural relevance of the building and, secondly, in the national and international tourist interest generated by its architect, Antonio Gaudí. In this vein, in 2010, El Capricho de Gaudí began to be exploited as a visitor center (see table 2).

Table 2.

How the Company creates value?	Turn an iconic and unique monument of the region into an experience
Advantages	Promote, preserve and maintain a unique modernist monument in the region. Gaudí brand - generates great national and international tourist interest
Position regarding the competition	Differentiation: an exclusive modernist monument in the region. It is one of Gaudí's first works, known worldwide and attracting tourists over the world

Source: Authors

The services that the visitor center offers are:

- Free and self-guided visits of both the exterior gardens and the interior of the building buying the ticket directly at the main door of the enclosure. The visit is supported by graphic material developed by the company -for example an informative video is projected in which the history of the building and its characters are explained-.
- Group visit service with special conditions and specialized guides.
- Special rates for school programs, with discounts and a visit adapted to each educational stage.
- Rental system for private and business events to use exclusively all the facilities of the enclosure: gardens, terraces and interior rooms.
- Shop and bookstore that has a collection of exclusive products of El Capricho de Gaudí, specially developed for this company and that are combined with Gaudí's generic products.

During this stage, the owners dedicated to preserving the building, keeping it in optimal condition and promoting the great work that Gaudí built 150 years ago. The visitor center closes only at Christmas, with a continuous schedule from March to October, which extends from July to September. This schedule is a sign of the property's commitment to visitors, given the high seasonality of regional tourism and, more specifically, tourism in Comillas, where most of the businesses are closed from October to May. However, El Capricho remains open throughout the year, which implies assuming high fixed costs. They also know that the position of El Capricho with respect to the competition contains disadvantages since it is a 100% private, self-sustaining structure without any type of grant, unlike the rest of the existing monuments placed in Comillas.

EL Capricho de Gaudí: Business Model Innovation Discussion

The descriptive analysis carried out in the previous section shows the radical transformation undergone by the business model that exploits this cultural tourism asset. The most relevant aspects of both stages and their comparison are summarized in table 3.

The analysis of this case study shows how the board of El Capricho de Gaudí has maintained great dynamism and constant concern for monitoring opportunities in the environment. The continuous adaptation to the environment can show the existence of a strong entrepreneurial management spirit that is fundamentally identified through two aspects: 1) capacity to resist failure and 2) development of the "locus of control" - degree in which the owners claim responsibility for what happens in their business and focus on their ability to modify the environment - which is clearly evidenced in the different stages that the building has gone through and which are discussed below. Fundamentally, the business model innovation implemented can be considered the result of a radical change in the value offered by the organization in response to changes in the environment.

Table 3. Business models under which the cultural asset has been exploited

<table>
<tr><th></th><th>El Capricho as a Restaurant (1989-2009)</th><th>El Capricho as a Visitor Center (2010-nowadays)</th></tr>
<tr><td>Business Model Description</td><td>The family group acquired the building in 1990 which continues with the restaurant services.
- Incremental modification of the business model by including French and Japanese food and a gift shop
-Internal and external reform of the building as a measure to improve customer service and exploit the three main business assets (exclusive quality cuisine, geographical situation and Gaudí architectural prestige).
-Business model failure in 2010 as a consequence of -the economic crisis that causes a decrease in the restaurant's activity.</td><td>Although the ownership is maintained, a radical business model innovation is implemented to find a new way to exploit the building, the main asset of the company.
The family started a new activity with guidelines and knowledge radically different from the restaurant. They are:
• Visits in different formats: group visits or night visits among other
• Rental the building for private or business events
• Store that offers exclusive products "El Capricho de Gaudí" and Bookstore
During this stage the building is emptied of all the catering equipment. A ticket office is installed where an entrance fee to visit the monument begins to be charged. The main purpose was to exploit two main business assets: a singular building by a renowned architect (Gaudí) and a tourist geographical situation</td></tr>
<tr><td>Main barriers and achievements</td><td><u>-Barriers</u>
• Constant losses.
• Tourist seasonality during winter months.
• Resistance to organizational change.
Failure suffered in the final stage as a restaurant, reform and investment.
<u>-Achievements:</u>
• Stage in which the restaurant generates interest.</td><td><u>-Barriers:</u>
• Period of experimentation at the beginning of the visitor center.
• The building remains closed during a decision-making process, undergoing media pressure (press sensationalism).
• Visitor complaints.
<u>-Achievements:</u>
• An average of 186,000 people a year visit it (70% national).
• Promote interest in culture within and outside the region.
• Economically sustainable business model.</td></tr>
</table>

Source: Authors

If we analyze the building's stage as a restaurant, it can be considered that the value that El Capricho de Gaudí created at that time was twofold. On one side, the Gaudí's modernist building, unique in Comillas and northern Cantabria, of great tourist interest. On the other side, the restaurant offered high-quality cuisine and the opportunity to eat in an emblematic place. The stage in which this value was maintained as valuable to customers was a phase in which national and international tourism was booming and in which the economic situation was favorable. This context allowed the organization to effectively manage the seasonality of this sector and maintain a high-quality culinary offer that provided sufficient business margins to guarantee business survival. However, as the context changed, the company had to face some challenges that caused the need to innovate its business model.

Is at that moment when activity ceases that the central role of management is shown in the innovation processes of business models (Demil & Lecocq, 2010). Management, faced with the problem of economic viability exacerbated by the severe economic crisis, initially resists radical changes in the business model due to risks associated with major modifications. This, together with a logic focused on maintaining a high-level restaurant, leads to a series of decisions that, in the medium term, definitely weighed down the restaurant's sustainability. However, once the closure of the restaurant is inevitable, the entrepreneurial spirit of the management is released. In this way, it is observed that the management is able to visualize, decide and carry out a radical innovation of the business model, taking into account the main factors that determine the exploitation of this cultural asset - geographic location, seasonality of demand and a high building's architectural value-. The strategic analysis of these factors leads to transform the restaurant business model, going from offering high-quality cuisine in an emblematic environment, to focusing on the exploitation of a cultural asset of great tourist interest. Thus the value of the new model business focuses on the cultural relevance of the building, its details and its architecture. The implementation and development of this new concept of value forced the management to overcome a series of barriers, among which the following stand out:

- Fear of the unknown and uncertainty: The building, as a company, had always been dedicated to high-level restoration, therefore when the business model was changed there was a total ignorance, in terms of customer service and prices of the visit.
- Resistance to change: managers needed to assume and assess the risk of radical innovation fighting the fear of the unknown.
- Comfort: The business model as a restaurant failed. Consequently the owners were forced to innovate and remodel its business model, so they moved away from their comfort zone to launch into a totally unknown new area: a cultural center that can be visited.

Overcoming these barriers required a continuous process of renewal and adaptation that made it possible to identify and transfer the value of the organization to the target audience based on internal and external demands. In order to better understand this process we answer the key questions posed by Teece (2010) in terms of business model innovation. These questions allow us to explain the guidelines followed by the organization in this last stage. The response of these questions are presented below:

1. What is the goal of the new business model?
   The main purpose of the new business model as a visitor center was to offer visitors an enriching experience, in which they learn while enjoying culture, for which a series of innovative activities is offered.
2. Which new activities are necessary to meet perceived needs?
   The new model required abandoning all well-known activities associated with high-level restoration for developing new activities related to the new model and which, at least, include: adoption of the basic decisions that determine the operation of the visitor center (hours, prices, among others); dissemination of the visitor center in the appropriate channels; implementation of new processes (reception, guidance, monitoring of results and satisfaction with the visit); workforce training for the new functions.
3. How might these activities be related to each other in novel ways?
   After the radical business model innovation the company board has implemented new activities. Some of the incremental innovations adopted were examples of how interaction between activities was leveraged to implement improvements. These included:
   a. Educational visits suitable for all audiences and perfectly valid for families. Visits should be a funny experience to know the Gaudí world. They also put at the visitor's disposal the "Gauditeca", a workshop for children, in which they work with the creativity of the little ones.
   b. Special activities and events that allow the visitor to get closer to the work, to be interested in it and its architect. An example of this is the so-called "Gaudí Expert" held on the occasion of the 162nd anniversary of the birth of Antonio Gaudí, in which a permanent exhibition of photos, drawings and images was inaugurated called "Gaudí, the origin", and in the that an enigma contest was prepared on the architect, with a series of winners that were published on its website.
   c. Night visits, opening for two weekends in August until 11:00 p.m.
4. How is value created through the new business model for each of the parties involved?
   The value of this new business model lied on the building relevance and the prestigious brand of its architect, Gaudí. On the other hand, the value that is created for customers is the satisfaction of their cultural concerns about Gaudí's work. The company tries to motivate the visitor and provoke a good memory: that the architecture and the history of the building, and in general the culture, be an experience worth paying for. In this vein, new activities or elements are

included constantly, with the intention of improving customer service (video, furniture, games for children, exhibitions ...)

For the company, the value generated derives from the ability to generate income with a lower cost structure that makes the Visitor Center economically viable and that allows the building to be maintained and a highly relevant cultural asset available to the public.

For the tourist destination in which it is located, it supposes a reinforcement of its cultural offer that contributes to greater visitor satisfaction as well as to attract new types of visitors who, otherwise, would not approach Comillas.

5. Who should carry out each of the activities that are part of the business model and what new management provisions could enable this structure?
   The human resources area of El Capricho de Gaudí is made up of a small number of trusted people with high labour flexibility -they can carry out different tasks throughout the day depending on the needs of every moment-. The activities that are part of the new business model are carried out in shifts by all the company's workers, both in customer service in the gift shop and ticket office, as well as the maintenance of the building and the premises as well as in the administration of the deal. During the summer months, there has been an innovation in terms of guided tours of the building, also carried out by some of the company's workers. Workers have received the necessary training when they have had to change the type of activity to be carried out. The company, when needed, has also used external technicians to carry out some activities (preparation of specific material for the store, production of videos, etc.)
6. What revenue model of the company will allow the company to appropriate part of the value created from the new business model?
   With the introduction of the new business model as a visitor center, the sources of income are twofold: 1) the entrance price paid by visitors to see the building and 2) the income obtained in the exclusive gift shop of the Gaudí brand.

The success of the radical innovation carried out in the business model is supported by the positive data regarding the number of visits they are receiving. Thus, the average number of visits in 2019, the last year available were 186.606 people per year, (of which 30% come from the international sphere). These data confirm that an economic model of sustainability has been found that offers greater stability for the future while confirming the promotion of interest in culture within the national context - a priority objective for management-. In fact, this building has now become one of Gaudí's buildings with the highest number of national visitors (70%).

However, although the radical innovation carried out is allowing the management to maintain a sustainable economic model, this success is the result of a constant and laborious phase of trial and error through which the management model has been refined over time (Sosna et al., 2010). In fact, such process continues to be carried out today as a constant activity that allows us to identify new ways of innovating in the current business model, evaluating it in a way that responds to the new needs of the environment faster. For example, in the current global pandemic situation, El capricho de Gaudí was the first building in recover activity after quarantine derived from COVID-19 health crisis, thus demonstrating again the innovative endevenour of this SMTE to survival in a time of crisis.

CONCLUSION

A business model starts from a generation and identification phase to later be developed and implemented in a way that generates value for both customers and the company. The achievement of this objective often implies a constant and continuous task of renovation and adjustment of the business model. Within the tourism sector this renewal process is a complex task as it requires the existence of a commitment and an entrepreneurial spirit that allows the organization to perceive risk, identify opportunities and adapt the company to the needs of the environment throughout its entire life.

Within this context, the main objective of this chapter was to analyze, through a longitudinal and in-depth analysis, the radical business model innovation experienced by a Cantabrian tourism experience-based company that exploits a protected cultural property called "El Capricho de Gaudí". Moreover, we tried to overcome main limitations fo business model innovation literature by offering a rich in wyde vision of the antecedents, context and process of business model innovation identifiying the main barriers faced in different moments, the critical role of managers' leadership and commitment, and the experimentation and exploration steps followed by the company. We also offered a deep analysis of the results. In a specific way, the metamorphosis process experienced by the business model of this cultural asset, which has been transformed from a restaurant to a cultural visitor center, has been described. This business model innovation has involved a laborious trial and error phase through which the business model has been refined with a relevant knowledge for learning. More specifically, this case study has evidenced that, in some cases, the resistance to organizational change (Chesbrough, 2010) led company managers to implement decisions that did not affect the existing business model and that were not sufficient to overcome their lack of profitability, thus contributing to the definitive collapse of the business model as a restaurant. However, the analysis of this case study also shows that once faced with the critical situation of having to cease the restaurant, the property conditions (family business, commitment to the building and the geographical environment) allow the manager's creative and innovative capacity to be released to lead a process of radical innovation in a context of very strong uncertainty and economic crisis that requires the company's ability to remain alert and receptive. In this vein, the radical innovation of this SMTE has exemplified that the company's ability to resist failure, to face and overcome resistance to change and to move away from the comfort zone, enabled to turn its old restaurant business model into a totally new one, a visitor center business model, that is allowing the development of a sustainable competitive advantage over time.

The case study presented also exemplifies that once the path of innovation has been started, it is verified how the evaluation, experimentation and innovation mechanisms are permanently maintained. In this way, the case analyzed shows how success in business model innovation is not something that should be taken for granted or a process that ends once it has been implemented, but rather requires a permanent attitude of questioning what exists and seeking improvements that can underpin its sustainability. This is due both to the learning process that takes place in the company and to the changes that can occur in the environmental conditions and that make the model or some business model elements being obsolete for exploiting new opportunities that can be identified.

In summary, this case study leads us to think that sustained success does not depend so much on the identification and implementation of a business model that may be adequate in a given business, but on the company's ability to continually modify it to maximize the value generated and its ability to appropriate a significant part of it as well as the ability to radically question, if necessary, the applied model. Within this context this chapter can contribute to business model research by addressing practical challenges in the tourism sector thus contributing to the limited evidence of examples of radical business model innovation in practice.

REFERENCES

Alegre, I., & Berbegal-Mirabent, J. (2016). Social innovation success factors: Hospitality and tourism social enterprises. *International Journal of Contemporary Hospitality Management*, *28*(6), 1155–1176. doi:10.1108/IJCHM-05-2014-0231

Ambrož, R., & Omerzel, D. G. (2018). Business model innovation in tourism: How to survive in highly competitive environments. *Academica Turistica-Tourism and Innovation Journal*, *10*(2), 175–183. doi:10.26493/2335-4194.10.175-183

Amit, R., & Zott, C. (2012). *Creating value through business model innovation*. Academic Press.

Andersen, P. N., & Andersen, J. B. (2012). *Green business model innovation in the tourism and experience economy*. Nordic Innovation.

Arbussà Reixach, A., Bikfalvi, A., & Marquès i Gou, P. (2017). Strategic agility-driven business model renewal: The case of an SME. *Management Decision*, *55*(2), 271–293. doi:10.1108/MD-05-2016-0355

Arcese, G., Valeri, M., Poponi, S. & Elmo, G. C. (2020). Innovative drivers for family business models in tourism. *Journal of Family Business Management*. doi:10.1108/JFBM-05-2020-0043

Bigelow, L., & Barney, J. (2021). What can strategy learn from the business model approach? *Journal of Management Studies*, *58*(2), 528–539. doi:10.1111/joms.12579

Bocken, N. M., & Samuel, W. (2014). A Literature and Practice Review to Develop Sustainable Business Model Archetypes. *Journal of Cleaner Production*, *65*, 42–56. doi:10.1016/j.jclepro.2013.11.039

Bogers, M., & Jensen, J. D. (2017). Open for business? An integrative framework and empirical assessment for business model innovation in the gastronomic sector. *British Food Journal*, *119*(11), 2325–2339. doi:10.1108/BFJ-07-2017-0394

Breier, M., Kallmuenzer, A., Clauss, T., Gast, J., Kraus, S., & Tiberius, V. (2021). The role of business model innovation in the hospitality industry during the COVID-19 crisis. *International Journal of Hospitality Management*, *92*, 102723. doi:10.1016/j.ijhm.2020.102723

Buhalis, D. (2019). Technology in tourism-from information communication technologies to eTourism and smart tourism towards ambient intelligence tourism: A perspective article. *Tourism Review*, *75*(1), 267–272. doi:10.1108/TR-06-2019-0258

Çakar, K., & Aykol, S. (2020). Case Study as a Research Method in Hospitality and Tourism Research: A Systematic Literature Review (1974–2020). *Cornell Hospitality Quarterly*, *62*(1), 21–31. doi:10.1177/1938965520971281

Cheah, S., Ho, Y. P., & Li, S. (2018). Business model innovation for sustainable performance in retail and hospitality industries. *Sustainability*, *10*(11), 3952. doi:10.3390u10113952

Chesbrough, H. (2010). Business model innovation: Opportunities and barriers. *Long Range Planning*, *43*(2-3), 354–363. doi:10.1016/j.lrp.2009.07.010

Christensen, C. M., & Raynor, M. E. (2003). *The innovator's solution: Creating and sustaining successful growth*. Harvard Business School Press.

Cranmer, E. E., Tom Dieck, M. C., & Fountoulaki, P. (2020). Exploring the value of augmented reality for tourism. *Tourism Management Perspectives*, *35*, 100672. doi:10.1016/j.tmp.2020.100672

Creswell, J. W. (1998). *Qualitative inquiry and research design: Choosing among five approaches*. SAGE.

Creswell, J. W., & Poth, C. N. (2018). *Qualitative inquiry and research design: Choosing among five approaches* (4th ed.). SAGE.

Crick, D., Chaudhry, S., & Crick, J. M. (2018). Risks/rewards and an evolving business model. *Qualitative Market Research*, *21*(2), 143–165. doi:10.1108/QMR-01-2017-0001

Cró, S., & Martins, A. M. (2017). Structural breaks in international tourism demand: Are they caused by crises or disasters? *Tourism Management*, *63*, 3–9. doi:10.1016/j.tourman.2017.05.009 PMID:32287750

De Montis, A., Ledda, A., Ganciu, A., Serra, V., & De Montis, S. (2015). Recovery of rural centres and "albergo diffuso": A case study in Sardinia, Italy. *Land Use Policy*, *47*, 12–28. doi:10.1016/j.landusepol.2015.03.003

Demil, B., & Lecocq, X. (2010). Business model evolution: In search of dynamic consistency. *Long Range Planning*, *43*(2-3), 227–246. doi:10.1016/j.lrp.2010.02.004

Desyllas, P., Salter, A., & Alexy, O. (2020). The breadth of business model reconfiguration and firm performance. *Strategic Organization*, 1–39.

Dorcic, J., Komsic, J., & Markovic, S. (2019). Mobile technologies and applications towards smart tourism–state of the art. *Tourism Review*, *74*(1), 82–103. doi:10.1108/TR-07-2017-0121

Druker, P. F. (1954). *The Practice of Management*. Harper and Row.

Fissi, S., Romolini, A., & Gori, E. (2020). Building a business model for a new form of hospitality: The albergo diffuso. *International Journal of Contemporary Hospitality Management*, *32*(1), 307–323. doi:10.1108/IJCHM-01-2019-0047

Flyvbjerg, B. (2006). Five misunderstandings about case-study research. *Qualitative Inquiry*, *12*(2), 219–245. doi:10.1177/1077800405284363

Foss, N. J., & Saebi, T. (2017). Fifteen Years of Research on Business Model Innovation: How far have we come, and where should we go? *Journal of Management, 43*(1), 200–227. doi:10.1177/0149206316675927

Foss, N. J., & Saebi, T. (2018). Business models and business model innovation: Between wicked and paradigmatic problems. *Long Range Planning, 51*(1), 9–21. doi:10.1016/j.lrp.2017.07.006

Gössling, S., & Michael Hall, C. (2019). Sharing versus collaborative economy: How to align ICT developments and the SDGs in tourism? *Journal of Sustainable Tourism, 27*(1), 74–96. doi:10.1080/09669582.2018.1560455

Gössling, S., Scott, D., & Hall, C. M. (2020). Pandemics, tourism and global change: A rapid assessment of COVID-19. *Journal of Sustainable Tourism, 29*(1), 1–20. do i:10.1080/09669582.2020.1758708

Gretzel, U., Sigala, M., Xiang, Z., & Koo, C. (2015). Smart Tourism, foundations and developments. *Electronic Markets, 25*(3), 179–188. doi:10.100712525-015-0196-8

Gustafsson, J. (2017). *Single case studies vs multiple case studies: A comparative study*. Academy of Business, Engineering and Science, Halmstad University.

Guttentag, D. (2015). Airbnb: Disruptive innovation and the rise of an informal tourism accommodation sector. *Current Issues in Tourism, 18*(12), 1192–1217. do i:10.1080/13683500.2013.827159

Han, D. I., Jung, T., & Gibson, A. (2013). Dublin AR: implementing augmented reality in tourism. In *Information and communication technologies in tourism 2014* (pp. 511–523). Springer. doi:10.1007/978-3-319-03973-2_37

Hjalager, A. M. (2010). A review of innovation research in tourism. *Tourism Management, 31*(1), 1–12. doi:10.1016/j.tourman.2009.08.012

Hjalager, A. M., Kwiatkowski, G., & Østervig Larsen, M. (2017). Innovation gaps in Scandinavian rural tourism. *Scandinavian Journal of Hospitality and Tourism, 33*(2), 1–17.

Hjalager, A. M., & Madsen, E. L. (2018). Business Model Innovation in Tourism: Opportunities and Challenges. The Sage Handbook of Tourism Management, 373.

Hoorani, B. H., Nair, L. B., & Gibbert, M. (2019). Designing for impact: The effect of rigor and case study design on citations of qualitative case studies in management. *Scientometrics, 121*(1), 285–306. doi:10.100711192-019-03178-w

Hyett, N., Kenny, A., & Dickson-Swift, V. (2014). Methodology or method? A critical review of qualitative case study reports. *International Journal of Qualitative Studies on Health and Well-being*, *9*(1), 23606. doi:10.3402/qhw.v9.23606 PMID:24809980

Jensen, & Prebensen, N. (2015). Innovation and value creation in experience-based tourism. *Scandinavian Journal of Hospitality and Tourism, 15*(1), 1-8,

Jones, P., & Comfort, D. (2020). The COVID-19 Crisis, Tourism and Sustainable Development. *Athens Journal of Tourism*, *7*(2), 75–86. doi:10.30958/ajt.7-2-1

Kounavis, C. D., Kasimati, A. E., & Zamani, E. D. (2012). Enhancing the tourism experience through mobile augmented reality: Challenges and prospects. *International Journal of Engineering Business Management*, *4*, 10. doi:10.5772/51644

Krajcik, V., Kljucnikov, A., & Rihova, E. (2019). Innovative Sharing Economy's Business Models in Tourism: Case of Airbnb in Prague. *Marketing and Management of Innovations*, *2*, 108–117. doi:10.21272/mmi.2019.2-10

Kraus, S., Clauss, T., Breier, M., Gast, J., Zardini, A., & Tiberius, V. (2020). The economics of COVID-19: Initial empirical evidence on how family firms in five European countries cope with the corona crisis. *International Journal of Entrepreneurial Behaviour & Research*, *26*(5), 1067–1092. doi:10.1108/IJEBR-04-2020-0214

Lanzolla, G., & Markides, C. (2021). A Business Model View of Strategy. *Journal of Management Studies*, *58*(2), 540–553. doi:10.1111/joms.12580

Laukkanen, M., & Patala, S. (2014). Analysing barriers to sustainable business model innovations: Innovation systems approach. *International Journal of Innovation Management*, *18*(6), 1440010. doi:10.1142/S1363919614400106

Lee, S. J. (2017). A review of audio guides in the era of smart tourism. *Information Systems Frontiers*, *19*(4), 705–715. doi:10.100710796-016-9666-6

Leitch, S., & Palmer, I. (2010). Analyzing texts in context: Current practices and new protocols for critical discourse analysis in organization studies. *Journal of Management Studies*, *47*(6), 1194–1212. doi:10.1111/j.1467-6486.2009.00884.x

Loureiro, S. M. C., Guerreiro, J., & Ali, F. (2020). 20 years of research on virtual reality and augmented reality in tourism context: A text-mining approach. *Tourism Management*, *77*, 104028. doi:10.1016/j.tourman.2019.104028

Massa, L., Tucci, C. L., & Afuah, A. (2017). A Critical Assessment of Business Model Research. *The Academy of Management Annals*, *11*(1), 73–104. doi:10.5465/annals.2014.0072

Mehraliyev, F., Chan, I. C. C., Choi, Y., Koseoglu, M. A., & Law, R. (2020). A state-of-the-art review of smart tourism research. *Journal of Travel & Tourism Marketing*, *37*(1), 78–91. doi:10.1080/10548408.2020.1712309

Mendoza, J. M. F., Gallego-Schmid, A., & Azapagic, A. (2019). A methodological framework for the implementation of circular economy thinking in higher education institutions: Towards sustainable campus management. *Journal of Cleaner Production*, *226*, 831–844. doi:10.1016/j.jclepro.2019.04.060

Mihalic, T., Zabkar, V., & Knezevic Cvelbar, L. (2012). A hotel sustainability business model: Evidence from Slovenia. *Journal of Sustainable Tourism*, *20*(5), 701–709. doi:10.1080/09669582.2011.632092

Mihardjo, L. W., & Alamsjah, F. (2019). Mediating Role of Co-creation Strategy on the Relationship between Business Model Innovation and Corporate Reputation: A Case Study on Indonesian Telecommunication Firms. *Journal of Technical Education and Training*, *11*(4), 67–76.

Morgan, T., Anokhin, S., Ofstein, L. F., & Friske, W. (2020). SME response to major exogenous shocks: The bright and dark sides of business model pivoting. *International Small Business Journal: Researching Entrepreneurship*, *38*(5), 369–379. doi:10.1177/0266242620936590

Morris, M., Schindehutte, M., & Allen, J. (2005). The entrepreneur's business model: Toward a unified perspective. *Journal of Business Research*, *58*(6), 726–735. doi:10.1016/j.jbusres.2003.11.001

Musa, S. M., & Mamun, S. A. (2020). Business Model Innovation as Strategic Renewal Process: A Case of China's Most Innovative Firm. *Global Business and Management Research*, *10*(2), 1–18.

Okumus, F., Altinay, M., & Arasli, H. (2005). The impact of Turkey's economic crisis of February 2001 on the tourism industry in Northern Cyprus. *Tourism Management*, *26*(1), 95–104. doi:10.1016/j.tourman.2003.08.013

Osiyevskyy, O., & Dewald, J. R. (2018). The pressure cooker: When crisis stimulates explorative business model change intentions. *Long Range Planning*, *51*(4), 540–560. doi:10.1016/j.lrp.2017.09.003

Paniccia, P., & Leoni, L. (2017). Co-evolution in tourism: The case of Albergo Diffuso. *Current Issues in Tourism*, 1–28.

Peric, M., & Wise, N. (2015). Understanding the delivery of experience: Conceptualising business models and sports tourism, assessing two case studies in Istria, Croatia. *Local Economy*, *30*(8), 1000–1016. doi:10.1177/0269094215604131

Pikkemaat, B., & Peters, M. (2006). Towards the measurement of innovation—A pilot study in the small and medium sized hotel industry. *Journal of Quality Assurance in Hospitality & Tourism*, *6*(3-4), 89–112. doi:10.1300/J162v06n03_06

Prescott, J. E., & Filatotchev, I. (2021). The Business Model Phenomenon: Towards Theoretical Relevance. *Journal of Management Studies*, *58*(2), 517–527. doi:10.1111/joms.12610

Randelli, F., Romei, P., & Tortora, M. (2014). An evolutionary approach to the study of rural tourism: The case of Tuscany. *Land Use Policy*, *38*, 276–281. doi:10.1016/j.landusepol.2013.11.009

Reinhold, S., Zach, F. J., & Krizaj, D. (2018). Business models in tourism – state of the art. *Tourism Review*, *74*(6), 1120–1134. doi:10.1108/TR-02-2018-0027

Riege, A. M. (2003). Validity and reliability tests in case study research: A literature review with "hands-on" applications for each research phase. *Qualitative Market Research*, *6*(2), 75–86. doi:10.1108/13522750310470055

Ritchie, B. W., & Jiang, Y. (2019). A review of research on tourism risk, crisis and disaster management: Launching the annals of tourism research curated collection on tourism risk, crisis and disaster management. *Annals of Tourism Research*, *79*, 102812. doi:10.1016/j.annals.2019.102812

Rosato, P. F., Caputo, A., Valente, D., & Pizzi, S. (2021). 2030 Agenda and Sustainable Business Models in Tourism: A bibliometric analysis. *Ecological Indicators*, *121*, 106978. doi:10.1016/j.ecolind.2020.106978

Runfola, A., Rosati, M., & Guercini, S. (2013). New business model in online hotel distribution: Emerging private sales versus leading IDS. *Service Business*, *7*(2), 183–205. doi:10.100711628-012-0150-1

Saebi, T., Lien, L. B., & Foss, N. (2017). What Drives Business Model Adaptation? The Impact of Opportunities, Threats and Strategic Orientation. *Long Range Planning*, *50*(5), 567–581. doi:10.1016/j.lrp.2016.06.006

Serravalle, F., Ferraris, A., Vrontis, D., Thrassou, A., & Christofi, M. (2019). Augmented reality in the tourism industry: A multi-stakeholder analysis of museums. *Tourism Management Perspectives*, *32*, 100549. doi:10.1016/j.tmp.2019.07.002

Sinkovics, R. R., & Alfoldi, E. A. (2012). Progressive focusing and trustworthiness in qualitative research: The enabling role of computer-assisted qualitative data analysis software (CAQDAS). *Management International Review*, *52*(6), 817–845. doi:10.100711575-012-0140-5

Sosna, M., Trevinyo-Rodríguez, R. N., & Velamuri, S. R. (2010). Business model innovation through trial-and-error learning: The Naturhouse case. *Long Range Planning*, *43*(2-3), 383–407. doi:10.1016/j.lrp.2010.02.003

Souto, J. E. (2015). Business model innovation and business concept innovation as the context of incremental innovation and radical innovation. *Tourism Management*, *51*, 142–155. doi:10.1016/j.tourman.2015.05.017

Spieth, P., Schneckenberg, D., & Ricart, J. E. (2014). Business model innovation–state of the art and future challenges for the field. *R & D Management*, *44*(3), 237–247. doi:10.1111/radm.12071

Stake, R. E. (2005). Qualitative case studies. In N. K. Denzin & Y. S. Lincoln (Eds.), *The Sage handbook of qualitative research* (3rd ed., pp. 443–466). SAGE.

Stamboulis, Y., & Skayannis, P. (2003). Innovation strategies and technology for experience-based tourism. *Tourism Management*, *24*(1), 35–43. doi:10.1016/S0261-5177(02)00047-X

Stamboulis, Y., & Skayannis, P. (2003). Innovation strategies and technology for experience-based tourism. *Tourism Management*, *24*(1), 35–43. doi:10.1016/S0261-5177(02)00047-X

Sundbo, J., Orfila-sintes, F., & Sorensen, F. (2007). The innovative behavior of tourism firmscomparative studies of Denmark and Spain. *Research Policy*, *36*(1), 88–106. doi:10.1016/j.respol.2006.08.004

Szromek, A. R., & Naramski, M. (2019). A Business Model in Spa Tourism Enterprises: Case Study from Poland. *Sustainability*, *11*(10), 2880. doi:10.3390u11102880

Teece, D. J. (2010). Business models, business strategy and innovation. *Long Range Planning*, *43*(2-3), 172–194. doi:10.1016/j.lrp.2009.07.003

Thomas, R., & Wood, E. (2014). Innovation in tourism: Re-conceptualising and measuring the absorptive capacity of the hotel sector. *Tourism Management*, *45*, 39–48. doi:10.1016/j.tourman.2014.03.012

Wiprächtiger, D., Narayanamurthy, G., Moser, R., & Sengupta, T. (2019). Access-based business model innovation in frontier markets: Case study of shared mobility in Timor-Leste. *Technological Forecasting and Social Change*, *143*, 224–238. doi:10.1016/j.techfore.2019.02.004

Yin, R. K. (2003). *Case Study Research: Design and Methods* (3rd ed.). Sage.

Yin, R. K. (2014). *Case Study Research: Design and Methods* (5th ed.). Sage.

Yin, R. K. (2018). *Case Study Research and Application: Design and Methods* (6th ed.). Sage.

Yung, R., & Khoo-Lattimore, C. (2019). New realities: A systematic literature review on virtual reality and augmented reality in tourism research. *Current Issues in Tourism*, *22*(17), 2056–2081. doi:10.1080/13683500.2017.1417359

Zeng, B., Carter, R. W., & De Lacy, T. (2005). Short-term perturbations and tourism effects: The case of SARS in China. *Current Issues in Tourism*, *8*(4), 306–322. doi:10.1080/13683500508668220

Zhang, W., Daim, T., & Zhang, Q. (2019). Exploring the multi-phase driven process for disruptive business model innovation of e-business microcredit: A multiple case study from China. *Journal of the Knowledge Economy*, *10*(2), 590–617. doi:10.100713132-017-0483-z

Zott, C., & Amit, R. (2010). Business model design: An activity system perspective. *Long Range Planning*, *43*(2-3), 216–226. doi:10.1016/j.lrp.2009.07.004

Zott, C., Amit, R., & Massa, L. (2011). The business model: Recent developments and future research. *Journal of Management*, *37*(4), 1019–1042. doi:10.1177/0149206311406265

ADDITIONAL READING

Afuah, A. (2014). *Business model innovation: concepts, analysis, and cases*. Routledge. doi:10.4324/9780203584583

Alvarez, M. D., Yuksel, A., & Go, F. (2016). *Heritage tourism destinations: preservation, communication and development*. CABI.

Foss, N. J., & Saebi, T. (2015). *Business model innovation: The organizational dimension*. OUP Oxford. doi:10.1093/acprof:oso/9780198701873.001.0001

Hall, C. M., & Williams, A. M. (2019). *Tourism and innovation*. Routledge. doi:10.4324/9781315162836

Leslie, D., & Sigala, M. (2005). *International cultural tourism: Management, implications and cases*. Routledge.

Raj, R., Griffin, K., & Morpeth, N. D. (2013). *Cultural tourism*. CABI. doi:10.1079/9781845939236.0000

Smith, M. K. (2009). *Issues in cultural tourism studies*. Routledge. doi:10.4324/9780203869857

KEY TERMS AND DEFINITIONS

Business Assets: Items of value owned by a company. Business assets span many categories. They can be physical, tangible goods, such as vehicles, real estate, computers, office furniture, and other fixtures, or intangible items, such as intellectual property.

Business Model: A business model is a "concise representation of how an interrelated set of decision variables in the areas of venture strategy, architecture, and economics are addressed to create sustainable competitive advantage in defined markets" (Morris, Schindehutte, & Allen, 2005: 727). It has six fundamental components: Value proposition, customer, internal processes/competencies, external positioning, economic model, and personal/investor factors.

Business Model Innovation: Business model innovation is the art of enhancing advantage and value creation by making simultaneous—and mutually supportive—changes both to an organization's value proposition to customers and to its underlying operating model.

Crisis: Situation in which something or someone is affected by one or more very serious problems.

Cultural Experience-Based Tourism Enterprises: Quite broadly, cultural experience-based tourism enterprises indicate a type of tourism that is different from more conventional tourism (Stamboulis & Skayannis, 2003) and is characterized with offerings with a relatively high degree of differentiation and intangible value perceived by the customers at a certain place and in a certain time. The basis for differentiating can be linked to elements such as specific themes, extreme contexts, novel and innovative arrangements offering exclusiveness, perceived experience value, specialized competence, multiple sense involvement and other elements that increase the perceived value for the customer.

Tourism Sector: The tourism industry is the total of all businesses that directly provide goods or services to facilitate business, pleasure and leisure activities away from the home environment.

Value: Value in business models is the worth in monetary terms of the technical, economic, service, and social benefits a customer company receives in exchange for the price it pays for a market offering.

Chapter 6
Effects of COVID-19 on Small Businesses in the Catering Sector:
Measures for recovery

Diego R. Toubes
https://orcid.org/0000-0001-7017-6659
University of Vigo, Spain

Noelia Araújo Vila
https://orcid.org/0000-0002-3395-8536
University of Vigo, Spain

Sara Fidalgo Estévez
University of Vigo, Spain

ABSTRACT

This study focuses on the effects of COVID-19 on independent restaurant operators in small cities. Proposals for recovery are assessed both in terms of business operations and service provision and in terms of the governance measures demanded by restaurateurs. Semi-structured open-ended interviews were conducted with 38 restaurateurs in Ourense (Spain) during the toughest months of the COVID-19 pandemic. This study provides key operational and governance measures for the recovery of the restaurant sector and confirms the need to generate a greater connection and empathy with the customer and to adapt to the new situation of instability and uncertainty by incorporating new business models. More effective governance measures are needed in line with increased public concern for sustainability, humanization, and environmental conservation.

DOI: 10.4018/978-1-7998-6996-2.ch006

INTRODUCTION

The COVID-19 pandemic has brought tourism activities to a halt in almost every country in the world. The most affected sub-sectors have been - and are likely to continue to be - those where measures to ensure physical distance from users are most needed. Restaurants, bars and cafés are included among these tourist activities, which are considered to be at risk. These businesses are being affected by the implementation of strong regulations that limit indoor capacity or even cancel it completely, having to serve customers in outdoor areas set up for this purpose.

The current and future implications of this way of operating and serving clients, besides changes in the response of the public and clients to health risks and the evolution of the health crisis, will provide guidelines for how the restoration sub-sector can evolve. In many cities, the way restaurants conduct business will likely change in the future. The hygiene protection measures that must be implemented to gain the confidence of customers now shape the urban environment, for example, by occupying outdoor spaces such as terraces and courtyards, or by advancing service hours.

According to the UN policy brief of July 2020 'COVID-19 in an Urban World', 90 per cent of infections occur in cities. Cities must adapt to the new scenario and possible change in the tourism paradigm. The debate that has arisen in recent years around the concept of the "smart city" in relation to its excessive focus on technology adds interest to the work. Criticism of the dehumanisation that the concept of "smart city" conveyed, may now have more resonance because of the need to humanize urban spaces, rather than adding spaces for control. The symbiosis of such aspects with the need for restaurants to be able to adapt and prosper in the new situation, in being more resilient, is the aim of this work.

This study focuses on the effects of COVID-19 on the independent restaurant sector in small cities -up to 100.000 inhabitants-. Independent restaurant operators often face different challenges than their chain restaurant counterparts. The relatively smaller scale of these businesses may limit their ability and business options to remain financially solvent. The purpose of this study is to explore and assess the perceptions and attitudes of independent restaurant operators regarding the operational and economic effects of the COVID-19 pandemic on their businesses at the end of 2020 and provide key operational and governance measures for recovery in the catering sector. Specifically, proposals for recovery are assessed, both in the area of business operations and service delivery, as well as on the governance measures that restaurateurs are demanding.

The results of this study are a contribution to the literature on crisis management in small firms in the restaurant industry. At the same time, they help independent restaurant operators to better understand the new scenario opened up by the COVID-19 pandemic in relation to new business models and consumer sensitivities.

THE NEW SCENARIO: HUMANIZING SMART CITIES

The recent outbreak of COVID-19 has demonstrated how cities are resilient to the emergency, for example, pre-existing ICT infrastructure has made it possible to easily switch to a virtual, or less face-to-face, mode of working (Kim, 2021). In this regard, COVID-19 has intensified the use of technology to deal with unexpected scenarios. The use of technology, e.g. in social networking and videoconferencing, has prevented social isolation, despite mandatory physical distancing for health reasons. As discussed in the previous section, measures have also been put in place in the foodservice sector for contact-free customer service and home service.

Before the pandemic, there was already a social and academic debate about how new technologies were affecting the lives of individuals in cities. At the beginning of the 21st century, smart cities (see key terms and definitions of the chapter) became a favourite concept all over the world, widely cited and discussed in numerous studies in different areas of knowledge, and promoted by smart digital industries, think tanks and consultants as well as media and thousands of digital start-ups. According to Angelidou (2015), smart city strategies play a decisive role in how cities will choose to take advantage of technology to favour the development of innovation networks, healthy societies and dynamic economies.

Yigitcanlar *et al.* (2019) performed a literature review about the challenges of smart cities in delivering sustainable outcomes. They found three major points: (a) smart city policies are characterized by heavy techno centricity; (b) smart city practices involve complexities, and; (c) smart city notions are conceptualized in an ad-hoc manner. Krivý (2018) proposed three main arguments against the concept of smart city: (1) it is incompatible with an informal character of the city, (2) it subjects the city to corporate power, and (3) it reproduces social and urban inequalities. These arguments show that achieving a goal of genuine sustainability is not straightforward to achieve in cities, and requires an anthropocentric approach to development practice and policy formulation, i.e., an approach more focused on the development of the individual, rather than on the development of technologies.

Society has evolved rapidly and the role of citizens in smart cities has changed. Governance models must also adapt and provide a human and social response to these changes (Hernàndez, 2021). According to UN reports, more than two-thirds of the world's population will live in cities, by 2030. Urban population growth has put governments in a serious challenge in various areas of urbanization. The ultimate goal of a smart city is to create a new urban management perspective that focuses on all aspects of urban real-life (Fakhimi, Khani, & Sardroud, 2021). There is a groundswell that encourages the move from smart cities to intelligent communities (Jung, 2019). Intelligent communities are collectively built from combinations of individual cognitive skills and information systems that operate in the physical, institutional, and digital spaces of cities (Komninos, 2006). Liugailaitė-Radzvickienė and Jucevičius (2014) noted that information communication technologies are important but only in the role of helping tools for more successful development. Hence, the focus should be on the comprehensive development of the individual in his environment, stead of the overuse of the technology of cities. Kunzmann (2020) maintains that the COVID-19 pandemic will be a welcomed lubricant for the innovative stakeholders of many smart-city development policies. If the goal is to plan genuinely sustainable communities, local political entities should take into account the human factor in their proposals.

IMPACT OF THE PANDEMIC IN CAFÉS, BARS AND RESTAURANTS

Background

Before the pandemic spread around the world, intense competition in the restaurant sector was one of the main challenges for businesses, mainly in countries or areas with a high density of bars and restaurants. Success in this environment requires brands to focus on a number of strategic imperatives (Langford & Weissenberg, 2018):

- *Embrace the experience.* They need to deliver a differentiated customer experience along with their meals. But differentiation is not so simple. It requires restaurants to execute on all dimensions of the customer experience: value, menu, quality, and the right managers and employees to tie it all together
- *Drive employee engagement.* A staff of friendly, hospitable employees was cited as the most important element needed for a positive experience at a restaurant.
- *Dominate delivery.* The shift to third-party digital distribution can translate to massive exposure and order volume.
- *Operational excellence and compliance.* Many restaurants are realizing the pressing need to augment store operations through better technology enablement, which reinforces operational excellence and enhances team member capabilities.

COVID-19 has put bars and restaurants under greater pressure, although it is not the first time in history that a virus of the coronavirus family has affected society and by extension the restaurant sector. Severe acute respiratory syndrome (SARS-CoV) was recognized as a global threat in mid-March 2003. The first known cases of SARS occurred in China's Guangdong province, and after five months of fighting, WHO declared the epidemic over on 5 July 2003. SARS-CoV resulted in 8,098 cases in 26 countries, with 774 deaths (WHO, 2004). Initially, it was a sanitary problem but its consequences affected in a very important way all the productive sectors, among them the restaurant and catering sector. Although previous global crises have had a major impact on the tourism industry, the present crisis has different outlines, as there was a mandatory lockdown which has never happened before.

Tse and Leo Sin (2006) examined the 2003 outbreak in Hong Kong restaurants from a crisis management perspective to illustrate how local restaurants could cope with such situations and develop recovery strategies. Their study suggested that the right strategy for dealing with such disasters is to react quickly to minimize the damage before it escalates exponentially and threatens the survival of the business. Physical crises cause social unrest, so restaurant operators must try to calm customers and negotiate with affected parties to achieve win-win resolutions. Tse and Leo Sin (2006) suggested a tiered approach to dealing with this type of crisis. First, the risk be assessed before a reconstruction plan can be drawn up. Then, immediately try to minimize operating costs and get financed at the lowest possible cost, for example, through forgivable loans issued by the government in emergency situations. Revenue optimization strategies should also be implemented to increase the value perceived by customers. The most obvious tactic is to improve the quality of food and service. In addition, restaurants may offer discounts or other forms of promotion to attract customers. Several restaurants also took steps to improve their public image through community philanthropy, with a portion of their sales being donated to SARS-related causes.

At Present

Faced with the current crisis, owners and managers of bars and restaurants are worried about the future of the sector. There is a common concern to all entrepreneurs in the restaurant business for the post-pandemic period and about the strategies and lessons learned for the future, about the measures that the governments should implement, i.e., fear of a lack of appropriate government measures, and the availability of working capital (Madeira, Palrão, & Mendes, 2020).

Despite the significant vulnerability to external shocks, restaurateurs have sought new opportunities based on the changing environment and new normality brought about by COVID-19. These opportunities include altering or adapting their business models to minimize customer uncertainty towards the restaurant industry. Clean safety message framing can be a highly efficient sales promotion strategy for small, independent restaurants, specifically during a pandemic such as COVID-19 (Kim, Bonn, & Cho, 2021). In addition, the rapid spread of COVID-19 has significantly increased delivery services. Because of the fear of infection, many people have chosen to stay home voluntarily and have resorted to delivery services to buy meals in restaurants. Within the restaurants, different measures of separation of spaces were tested, such as partitions between tables or placement of mannequins, with a greater preference for the first measure than the second (Taylor, 2020). In a macro-study conducted with sales data from 86,507 small and medium-sized catering enterprises, in nine cities in mainland China, Kim, Kim and Wang (2021) recommend sharing information on the current effort of the catering industry and measures taken to cope with financial difficulties amid the COVID-19 pandemic. For example, offering contactless delivery service, or partnering with mobile delivery platforms.

The pandemic has accelerated the process of business automation and the trend towards less contact with customers, for example, through the adoption of QR codes for customer registration or menu consultation. In some cases, they have been adopted because of regulatory requirements, but in other cases, it has been the need to offer customers the level of security they were demanding. The situation presents an opportunity for automation measures and the development of robotic applications that enhance tourist experiences and the emergence of new 'high-touch' employment opportunities for hospitality workers (Zeng, Chen, & Lew, 2020).

Malbec (2020, 2020a) points out some changes or innovations and new security measures in business models. Among those proposed are the following:

- Payment Solutions. Less "dangerous" contactless payment systems will be promoted. Banknotes are losing their use because they are major carriers of germs, and cards are also looking for alternatives because a virus can remain on them for 72 hours.
- Business Model Changes: Delivery & Customer Pickup
- Fine Dining & Table Service Menus. Digital menu, without paper or plastic, displayed on a device, or through an application shared with the customer. Enable payment from the table.
- Employee-Facing Restaurant Technology. Request orders by voice, avoiding the use of touch devices. Offer electronic receipts instead of paper.
- Daily Health Checks (temperature).
- Checklists & Alerting. Access registration with QR code at the entrance of the establishment and management of alerts through the COVID Radar application, in case of infection of a customer.

- Supply Chain Visibility. Ensure the safety of the chain from the kitchen to the customer's home. Food safety throughout the service chain before the customer picks up or receives their food.
- Greater emphasis on Food Safety Scoring and Cleanliness Standards. Hygiene and cleaning measures on any frequently used device, keyboard or machinery. Wipes, gels, gloves, masks, etc. should be included in the inventories.
- Safety over Efficiency. Anti-Microbial POS Screens and other device surfaces. Elimination of salt and pepper shakers or other common or shared dispensers. Sanitized cutlery, glasses and plates on the table or with a seal.
- Restaurant Design Changes. Increase space between tables. Improve access to hygiene products (wipes and gels on tables and in access areas).

Client risk reduction strategies aim to prepare for future external shocks caused by such disease-related risk situations.

The Impact on the Sector

The impact of COVID-19 on the global restaurant industry has been enormous (Dube, Nhamo & Chikodzi, 2020). In many countries around the world, this sector has been disproportionately negatively affected (Bartik *et al.*, 2020). Patrons, mainly indoor service, were reduced to zero in many countries, as governments around the world enacted social distancing measures, mobility restrictions and lockdowns. COVID-19 also led to an unprecedented loss of employment and revenue, resulting in millions of jobs and billions of euros in potential revenue lost (Dube, Nhamo, & Chikodzi, 2020).

Gunay and Kurtulmuş (2021) analysed the impact of social distancing measures in the US service sector (hotels, entertainment, restaurants and airlines). Their study suggested that the pandemic initially affected mainly the entertainment and airline sectors, with a gradual deterioration in the hotel sector, mainly small businesses. As time goes by, the impacts of the crisis on the restaurant sector are becoming more evident. For example, Panzone, Larcom and She (2021) estimated that in the period March to August 2020, COVID-19 restrictions in the UK accounted for a loss of £25 billion in food and beverage service turnover.

Social distancing requirements reduce the capacity of hospitality operations, such as restaurants, to be able to host guests while the most extreme measures require the closure of operations to prevent social contact (Hall, Scott, & Gössling, 2020). Public perception of the risk of infection in restaurants is high. The prominence of the virus systematically increases the preference for private dining facilities (Kim & Lee, 2020). Byrd *et al.* (2021) note that the risk perceptions of consumers varied with food price, gender, and being in a high-risk category of COVID-19. And Brewer and Sebby (2021) find out that the menu's visual appeal and informativeness, and the perception of COVID-19 risks, impact on consumer purchase intentions; this causal relationship was significantly mediated by consumers' desire for food and their perceived convenience of online food ordering.

Despite the strong impact that the restaurant sector has received, some studies also point to greater resilience. In some locations, individuals, businesses and communities will seek to transform and change the way in which tourism and hospitality are conducted (Nunn, 2020). Sobaih, Elshaer, Hasanein and Abdelaziz's (2021) study explored the direct impact of small hospitality business resilience on sustainable tourism development, as well as the indirect impact through performance. A survey of small hospitality businesses owners and managers in Cairo (Egypt) concluded that the type of business has a significant effect on outcomes, with restaurant owners showing more resilience than their hotel counterparts.

CONTEXT OF THE PANDEMIC ON THE CATERING BUSINESS IN SPAIN

The first case of COVID-19 detected in Spain was on January 31, 2020. It was a German tourist visiting La Gomera (Canary Islands). The second case became known on February 10, a British tourist in Mallorca. On February 12, the Mobile World Congress (MWC) in Barcelona was cancelled due to the risk of contagion of the participants. With this decision, it is estimated that the city of Barcelona lost 500 million euros in hotels and restaurants (Millán, 2020). On February 25, the third positive result was announced, an Italian tourist staying at the H10 hotel in Adeje (Tenerife), which led to the quarantine of the 700 tourists staying at the hotel.

The impact of the pandemic mainly affected those activities where individuals gather in confined spaces, stay for long periods of time, and mask use is not permanent. All these characteristics are found in bars, cafés and restaurants. The hospitality segment is one of the most important segments of the service sector in Spain. According to KPMG (2019) accommodation, food and beverage sectors employs 1,587 million workers in 2017, 7.1% of the total number of active workers, and turned over 129.45 billion euros, 6.5% of Spain's GDP.

Royal Decree 463/2020, articles 10 and 11, established the first measures aimed at the catering sector. Under the heading "containment measures in the field of commercial activity, cultural facilities, establishments and recreational activities, hotel and catering activities, and other additional activities", the suspension of hotel and catering activities was decreed, allowing only the provision of home delivery services. Article 11 established a series of measures to limit civil and religious ceremonies that had a great impact on the catering sector, such as weddings, baptisms and communions.

Following the situation of lockdown and temporary restraining orders in Spain on 13 March 2020, the sharpest deterioration in the profitability of companies in 2020 occurred in the hotel, restaurant and leisure industry, which is also the sector that, before the crisis, started from the lowest average level of profitability, 2.4%. As a result of the pandemic, this level would fall sharply to -12.4%. Moreover, in this sector, 75% of the companies went into negative profitability in 2020 (Bank of Spain, 2020). The current hospitality industry shows unequal capacities to cover fixed costs when comparing accommodation and catering businesses: 42% of companies do not have the liquidity to cover expenses beyond 25 days, and another large percentage would need to make serious adjustments (Bain and Ernst & Young, 2020).

On the other hand, the closure of restaurants, bars and cafés meant serious economic damage for those who provide their services on a lease basis (Guerra, 2020). For this reason, some voices advocated adapting to these new circumstances by reaching an agreement that satisfies both parties and in which the following options can be incorporated: the extension of the lease terms, the temporary suspension of rent payment, and the adjustment of the price based on the current situation and the partial cancellation of the agreed amount (López-Medel, 2020). It is uncertain how long many businesses will be able to last with no turnover or with revenues well below normal. In Spain, various measures were implemented to support the sector, which are discussed below.

Support and Protection Act

Some support and protection act were applied to alleviate the effects of the standstill in industry and business activities. One of the most widely used by catering companies was to claim coverage of the wages for employees on temporary leave ('furlough'), which is called ERTE in Spanish law. It is a procedure by which a company in an exceptional situation obtains authorization to suspend work contracts or reduce working hours on a temporary basis. That is to say, the worker's temporary leave is limited in time since, after a period of time, he/she returns to his/her job. During the mandatory closure decreed by the Spanish government during the period of confinement, a large number of bars and restaurants used this procedure to save the complicated situation and keep the workers' jobs. In the first 180 days the wage is reduced to 70% of the total and, in the event of loss of employment, the days consumed by the ERTE are deducted from the period with the right to payment, i.e., the workers consume their unemployment benefits while the ERTE is being applied.

Furloughs have been a tool especially used in sectors subject to partial closures such as catering and retail trade. The main requests and demands of the employers are aimed at obtaining a temporary extension of the ERTEs or the exemption of the payment of Social Security for workers. The bad start of 2021 with the rise of the third wave and the tightening of social distancing measures led the government to authorize successive extensions in the application of the ERTEs, the last one until May 31, 2021 (ESADE, 2021). In February 2021, the figure for registered unemployment in Spain was over four million people, to which should be added the almost 900,000 workers on furlough (Ministry of Labour and Social Economy, 2021), with a high percentage of these jobs belonging to the hotel and catering and commercial sector (ESADE, 2021).

In addition to the flexibility in the application of the ERTEs, most of the measures requested by hoteliers and industry associations are aimed at reducing costs and ensuring the liquidity of companies in a period of high uncertainty. Among the various measures, the following may be noted:

- Exemption or reduction of municipal fees and taxes that apply to establishments, such as the use of public space in terraces and sidewalks, garbage collection taxes, the economic activities tax (named IAE in Spain) which directly taxes the performance of any type of economic activity, and the amount is constant regardless of the size of the business or turnover, and a temporary waiver of the real estate tax (named IBI in Spain).
- Exempt the self-employed from contributions, energy supply taxes and financial obligations to landlords during the time of closure or reduced activity caused by the pandemic.
- Regarding the relationship with banks, freezing of loan and mortgage obligations, renegotiation of credit policies and greater flexibility in financing.
- Finally, direct and non-refundable aid measures are also proposed for businesses to stimulate recovery.

Regarding the operational aspects of the business, Berbel and Reyes (2020) conducted a survey among 510 operators in the sector on the actions that are being employed to deal with the crisis. The answers can be grouped into two broad areas, building trust and the business model. It is possible to generate confidence in customers when a series of aspects are taken care of, among which are: hygiene protocols, obtaining hygiene and safety certificates, strict compliance with legislation on the minimum distance between tables and the capacity, the use of protective elements in the establishment, such as screens, disinfectant gel dispensers, temperature control when entering the venues, extension of the terrace and giving visibility to the kitchen. The other group of actions refers to changes in the business model, including the following: take away and home delivery, the organisation of online ordering, management aimed at reducing waiting times (shift management), optimisation of resources, e. g. dispensing like food shops (some restaurants already offer products, preparations, utensils and recipes), greater use of discount vouchers, promotions and offers.

RESEARCH METHOD

The research focuses on the operational problems of small businesses during the pandemic, with possible governmental remedial solutions.

Context and Population of the Study Area

In addition to the ordinary food service provided by bars and restaurants, from a social and cultural perspective, in Spanish society, the bar is a meeting and gathering place. The "bar culture" is widespread throughout Spain. Bars, restaurants and nightlife venues have been part of Spanish society and the way we relate to each other. In mid-2020, a social movement in support of the restaurant industry emerged directly from the platform *Juntos por la Hostelería* (together for the catering industry), which aims to have bars and restaurants considered a World Heritage Site.

This "bar culture" is visible in the high number of catering businesses, cafés and bars open, and in the fact that a large part of the population works in this field. At the beginning of 2021, restaurants, bars, cafés and pubs employed 1.3 million people in Spain and had a turnover of around 94,000 million euros, which accounted for 4.7% of the national GDP (CEHE, 2021). In 2020 there were 253,544 establishments operating in the restaurant sector in Spain. Table 1 shows the number of establishments in Spain according to the three categories: bars, restaurants and food and catering businesses.

Table 1. Distribution of catering establishments in Spain in 2020

Type of Establishment	Total	Percentage
Bars	167,955	66,2%
Restaurants and cafés	70,747	27,9%
Meals for events and catering	14,852	5,9%

Source: INE (2021)

The city of Ourense, a provincial capital city of the autonomous community of Galicia, in the northwest of Spain, was chosen for the analysis. Ourense had 105,643 inhabitants by 2020 (IGE, 2021). The population of the province has an average age of 50.8 years, the second-highest in Spain after Zamora (51.2 years) (INE, 2020). The productive and employment structure of the city of Ourense depends to a large extent on the catering sector. In 2020 Ourense had 2,311 open restaurant businesses, a very high number compared to the total population. In 2020, the population density for each type of restaurant business was 187 inhabitants for Spain, 160 in Galicia and 132 inhabitants for each restaurant/bar in the province of Ourense (INE, 2021). The capital leads the Galician ranking in terms of the number of bars per inhabitant; with 429 establishments, the city has an average of four bars per thousand inhabitants (Vázquez, 2017). Table 2 shows the catering businesses open in Spain, Galicia and Ourense.

Table 2. Catering businesses in operation

	2017	2018	2019	2020
Spain	253.344	254.496	252.011	253.554
Galicia	17.727	17.702	17.393	16.876
Ourense	2.399	2.401	2.362	2.311

Source: INE (2021)

The predominant business profile of the catering sector in the study area is that of a microenterprise. Figure 1 shows the distribution of the volume of open businesses by the number of employees (INE, 2021). It can be seen that 92.7% of the businesses do not exceed five employees, and the most frequent type of business is 1 or 2 employees (44%).

Figure 1. Restaurant business by number of employees in Ourense. Year 2020

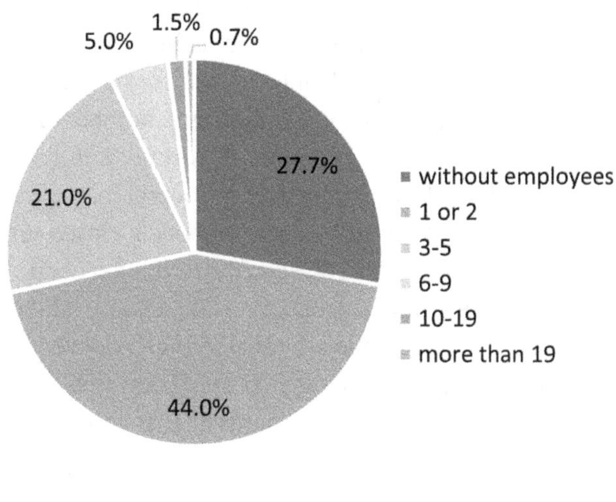

The hospitality sector in the province of Ourense is one of the most affected in Galicia (Nespereira, 2021). In March 2020, the number of workers registered with Social Security in the province was 5,235 hired and 2,564 self-employed, in January 2021 the number of contracts fell by 19.9% and 3.7% respectively, to 4,197 hired employees and 2,470 self-employed.

Research Sample and Procedures

Qualitative research is carried out to obtain the expected results. Interviews with semi-structured questionnaires have been used as a method of collecting business data. Semi-structured interviews offer an acceptable degree of flexibility, while maintaining sufficient uniformity to achieve interpretations consistent with the purposes of the study (Diaz-Bravo *et al.*, 2013). The data were analysed in order to explore the complexity and fragmented nature of the business and social context (Merriam, 2009).

We selected individuals who met the requirements of the target population, in our case, managers and owners of small catering businesses. The study sample was selected in a non-probabilistic and intentional way. The final sample consisted of 38 bars and cafés, all of the selected business were micro-enterprises. As it is a non-probabilistic sampling and no statistical inference is made we consider the requirements for representativeness of the sample to be sufficient to carry out the interviews.

A questionnaire was developed to analyse the impact of the COVID-19 pandemic on operations in hospitality, bar and restaurant businesses, and ultimately to evaluate their response and forecasts for future operations in the sector (see Appendix 1). The survey consisted of four sections. In the first section, respondents were asked about the business details, such as year of the start of the activity, maximum capacity, number of employees, existence and availability of a terrace and/or courtyard, and if yes, the number of tables. In the second section, the current impact of COVID-19, respondents were asked three questions about changes brought about by the pandemic in three aspects of their business, 1) changes in the service, 2) reduction of staff (number of employees in a temporary layoff (furlough), number of permanent layoff and duration of the furlough, and 3) reduction of working hours. The third section focuses on the forecasts for the future of bar and restaurant managers and owners. In this section they were asked five questions, the first one a closed-answer question - *are you considering closing the business? -, and* the following open-ended questions about (1) expected reductions in the capacity of the establishment to continue in the future, (2) expected changes in the way the service is developed, (3) changes in client schedules, and (4) expected changes in staff recruitment. The last section, called proposed solutions, sets out eight measures to deal with the current situation.

These measures have been taken from those that are currently in operation and from proposals demanded by hotel and catering associations or identified in academic studies on the subject. Finally, an open-ended question is posed to allow respondents to indicate other measures that do not appear in the survey.

A total of 38 interviews were carried out with restaurateurs. Most of those interviewed were the business owners or the main people in charge. The interviews were conducted face-to-face at the establishment itself in December 2020 and January 2021. During this period, restrictions on the service of restaurants were in force, such as the prohibition of serving meals inside the establishment, and the mandatory closing of the business at 6 pm.

RESULTS

The results of the interviews in the four sections into which the questionnaire was divided are presented below.

Section 1. Establishment Details

In general, the year of opening of the business in the sample is recent. Almost half of them are less than ten years old. The establishments are small in size. The average capacity of the establishments in the sample is 74 people. Figure 2 and 3 show the distribution of the sample by year of start and maximum capacity.

Figure 2. Start year of the businesses

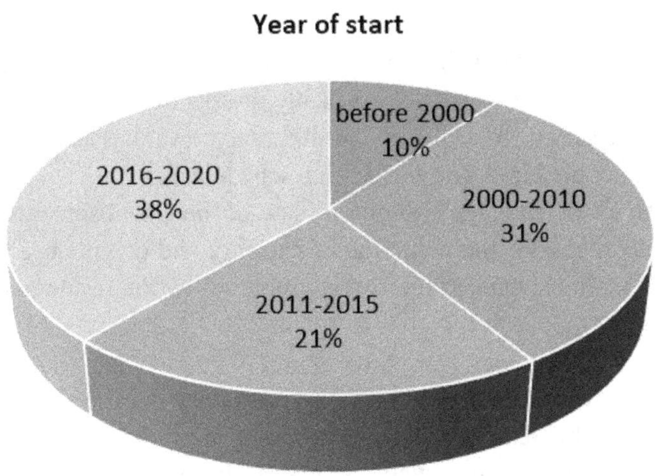

Figure 3. Maximum capacity of the businesses

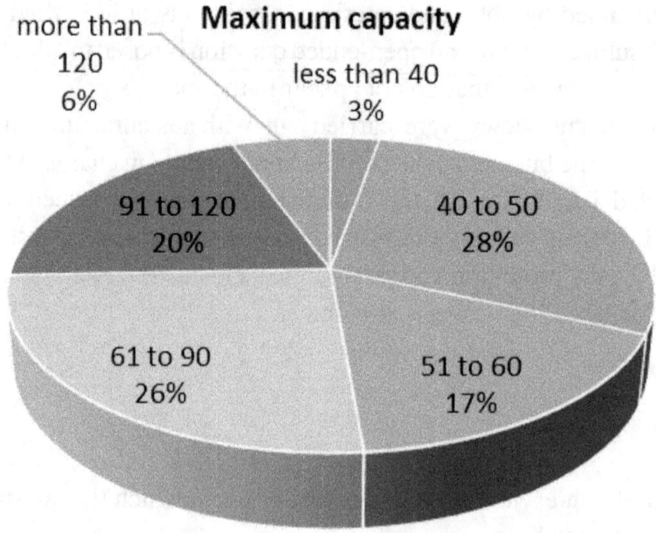

Of the 38 establishments in the sample, 32 have a terrace and 6 do not. On the other hand, 32 say they do not have a courtyard and 6 do. This is a mere coincidence of figures since the establishments without a terrace do not have a courtyard either. The establishments with a terrace add up to 233 tables and the establishments with a courtyard add up to 81 tables. The 38 establishments in the sample have 154 permanent employees, 6 temporary employees and 4 self-employed workers. Table 3 with this information broken down is included in appendix 2.

Section 2. Current Impact of COVID-19

This section asks respondents about changes in business operations that have occurred due to the impact of COVID-19. Specifically, it asks about the changes that have been made in the service to deal with the situation, how it has affected the staff and the number of people hired, and the service schedules.

In addition to the changes imposed by law in the establishments, such as the mandatory use of masks, the reduction of capacity and compliance with hygiene protocols, the main changes in the service reported by the restaurateurs were the following:

- Temporary closure. In addition to mandatory closures due to state of alarm or health reasons, fourteen respondents indicated that they decided to temporarily close the business, mainly because revenues did not cover operating expenses. This measure is indicated regardless of the size of the establishment. Two restaurants decided to close at noon.

- Takeaway and delivery service. Up to thirteen establishments report that they began delivering meals (from the front door when it was not possible inside the establishment) and hiring delivery services. A few interviewees stated that they took on the delivery service with their own staff.
- Reduction of working hours, mainly at night. Eleven respondents indicate that they changed schedules by reducing late-night service. The change in customer behaviour is one of the reasons behind this decision.
- Terrace service. The restrictions on consumption inside the establishments led businesses to look for customers on the terraces. Five businesses stated that they have expanded their terrace space. The capacity limitation on the terraces was set by law (50% of the maximum capacity in normal conditions on the terrace and 30% inside). The low winter temperatures - between 6ºC and 15ºC on average in the city - did not help customers to stay outdoors for long periods.

The pandemic has had a very strong impact on employment. The 38 establishments add up to 26 dismissals of permanent staff and 122 furloughs. Appendix 2 includes Table 4, which breaks down for each establishment, the number of furloughs, the months of duration, the permanent layoffs and the reduction in working hours, specifying the number of hours.

Section 3. Foresight for the Future

This section asks about future foresight for the business. The first question asks whether the restaurateur is thinking of closing the business permanently. Economic difficulties are pressing for many of them. Eight answered yes, nine said they were considering a partial seasonal closure (other than holidays), and the rest (21) said no. A summary of the responses is shown in Table 5.

Of the 21 businesses that do not plan to close, nine do not expect to reduce capacity in the future, five expect the reduction to vary between 20 and 50 per cent, and the remainder did not answer this question. Of the eight businesses that are thinking of closing, seven believe that in the near future the capacity of the establishments will be reduced by 20 to 50 per cent, and only one does not expect it to be reduced. Of the nine facilities considering partial, seasonal closure, five expect no future reduction in maximum capacity, three expect a 50% reduction, and one expects a 30% reduction.

Table 3. Expected changes in the future of the business

Are you thinking of closing your business?	*In the future, what changes do you foresee...*		
	...in the service?	*...in customer schedules?*	*...in staff recruitment?*
Yes, totally (8)	Home delivery and takeaway (5) None* (2) Terrace service (1)	Loss of customers at certain times slots (4) None (2) Earlier dinners (1) Cancel dinner service (1) Less customer turnover due to protocol compliance (1)	Reduction of staff (7) Reduction of working hours (1)
Partial (seasonally) (9)	Home delivery and takeaway (4) None (3) Earlier drink service (1) Distances and hydro alcoholic gel (1)	Loss of customers at certain times slots (6) None (3) Earlier dinners (3)	Reduction of staff (5) Recruitment of temporary staff (3) None (2) Reduction of working hours (1)
No (21)	Home delivery and takeaway (8) None (6) Terrace service (4)	Widespread loss of customers (7) Earlier service, loss of evening customers (7) None (5)	Back to normality (9) Reduction of staff (8) Reduction of working hours (1) Recruitment of temporary staff (1)

Note: in brackets the number of responses
*«None» means that respondents believe we will return to pre-pandemic normalcy

Section 4. Remedial Measures

In the last section, the interviewees were asked to indicate what, in their opinion, are the three most effective measures to deal with the current situation. Eight measures are proposed, with the possibility of contributing further measures. Table 6 shows the results.

The most widely advocated measure was to receive aid to cover fixed costs during compulsory closures. This measure was chosen by 27 of the 38 businesses. Exemption from Social Security payments for employees on furlough is the second most frequent option. Both of these measures include the idea of "pay-as-you-go", i.e. not requiring companies to pay taxes and fees if, for health and safety reasons, restrictions are imposed on their operations and they are unable to operate normally.

Table 4. Governance measures for recovery in the restaurant sector

Measures			1º	2º	3º	Unnumbered
Aid to cover fixed costs during the closure of the establishment	27	25,5%	6	5	3	13
Social Security exemption for employees on furlough	17	16,0%	1	4	4	8
Extension of the furlough period (ERTEs)	14	13,2%	4	2	2	6
Exemption from payment of the Economic Activities Tax (IAE)	11	10,4%	6	2		3
Fees exemption for the use of terraces	11	10,4%	1	3	2	5
Extension of terrace licenses	9	8,5%		2	5	2
Exemption from municipal taxes (e. g. garbage collection)	6	5,7%		4	2	
Tax flexibility	5	4,7%	3			2
Do not charge for services that cannot be used, e.g. tables, bars	2	1,9%			1	1
Social Security exemption for self-employed employees	1	0,9%	1			
Tax deferment	1	0,9%			1	
Increase capacity allowed	1	0,9%			1	
Exemption from the six months required to dismiss an employee after the period of furlough	1	0,9%			1	

Note: respondents who did not rank the answers were reflected under the label "unnumbered".

DISCUSSION AND CONCLUSION

The crisis caused by COVID-19 has particularly affected cities. The higher population density and the constant movement of citizens are an ideal breeding ground for the spread of the virus. Most businesses in the hospitality sector - bars, restaurants and cafés - are located in cities and have been among the hardest hit by the pandemic. Some businesses have had to close and others have survived. The restaurateurs interviewed in this study face a future of high uncertainty in which re-opening is expected to be slow and difficult, in fact, 20 per cent of those interviewed are thinking about closing.

However, this qualitative research also highlights the resilience of these establishments as they have been able to restructure and implement new business models in a very short time. Although restaurateurs were very concerned about their financial viability, the adaptation of businesses to the complicated situation has been astonishingly effective, by successfully integrating the complexities that have arisen (Yigitcanlar *et al.,* 2019).

Most respondents do not expect business activity to return to normal quickly. Some owners commented on the difficulty of meeting customer separation requirements in restaurants, given the limited space in the establishments. On the other hand, they also find it difficult to comply with timetable regulations or health-security guidelines due to continuous changes and updates. Businesses more dependent on tourism, i.e. more dependent on out-of-town clientele, such as higher-end restaurants, were pessimistic about a quick return of customers due to the perimeter entrance closure imposed on the city. Other businesses that cater primarily to the local customer and do not depend so much on tourism, such as bars and cafés, see the future with more optimistic prospects.

Some trends in the catering sector detected in this study on the supply side confirm the desirability of adapting to the needs of customers, something already identified in other studies (Ballarín, 2020), for example, by increasing takeaway or home delivery options and making it easier for customers to consume in open spaces. Business development will require improving processes and procedures in the management of the business to achieve greater profitability and changes in the internal communication channels with employees and collaborators to generate ideas. A growing penchant for unique dining experiences, authentic local menus and convenience is driving customers away from traditional venues and towards exploring other alternatives such as gastro-bars, restaurants gourmet pop-up food kitchens, food trucks, restaurants in premium grocery stores, subscription-based meal kits, and even privately hosted meals enabled by the sharing economy. Traditional players should consider the risks of sticking with traditional strategies (Langford & Weissenberg, 2018).

Government assistance as an element of support for small restaurant businesses and their employees is necessary for many businesses to weather a long-term macroeconomic downturn. The main measures preferred by the business consulted in this study include, aid to cover fixed costs during the closure of the establishment, the Social Security exemption for employees on furlough, and the extension of the furlough period (ERTEs). However, governance models in the new situation should not be limited to providing a series of aids or reducing taxes and duties. Society and small businesses have changed, with a growing sensitivity towards values related to sustainability, respect for the environment, protection and humanisation towards what surrounds us, and it is a good opportunity for planning and development policies in cities to adapt accordingly to these new scenarios (Hernàndez, 2021).

The need for restaurants to provide service in outdoor spaces has led to occupying sidewalks, squares and terraces, with significant amounts of money being invested in the fitting out of these spaces. As a result, eating outside has been a hit in many cities, and it will surely provide a new channel of income for restaurateurs once the pandemic is over. Local governments should capture current trends in this regard, such as the installation of parklets and the Design for Distance (Gunts, 2020), for example, by promoting contests between urban designers and planners to create innovative designs to bring life back to public spaces. This will not be easy, since the income of the public administrations of small towns has been decreasing accordingly with the decrease in the tax burden on the city's businesses. While national governments are generously distributing a considerable amount of money to safeguard national and regional economies from ruin, cities and municipalities will suffer from the decline of tax revenues (Kunzmann, 2020).

The small size of the sample used can be considered a limitation, moreover, this study was carried out in a city with little international tourism, making it difficult to generalize the findings of this study to other destinations and cities where the dependence on foreign or non-local clients is greater. In future research, it would be advisable to continue this cross-sectional study with a longitudinal study, exploring the business's responses to the crisis and the changes that have occurred in a year. The interviews were conducted during the lockdown period caused by the pandemic, which is certainly a factor that adds interest to the study, although it is probable that at other times of lesser uncertainty the answers would have been different. On the other hand, it would be of great interest to advance the research from a demand-driven perspective. The impact of the crisis will generate much more demanding customers, with a greater sensitivity to food safety, so the factor of trust between the customer and the restaurateur must be strengthened in order to generate greater connection and empathy with the customer and to adapt to new situations of vulnerability and instability.

REFERENCES

Angelidou, M. (2015). Smart cities: A conjuncture of four forces. *Cities (London, England)*, *47*, 95–106. doi:10.1016/j.cities.2015.05.004

Bain and Ernst & Young. (2020). *Impacto de COVID-19 en hostelería en España.* https://www.ey.com/Publication/vwLUAssets/ey-bain-company-impacto-de-covid-19-en-hosteleria-en-espana/$FILE/ey-bain-company-impacto-de-covid-19-en-hosteleria-en-espana.pdf

Ballarín, E. (2020). *Tendencias de la hostelería post coronavirus.* Live Horeca Live. Retrieved from https://www.smarttravel.news/10-tendencias-explican-sera-la-hosteleria-post-coronavirus/

Banco de España. (2020). *Artículos analíticos. Boletín Económico 4/2020. El impacto de la crisis del COVID-19 sobre la situación financiera de las empresas no financieras en 2020: Evidencia basada en la central de balances. Banco de España*. Retrieved from https://www.bde.es/bde/es/secciones/informes/boletines/relac/Boletin_Economic/Informes_trimest/

Bartik, A. W., Bertrand, M., Cullen, Z. B., Glaeser, E. L., Luca, M., & Stanton, C. T. (2020). *How are small businesses adjusting to COVID-19? Early evidence from a survey (No. w26989)*. National Bureau of Economic Research. doi:10.3386/w26989

Berbel Giménez, G., & Reyes Gómez, J. D. (2020). Impacto de la COVID-19 en la restauración. Estadísticas, visión de expertos y de clientes. In *F.J. Bauzá Martorell y F.J. Melgosa Arcos (Dirs.), Turismo Post COVID-19. El turismo después de la pandemia global. Análisis, perspectivas y vías de recuperación* (pp. 127–146). Aquilafuente. Ediciones Universidad de Salamanca.

Brewer, P., & Sebby, A. G. (2021). The effect of online restaurant menus on consumers' purchase intentions during the COVID-19 pandemic. *International Journal of Hospitality Management*, *94*, 102777. Advance online publication. doi:10.1016/j.ijhm.2020.102777

Byrd, K., Her, E., Fan, A., Almanza, B., Liu, Y., & Leitch, S. (2021). Restaurants and COVID-19: What are consumers' risk perceptions about restaurant food and its packaging during the pandemic? *International Journal of Hospitality Management*, *94*, 102821. doi:10.1016/j.ijhm.2020.102821

CEHE. (2021). *Confederación Empresarial de Hostelería de España*. Retrieved from https://www.cehe.es/index.html

Díaz-Bravo, L., Torruco-García, U., Martínez-Hernández, M., & Varela-Ruiz, M. (2013). La entrevista, recurso flexible y dinámico. *Investigación en Educación Médica*, *2*(7), 162–167. doi:10.1016/S2007-5057(13)72706-6

Dube, K., Nhamo, G., & Chikodzi, D. (2020). COVID-19 cripples global restaurant and hospitality industry. *Current Issues in Tourism*, *1-4*. Advance online publication. doi:10.1080/13683500.2020.1773416

ESADE. (2021). *Informe económico y financiero*. 1er Semestre 2021, 28. Esade Center for Economy Policy. Retrieved from http://itemsweb.esade.es/wi/Prensa/InformeEconomicoFinancieroEsade28.pdf

Fakhimi, A. H., Khani, A. H., & Sardroud, J. M. (2021). Smart-city infrastructure components. In J. R. Vacca (Ed.), *Solving Urban Infrastructure Problems Using Smart City Technologies* (pp. 17–54). Elsevier. doi:10.1016/B978-0-12-816816-5.00002-4

Guerra, R. P. (2020). Algunas notas sobre el derecho administrativo del turismo: COVID-19. *Revista General de Derecho Administrativo, 54*(4).

Gunay, S., & Kurtulmuş, B. E. (2021). COVID-19 social distancing and the US service sector: What do we learn? *Research in International Business and Finance, 56*, 101361. Advance online publication. doi:10.1016/j.ribaf.2020.101361

Gunts, E. (2020). Design for Distancing aims to bring life back to public spaces left dormant in lockdown. *The Architect's Newspaper*. Retrieved from https://www.archpaper.com/2020/07/design-for-distancing-initiative-aims-to-bring-life-back-to-public-baltimore-spaces/

Hall, C. M., Scott, D., & Gössling, S. (2020). Pandemics, transformations and tourism: Be careful what you wish for. *Tourism Geographies, 22*(3), 577–598. doi:10.1080/14616688.2020.1759131

Hernàndez, C. A. (2021). The role of citizens in smart cities and urban infrastructures. In J. R. Vacca (Ed.), *Solving Urban Infrastructure Problems Using Smart City Technologies* (pp. 213–234). Elsevier. doi:10.1016/B978-0-12-816816-5.00010-3

IGE. (2021). *Padrón municipal de habitantes*. Statistics Galician Institute. Retrieved from https://www.ige.eu/web/mostrar_actividade_estatistica.jsp?idioma=es&codigo=0201001002

INE. (2020). *Edad media por comunidad autónoma y provincia, sexo y tamaño de municipio*. Statistics National Institute. Retrieved from https://www.ine.es/jaxi/Datos.htm?path=/t20/e245/p04/provi/l0/&file=0tamu005.px

INE. (2021). *Empresas por CCAA, actividad principal (grupos CNAE 2009) y estrato de asalariados*. Statistics National Institute. Retrieved from https://www.ine.es/jaxiT3/Datos.htm?t=39372

Jung, J. G. (2019). Megacities, utopian cities and planning our cities for the future. In S. Albert (Ed.), *Innovative Solutions for Creating Sustainable Cities* (pp. 269–289). Cambridge Scholars Publishing.

Kim, H. M. (2021). Smart cities beyond COVID-19. In H. M. Kim, S. Sabri, & A. Kent (Eds.), *Smart Cities for Technological and Social Innovation* (pp. 299–308). Academic Press., doi:10.1016/B978-0-12-818886-6.00016-2

Kim, J., Kim, J., & Wang, Y. (2021). Uncertainty risks and strategic reaction of restaurant firms amid COVID-19: Evidence from China. *International Journal of Hospitality Management, 92*, 102752. Advance online publication. doi:10.1016/j.ijhm.2020.102752 PMID:33223596

Kim, J., & Lee, J. C. (2020). Effects of COVID-19 on preferences for private dining facilities in restaurants. *Journal of Hospitality and Tourism Management*, *45*, 67–70. doi:10.1016/j.jhtm.2020.07.008

Kim, K., Bonn, M. A., & Cho, M. (2021). Clean safety message framing as survival strategies for small independent restaurants during the COVID-19 pandemic. *Journal of Hospitality and Tourism Management*, *46*, 423–431. doi:10.1016/j.jhtm.2021.01.016

Komninos, N. (2006). The architecture of intelligent cities. *Intelligent Environments*, *6*, 53–61.

KPMG. (2019). *La gastronomía en la economía española. Impacto económico de los sectores asociados.* KPMG. Retrieved from https://assets.kpmg/content/dam/kpmg/es/pdf/2019/01/gastronomia-en-economia-espanola.pdf

Krivý, M. (2018). Towards a critique of cybernetic urbanism: The smart city and the society of control. *Planning Theory*, *17*(1), 8–30. doi:10.1177/1473095216645631

Kunzmann, K. R. (2020). Smart Cities After Covid-19: Ten Narratives. *The Planning Review*, *56*(2), 20–31. doi:10.1080/02513625.2020.1794120

Langford, G., & Weissenberg, A. (2018). *2018 Travel and Hospitality Industry Outlook*. Deloitte Center for Industry Insights. Retrieved from www.deloitte.com/us/travel-hospitality-trends

Liugailaitė-Radzvickienė, L., & Jucevičius, R. (2014). Going to be an intelligent city. *Procedia: Social and Behavioral Sciences*, *156*, 116–120. doi:10.1016/j.sbspro.2014.11.131

López-Medel Báscones, J. (2020). *La cláusula rebus sic stantibus. Modificación o resolución de contratos ante la crisis del coronavirus*. Retrieved from: https://www.abogacia.es/actualidad/opinion-y-analisis/la-clausula-rebus-sic-stantibus-modificacion-o-resolucion-de-contratos-ante-la-crisis-del-coronavirus/

Madeira, A., Palrão, T., & Mendes, A. S. (2020). The Impact of Pandemic Crisis on the Restaurant Business. *Sustainability, 13*(1), 40. . doi:10.3390u13010040

Malbec, T. (2020). *Restaurant Technology in the Post-COVID-19 World.* Hospitality Technology. Retrieved from https://hospitalitytech.com/restaurant-technology-post-Covid-19-world

Malbec, T. (2020a). *Restaurant Business Model Changes in a Post COVID-19 World*. Hospitality Technology. Retrieved from https://hospitalitytech.com/restaurant-business-model-changes-post-Covid-19-world

Merriam, S. B. (2009). *Qualitative research: A guide to design and implementation*. Jossey-Bass.

Millán, S. (2020). La cancelación del Mobile por el coronavirus provoca fuertes pérdidas económicas. *Cinco Días*. Retrieved from https://cincodias.elpais.com/cincodias/2020/02/12/companias/1581541189_908014.html

Ministry of Labour and Social Economy. (2021). *Resumen datos de empleo*. Retrieved from https://www.mites.gob.es/ficheros/ministerio/estadisticas/documentos/RUD.pdf

Nespereira, A. (2021). La hostelería de la provincia es la más tocada de Galicia. La Región, 6.

Nunn, J. (2020, April 14). Restaurants will never be the same after coronavirus – but that may be a good thing. *The Guardian*. Retrieved from https://www.theguardian.com/commentisfree/2020/apr/14/coronavirus-restaurants-pandemic-workers-communities-prices

Panzone, L. A., Larcom, S., & She, P. (2021). Estimating the impact of the first COVID-19 lockdown on UK food retailers and the restaurant sector. *Global Food Security*, *28*, 100495. Advance online publication. doi:10.1016/j.gfs.2021.100495

Sobaih, A. E. E., Elshaer, I., Hasanein, A. M., & Abdelaziz, A. S. (2021). Responses to COVID-19: The role of performance in the relationship between small hospitality enterprises' resilience and sustainable tourism development. *International Journal of Hospitality Management*, *94*, 102824. Advance online publication. doi:10.1016/j.ijhm.2020.102824

Taylor, S. Jr. (2020). The socially distant service scape: An investigation of consumer preference's during the re-opening phase. *International Journal of Hospitality Management*, *91*, 102692. Advance online publication. doi:10.1016/j.ijhm.2020.102692 PMID:33012948

Tse, A. C. B., & Leo Sin, S. S. (2006). Crisis management and recovery: How restaurants in Hong Kong responded to SARS. *International Journal of Hospitality Management*, *25*(1), 3–11. doi:10.1016/j.ijhm.2004.12.001 PMID:32287848

Vázquez, M. (2017). Ourense tiene más bares por habitante que las otras tres capitales gallegas. *La Voz de Galicia*. Retrieved from https://www.lavozdegalicia.es/noticia/ourense/a-gudina/2017/06/18/ourense-bares-habitante-tres-capitales-gallegas/0003_201706O18C5991.htm

WHO. (2004). *WHO guidelines for the global surveillance of severe acute respiratory syndrome (SARS)*. Updated recommendations, October 2004. World Health Organization. Retrieved from https://www.who.int/csr/resources/publications/WHO_CDS_CSR_ARO_2004_1.pdf

Yigitcanlar, T., Kamruzzaman, M., Foth, M., Sabatini-Marques, J., da Costa, E., & Ioppolo, G. (2019). Can cities become smart without being sustainable? A systematic review of the literature. *Sustainable Cities and Society*, *45*, 348–365. doi:10.1016/j.scs.2018.11.033

Zeng, Z., Chen, P., & Lew, A. A. (2020). From high-touch to high-tech: COVID-19 drives robotics adoption. *Tourism Geographies*, *22*(3), 724–734. doi:10.1080/14616688.2020.1762118

ADDITIONAL READING

Farrer, J. (2020). How are Tokyo's independent restauranteurs surviving the pandemic? *Asia-Pacific Journal: Japan Focus*, *18*(18), 1–13.

Johnson, T. P., Lu, Z., Tolomiczenko, G., & Gellatly, J. (2008). SARS: Lessons in strategic planning for hoteliers and destinations marketers. *International Journal of Contemporary Hospitality Management*, *20*(3), 332–346. doi:10.1108/09596110810866145

McKercher, B., & Chon, K. (2004). The over-reaction to SARS and the collapse of Asian Tourism. *Annals of Tourism Research*, *31*(3), 716–719. doi:10.1016/j.annals.2003.11.002 PMID:32572281

Pine, R., & McKercher, B. (2004). The impact of SARS on Hong Kong's tourism industry. *International Journal of Contemporary Hospitality Management*, *16*(2), 139–143. doi:10.1108/09596110410520034

KEY TERMS AND DEFINITIONS

ERTE: It is the acronym used in Spanish legislation for the furlough regulation procedure.

Furlough: A period of time when a company stops employing someone, usually temporarily, because the company does not have enough money or enough work.

Governance: Governance refers, therefore, to all processes of governing, whether undertaken by a government, market, or network, whether over a family, tribe, formal or informal organization, or territory, and whether through laws, norms, power, or language. Governance differs from government in that it focuses less on the state and its institutions and more on social practices and activities.

Humanizing: In relation to cities and landscape, adapt them to human nature or use.

Smart City: It is the holistic vision of a city that applies ICT for the improvement of the quality of life and accessibility of its inhabitants and ensures a sustainable economic, social, and environmental development in permanent improvement.

APPENDIX 1

Questionnaire. Impact of COVID-19 on the Catering Sector

Establishment Details

1. Year of start of activity:
2. Maximum capacity:
3. Terrace. Yes (number of tables: x); No4. Courtyard/Backyard. Yes (number of tables: x); No 5. Number of employees. Permanent: Temporary:

Current Impact of COVID-19

1. Changes in the service (more terrace service, "prepare and take away", opening hours, temporary closure, etc.)
2. Reduction of staff
    a. Number of employees on furlough (ERTE):
    b. Duration of the furlough:
    c. Number of permanent layoff:
3. Reduction of working hours:

Foresight for the Future

1. Are you considering closing the business? No; Yes; Partial (low season)
2. Do you anticipate that the capacity of the establishment will be reduced by law in the future? In what percentage?
3. What changes do you foresee in the service? For example, increased terrace service, home service, none, etc.
4. What changes do you foresee in customer schedules? For example, earlier dinners, loss of clients in certain time slots, none, etc.
5. What changes do you foresee in staff recruitment? For example, more temporary staffing, downsizing, we'll go back to business as usual, etc.

Remedies Proposal

In your opinion, what would be the THREE most effective measures to deal with the current situation? (Indicate 1st, 2nd or 3rd in the selected box)

- Exemption of fees for the use of terraces
- Extension of terrace licenses

- Exemption from payment of the Economic Activities Tax (IAE)
- Extension of the furlough period (ERTEs)
- Exemption from municipal taxes (e. g. garbage collection)
- Tax flexibility
- Aid to cover fixed costs during the closure of the establishment
- Exemption from payment of Social Security for employees on furlough
- Others (open answer):

APPENDIX 2

Table 5. Profile of the small businesses in the sample (restaurants, bars, cafés)

Start of Activity	Maximum Capacity	Terrace Tables	Courtyard Tables	Permanent Employees	Temporary Employees	Self-employed
1986	150	14	40	8		
1990	40	7		3	1	
1993	120	9		6		
2000	54	4		3		2
2002	45	4	2	2	0	
2003	70	15		4		
2003		7				2
2005	100	5		3		
2006	90	13	12	7	2	
2006		6		2		
2007	90	10		1		
2008	198			12		
2008	120	6		1	0	
2008	60	4		7,5		
2009	40	4		4		
2011	100			5		
2012	64	4		3		
2012	100	4		5		
2012	55	4		5		2
2013	65	10		6	2	
2014	44	20		5		

continues on following page

Table 5. Continued

Start of Activity	Maximum Capacity	Terrace Tables	Courtyard Tables	Permanent Employees	Temporary Employees	Self-employed
2015	120			4	0	
2015	55	6		1	0	
2016	50	8	10	6	2	
2017	100			7		
2017	60			3		
2018	40	10	10	3		
2018	80	7		1		
2018	50	7		3	1	
2018	65	6		4		
2018	75	4		3	0	
2018	40	3		3		
2018	20	6		1		
2019	50	4		2	1	
2020	80	4		10		
2020		10	7	3		
2020	54	8		3		
2020	50			2		

Table 6. Furloughs, permanent layoffs and reduction of hours

Furloughs	Furlough Period (months)	Permanent Layoff	Reduction of Working Hours
7	6		
6	6		
6	9	1	
5	5		4 hours
4			2 hours
3	11	1	1 hour
3	4	1	yes
3		1	
3	10		
3	March-current		
2	3		yes
2	3		
2	6	1	
2		1	yes
2	8		
2	March-current		yes
2	6	2	1 hour
2	10		
2	5	1	1 hour
2		1	
2	March-current		5 hours
2	4		
1	3	3	1 hour
1	5	4,5	
1	March-current		3 hours
1	4		
1	8		
1	5		
			yes
			yes
		1	
	1	4	
		1	
		3	

Chapter 7
The Effects of COVID-19 on the Rent-a-Car Industry:
A Study Case Based on Argentina

Maximiliano Emanuel Korstanje
https://orcid.org/0000-0002-5149-1669
University of Palermo, Buenos Aires, Argentina

ABSTRACT

Doubtless, COVID-19 has accelerated an economic financial crisis since 2008, affecting seriously not only the tourism industry but the global commerce. Governments have adopted different positions and programs to mitigate the economic aftermath of COVID-19. As never before in its history, tourism has been placed between the wall and the deep blue sea. Although the interests and studies evaluating the impact of COVID-19 have captivated the attention of countless scholars, less attention was given to the rent-a-car industry, which occupies a central position in the tourist system. As substitute competitors of train, bus, and airplanes, the rent-a-car organizations seem to be a quintessential actor of the tourist system. Of course, because we live in a world without tourists, empirical-based studies do not abound. To fill such a gap, the present chapter describes the economic downturn of a rent-a-car organization giving a firm empirical case. Although illustrative to some extent, the obtained outcome cannot be extrapolated to other universes or samples.

INTRODUCTION

By the end of 2019, probably months earlier, China went through a new virus outbreak in the city of Wuhan. Although originally the situation was trivialized by Chinese

DOI: 10.4018/978-1-7998-6996-2.ch007

authorities, no less true was that the virus spread rapidly to the four continents affecting not only global commerce but also forcing governments to close their borders and airspaces. With the benefits of hindsight, Coronavirus disease became an unparalleled disaster (pandemic) so that the world was not prepared. To date, CO-SARS-v2 (known as COVID-19) infected almost 23.472.214 persons causing 815.262 fatal victims (last updated on 08-24-2020). The most affected countries include the US (180.788 victims) followed by Brazil (114.913), Mexico (60.480), India (58.546), the UK (41.433), Italy (35.441), France (30.528), and Spain (28.870). Even if the virus originally wreaked havoc the public health system in Spain, Italy and France, a similar state of the crisis involved other countries such as the US, Germany and the Netherlands. At first glance, the WHO (World Health Organization) failed to adopt and impose a standardized plan of contingency while each nation followed their methods of prevention. As a result of this, nations –no matter their economic maturation- opted for two clear-cut decisions. Some countries like the US, Mexico, Ecuador, Brazil and the UK engaged to protect the domestic market adopting soft measures while others imposed a strict lockdown, lasting for weeks or months (China, Japan, Argentina, South Korea, Australia). Quite aside from the discrepancies revolving around the correct –if not most efficient- course of action to prevent the pandemic, the global economic downturn teetered the entire world on the brink of a financial collapse. This opens the doors to a new paradox: in a hyper-mobile world, the modern means of transportation that often connected far-away countries in hours seems to be the most efficient carrier for the virus dissemination. This point has particularly led to tourism scholars to express a much deeper concern about the future of tourism. Echoing Adrian Franklin (2007), scholarship enthralled the figure of the tourist as the only valid source for scientific knowledge, relegating other voices to a peripheral position. In so doing, tourist-centricity monopolized the methods and epistemology of tourism research as well as the content to be published in leading journals. Without going into further details, COVID-19 offers a similar dilemma regarding how to study tourism in a world without tourists. Today scholars who urged to write and publish about COVID-19 debates between two opposed poles; those interested in developing new methodological toolkits to understand tourism in a context of new normalcy (Wen et al. 2020; Miles & Shipway, 2020; Zenker & Kock, 2020; Verma & Gustaffsson, 2020), and those voices who enthusiastically hoped for the opportunities to create more resilient and sustainable destinations. De-growth tourism, for them, exhibits an interesting point of departure for the next decades (Gossling, Scott, & Hall, 2020; Gallego & Font 2020; Higgins-Desbiolles, 2020; Prayag, 2020; Galvani, Lew & Perez, 2020; Crossley, 2020). Despite their eloquence and the high-quality texts, these studies are based on a speculative spirit, which is characterized by the lack of empirical evidence or information or follows theoretical insights about good practices towards more sustainable tourism. What is more important, tourism-based research –though widely developed in these decades- has systematically overlooked the role played by rent-a-car subsector. As stated, tourists, as well as their perception, fears, hopes and experience, have been methodologically situated as the main object of study in tourism research. To fill this gap, the present chapter explores empirically the real economic and socio-cultural effects in a rent-a-car company which is geographically located in Argentina.

CONCEPTUAL BACKGROUND

History of Rental-Cars

At a closer look, the history of the rent-a-car industry goes in a similar direction. Although the specialized literature in rent-a-car subsector does not abound (Korstanje 2011), it is important to delineate the first steps of rental cars organization in the US. The historical roots of rental cars date us back to Germany (1906). The company Sixt Authofharen und selbsfahrer [Sixt Car Cruises and Self drivers] was established by the end of 1913. The model was rapidly adopted by Joe Saunders in Omaha, Nebraska by 1917. His innovation led to expanding to almost 50 cities in the next years importing what American historians dubbed as "Saunders system". The company was finally bought by John Hertz in 1923. Meanwhile, the UK Godfrey Davies (a young visionary man) borrows Saunders system to create a company which will become in Europcar. The end of WWII set the pace to new transport means which revolutionized the tourism industry while replacing the trolley and horse. Today, this subsector offers plenty of car classes which are allocated throughout the globe, in cases, at least as an ideal option, substituting mass transport forms as the commercial airlines, the train or buses. In Argentina the history of the first rental companies are not pretty different than the rest of the world. Hertz Argentina, Localiza and Avis were the first pioneering companies who operated since 80s decade. In the successive decades, new minor competitors will take the opportunity to invest in Argentina as Cactus, Thrifty, Europcar, Budget, and Sixt (only to name a few). Today the market is centralized by the three main organizations totalizing almost 70% of sales volume. We have focused our study in one of these three main companies though because of professional secrecy we are unable to reveal the real name of the studied company.

As the previous argument is given, COVID-19 has doubtless accelerated an economic-financial crisis since 2008, affecting seriously not only the tourism industry but the global commerce. Even if the interests and studies evaluating the impact of COVID19 have captivated the attention of countless scholars, less attention was given to the rent-a-car industry, which occupies a central position in the tourist system.

As substitute competitors of train, bus, and airplanes, the rent-a-car organizations seem to be a quintessential actor of the tourist system. Of course, because we live in a world without tourists, empirical-based studies do not abound for these days.

How to Read This Chapter

The first section synthesizes a short introductory discussion about the problem of COVID-19 and its direct and durable aftermaths on the industry of tourism and hospitality. The section proffers an interesting discussion on the role of automobile as an emerging (alternative) vehicle of transport just after the WWII end of the inter-war period. In this respect, countless global risks which range from terrorism to virus outbreaks are gradually changing the nature of tourism towards more globalized (hybridized) forms. The second section dissects the origin of the automobile and how the expansion of consumption, which is enrooted into a cosmopolitan character, founded the pillars of global capitalism. It is tempting to accept that cars, though undermined by scholars in tourism research, are a key factor of global capitalism. We explore the effects of viruses in the tourism industries as well as the future of the industry in a post-Covid19 context. The last section is a snapshot of the economic effects of the pandemics in a well-renowned rent-a-car company that covers rentals Argentina-wide.

DEBATE

The turn of the century has brought some durable effects on the tourism industry. The rise and multiplication of countless global threats ranging from terrorism to natural disasters and today a pandemic have systematically placed the tourism industry in jeopardy. In fact, climate change seems to be a challenge posed to the humanity which forces governments to set a new agenda for the years to come (Blake & Sinclair, 2003; Ritchie 2004; Santana 2004; Laws, Prideaux & Chon, 2007). As Professor Raoul Bianchi (2006) puts it, the turn of the century inaugurated a new world where the concept of security and safety were radically transformed. The tourism industry faced several crises which led policy makers and experts to adopt the risk perception theory as the main paradigm (Fuchs & Reichel 2006; Yuksel & Yuksel, 2007; Lepp & Gibson, 2008). Studies in risk perception theory emulated potential scenarios to identify and eradicate potential variables or factors that may very well place the industry in jeopardy. Hence, the understanding of tourist´s motivation was of paramount importance to protect the organic image of the destinations (Reisinger & Mavondo 2006; Floyd et al 2004). Risk perception theory –as a catch-all model-looked to encompass different fields and matters oscillating from terrorism (Floyd et al. 2004; Reisinger & Mavondo 2005; Seabra, Abrantes & Kastenholz, 2014; Wolff & Larsen 2014) to natural disasters (Matyas et al, 2011; Park & Reisinger, 2010; Jonas et al 2011), without mentioning virus outbreaks (McKercher & Chon, 2004; Monterrubio 2010), and local crime without mentioning political violence (Baker, Page & Meyer, 2003; George 2010). Although the theory preliminarily obtained promising results, the radicalization of some terrorist cells, accompanied with the rapid deterioration of ecology which created an unparalleled ecological crisis has led to scholar to calibrate new strategies to adapt the tourist destinations to much complex world. Literature in post-disaster and post-conflict destination theory toys with the belief that the precautionary logic, which is present in risk perception models, does not suffice to keep safe the destination in a hyper-changing (if not uncertain) environment. Therefore, the figure of resiliency emerges as a new concept tailored and adapted to risky conditions (Robinson & Jarvie 2008; Mair, Ritchie & Walters, 2016; Amujo & Otubanjo 2012).

As the previous backdrop, post disaster studies emphasized on the needs to develop more resilient institutions and protocols while employing tourism as a mechanism that helps to accelerate the post-recovery timeframe (Seraphin, Butcher & Korstanje 2017; Seraphin et al 2018; Seraphin, Korstanje & Gowreesunkar 2020). In a nutshell, risk perception theory was gradually substituted by far-reaching models that accept the risk as a quintessential feature of the tourist system (Tarlow 2014). This seems to be exactly what happens with the conceptual models applied to the effects of global pandemic on the tourism industry. The failure of risk perception theory to make safer destination has ushered policy makers on the needs of introducing adaptation as their main buzzword (Tarlow, 2014).

Mobilities, Virus and Tourism

The evolution of modern tourism as a growing industry is threefold. The technological breakthrough associated with the reduction of working hours as well as the rise of the welfare state which gave to workforce new benefits and opportunities towards social upward. In fact, mobilities and tourism seem to be inevitably entwined (Sheller & Urry, 2004; Hannam, Sheller & Urry, 2006; Korstanje 2018). Over recent years, scholars reached a consensus that in such a process the car substituted other classic means of transport as the horse, or trolleys (Packer, 2008). The WWII end cemented the authority of new transport means as the train, the commercial aeroplanes, and cars which rapidly connected distant cities and areas into a global system. There is a clear correlation between war, production and transport mean. As John Urry (2006) eloquently observed, for some reason, sociology rarely introduced the car as an object of study or debate over recent years. Automobility exhibits a relevant aspect of modern life and the global city. It dominates not only modern politics but also the economy. The car not only accelerated the means of production, causing a real revolution but installed this machine as a cult object for drivers. The act of driving was more than the will of displacing, it was related to individualism. The power of driving is based on the needs of exercising the own autonomy other classic means of transport offered. The car gradually became a sign of status and distinction. Thousands of tourists visited Ford factories to gaze the assembly lines in the US and Europe. Modernity disposed and commoditized the economic production process as a spectacle in itself. In so doing, the car situates as a symbol of modernity. The history of rent-a-cars goes in a similar direction. It associates mainly to the opportunities of new business in the US and Europe, above all after the end of WWII. The technological advances in the fields of mobilities linked to the rise of new transport means paved the ways for the rise of a new growing industry worldwide, the rent-a-car system (as it was explained in the introductory chapter). Capitalism and renting system has been inextricably intertwined.

Viruses have affected, rattled, and molded the history of humankind. And of course, tourism is not an exception. Since 2001 some virus outbreaks have affected the tourism industry. The question of whether travel behavior was certainly centered on the possibilities of a destination to offer a pleasurable experience radically shifted to the quest of safer destinations as the main criterion in the decision-making process. Security and safety occupied a central position in the business talks as well as in the tourism management policies (Hall, Williams & Lew 2004; Brouder et al 2020). In 2002, a new virus known as Severe Acute respiratory syndrome (SARS) originated in Foshan, China threatened to be expanded to Western countries when it was contained by Chinese authorities. However, over 8.000 persons were infected in more than 30 countries and 770 deceased. In 2009, Mexico and the US witnessed an outbreak of Swine influenza (H1N1) paralyzing part of the international flights and global commerce worldwide. Some estimate numbers show that almost 284.000 persons died in the world because of H1N1. In 2012, a viral respiratory infection (Middle East respiratory syndrome MERS) caused mainly by MERS Coronavirus grinds the planet to a halt. Last but not least, the frightening Ebola virus disease arrives at the US in 2014 alerting all the health authorities in the Western hemisphere. In almost all cases, the effects and reactions of the government were the same. These viruses disseminate suddenly through modern transport means and tourism industry which are fertile ground for high infectivity (McKercher & Chon 2004; Monterrubio 2010; Solarin 2015; Novelli et al. 2018). Having said this, governments moved almost all in the same direction, which means closing international borders, imposing lockdowns, and closing airspaces. As Gossling, Scott & Hall (2020) note, the advances of viruses evince a complex reality which is inscribed in the unsustainable forms of mass consumption, drawing a parallel between the current COVID19 crisis and climate change, but at the same time, the crowded conditions of the large cities. Authors elaborate a careful diagnosis of the economic impact in the different subsectors but they overlook to the rent-a-car companies (Gossling, Scott & Hall, 2020).

Similar concerns accompany Higgins-Desbiolles´ paper entitled as Socialicing tourism for social and ecological justice. For her viewpoint, COVID19 brings the problem of sustainability into the foreground. The resulted crisis allows laying the foundations towards a new (so to speak, more sustainable) capitalism. The tourism industry should adapt to economic changes and a de-growing process for a more responsible form of consumption. COVID19 not only interrogates to the neoliberal agenda and the material injustices, which are proper of capitalism but also challenges the premise of the economic-based paradigm which appears to be finely oriented to profit-maximization (Sigala 2020; Carr 2020). Gretzel et al (2020) and Wen et al (2020) open –from different stances- an interesting discussion in this entry regarding the use of technology to forge new methodologies in the constellations of tourism research. Coronavirus disease calls attention to emerging practices of tourism ingrained in virtual consumption. It is safe to admit virtual or e-tourism are direct consequences of the economic crisis accelerated by COVID19. What is more important to discuss, the digital revolution allows a re-elaboration of transformative research. As earlier mentioned, despite their eloquence, the hot-released publications –with foci in COVID19- are abstract lucubration, probably based on the urgency of saying something in days of uncertainty, but lacking of an empirical basis. In fact, this happens because of two main reasons. On one hand, scientists know little about the real effects of COVID19 in the body, and what is more accurate we do not know how this story will finally end. On another, tourism research has been centered on an economic-based paradigm which prioritizes the tourist as the main agent of the system. Methodologically, this tourist-centricity leads epistemologists to marginalize other actors and voices towards a peripheral position. In this vein, Fransesc Romagosa (2020) debates critically the social consequences of the pandemic in society, alerting on the problems of neglecting the "Other" because of fear and mistrust. He argues convincingly that the crisis has far-reaching consequences which make very hard to punctuate realistic predictions. As he writes: no-one knows exactly the future of the tourism industry or what would happen at a short-run. After all, COVID19 begs a more than interesting question: how do scientists study tourism in a world without tourists?

To respond this above-mentioned answer, we shall combine a method of triangulation (using quantitative and qualitative methods) to give an exploratory approach of how COVID19 and the current global crisis affect a rent-a-car company in Argentina.

THE MARKET OF RENT-A-CARS IN ARGENTINA

Because of confidentiality issues, the real name of the organization will be changed. We have been provided of statistic information vital for business and competition, but in any case we have facilitated to the costs or budget details. The statistic information was limited to volume of sales and inter-location businesses. The gathered information was complimented with a formal and informal interviews conducted from Jan to August 2020. In addition, almost 45 interviews were recorded and supported via Zoom and Skype. The main method of sampling was snowball, which means each interviewee recommended to the next one just after the interview ends. The gathered information was contrasted with statistical information provided by top ranked managers.

The company has more than 30 years of experience in the rent-a-car market and situates as a leading global organization. The main segments of renters are foreigner tourists and domestic tourists. The foreigner renters come from Brazil, Chile, the US, and Europe while domestic renters are city-dwellers from Buenos Aires, Cordoba and Rosario. Officially, 24 locations offer different car classes to different renters. The process of renting a car is very simple the class is previously reserved in order for the car to be picked up as specified. The rental length varies according to the driver´s needs and the vehicle can be dropped out at another than picked up location –often by paying a one way fee. Over the recent years, renting a car substituted as an alternative way to drive freely when the aero commercial industry does not work. For the sake of clarity, locations may be grouped in two groups: industrial and touristic locations. Industrial locations are primarily based on private corporations and factories while tourist locations focus on foreign or domestic tourists as main segments.

Given the advance of the pandemic, on 20 April of 2020 President Alberto Fernandez declared the state of emergency imposing a strict lockdown which closed different commercial activities, subsectors and industries. Although the lockdown remains even to date, it gradually evolved towards different facets. The government decided to bail out countless companies, above all the tourist sector, which was seriously compromised by the economic crisis generated by the pandemic. The sales of small size tourist companies slumped down considerably tenfold. This scenario leads Tourism ministry Matias Lammens to bid for financial aid programs to assist the tourist sector. With the airspace completely closed, hotels in the brink of collapse, and borders controlled by army forces, Argentina went through in the worse economic downturn of its history. In view of this mayhem, this chapter describes the turbulences, problems and limitations a rent-a-car company experienced during the pandemic. We try to validate or refute four types of different hypotheses.

H1. The pandemic accelerated a serious decline in the sales of the company.
H2. All the locations show a similar decline in their sales.
H3. In view of the absence of the aero commercial activity, the rent-a-car situated as an option to travel to other cities.
H4. Domestic drivers or renters replaced to foreign renters.

In the next section, we discuss the ebbs and flows of a well-famous rent-a-car organization as well as the negative impact of COVID19 and the restrictive measures of the government to intervene in the market. As a result, we combine qualitative methods (interviews conducted through digital platform) and quantitative ones (characterized by internal statistics we have been kindly authored to make public). Although we asked for the budget details, as well as the preliminary costs model but it was declined because of the revelation of these ciphers may compromise seriously the company. As stated, the obtained outcomes should be understood within the sample and universe they represent, and they cannot be extrapolated to other cases.

Outcomes and Discussions

It is important not to lose the sight of the fact that sales are not only reliable but credible indicators of business performance but gives a certain insight into the potential businesses in the short run. As discussed, in the tourism industry, which is a sensitive sector respecting to disrupting events, it is important to add, tourist often select safe avoiding unsafe destinations. Security and safety seem to be the tugs of the war of the industry.

Sales

As table 1 shows, sales have invariably collapsed because of COVID19. On January of 2019, the company amassed almost ARS 77.213.334 (including VAT and other taxes), an amount which notably increased to ARS 113.429.941 in February of the same year. In the low season, which means the months of April, May and June, the sales declined to ARS 51.706.987, ARS 43.775.670, and 38.024.031 respectively. In April of 2020 –next to the imposition to the lockdown- this company had a turnover of ARS 9.712.386. The level of sales keeps for May for ARS 9.341.356 but declines even more for June (ARS 6.473.256) and July (ARS 6.361.175). At a closer look, at the time of comparing April 2019 and April 2020, we understand the problem with clarity. In April 2019 the sales were ARS 51. 706.987 while in 2020 the sales declined to ARS 9.712.386. This authorizes us to say sales were down 81%. The same tendency can be observed in the subsequent months of May, June and July. In this period the company reduced its fleet size, selling more than

500 vehicles to pay wages and salaries. In the meantime, the rewarding program for sales was suspended until further notice. The number of closed rental agreement which exhibits the number of rental of the company was reduced from 5.905 in April 19 to 92 in April 2020 (98% of the total activity). The tendency replicates in the months of May, June and July.

Table 1. Indicators and sales

Months	Sales	Closed Ra	Opened Ra	DBR Days	Avg Days	Driven kms	AVG kms
ene-19	77213334,92	6303	6232	34170,72	5,42	5443147	863,58
feb-19	113429941	7316	7655	36299,25	4,96	6044503	826,20
mar-19	74674390,76	4755	3995	23250,3	4,89	3969989	834,91
abr-19	51706987,16	5801	5905	23117,37	3,99	4072941	702,11
may-19	43775670	5081	5137	18991,77	3,74	3316364	652,70
jun-19	38024031,1	5171	5556	17460	3,38	3023765	584,75
jul-19	62345948,06	6156	6866	27305,67	4,44	4173966	678,03
ago-19	68362898,31	7006	7043	28481,35	4,07	4474323	638,64
sep-19	65957098,35	6644	6821	27349,1	4,12	4468094	672,50
oct-19	68548389,12	6906	7465	26892,18	3,89	4927580	713,52
nov-19	79355024,21	7596	8026	30426,85	4,01	5607464	738,21
dic-19	80572668,29	6981	7668	30221,7	4,33	5264876	754,17
ene-20	23306241	7328	7478	40092,46	5,47	6537841	892,17
feb-20	13474781,2	7316	7655	36300	4,96	6044503	826,20
mar-20	74713350,11	4755	3995	23252,2	4,89	3969989	834,91
abr-20	9712386	106	92	2822	26,62	291653	2751,44
may-20	9341356,04	119	123	1941,44	16,31	197950	1663,45
jun-20	6473256,14	183	186	2063,1	11,27	313081	1710,83
jul-20	6361175,08	120	139	1865,06	15,54	175010	1458,42

Source: self-elaboration, 2020

What is more important, the average of days per RA (contract agreement) is notably increased during the pandemic because of two main reasons. On one hand, there are fewer rentals than in other months but at the same time, (on another) the rentals now are used for working and not for leisure prepuces. In April 2019 the average of rental days was 3.99 days per rental but in April 2020 it raised 26.62 days. The same happens with the average of driven mileages. This indicator is formed by the sum of all rentals and all driven mileages in a period of time. In April 2019 this amount was 702.11 mileages per RA while it radically changed in April 2020 to 2.751.44 per RA. These results confirm that renters are now using the vehicles to move to other states only for working. In view of this H1 and H3 are clearly confirmed.

Figure 1. Evolution of sales

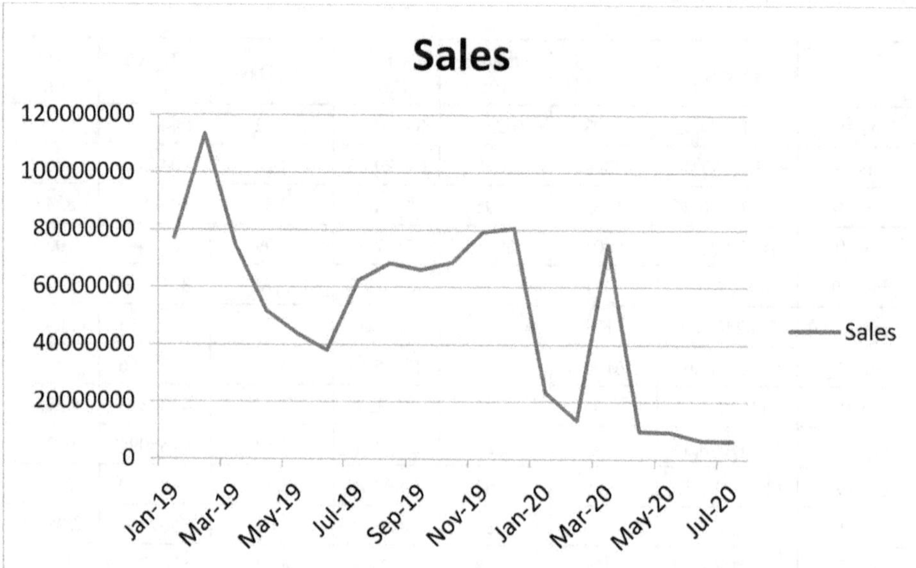

At the time of catching a glimpse of the number of reservations created in the month the downward trend seems to be evidently clear. In January 2019, 8.661 new bookings were created, followed by the months of February (4.800), March (8.299), April (7.474), and May (8.036). Although we find a decline in February, respecting January, it was resulted from the economic crisis and the process of devaluation that whipped the presidency of Mauricio Macri in February of 2019. Quite aside from this little discrepancy, these numbers are radically changed for the months of April 2020 (333 new bookings), May (332) (For further details see Table 2).

The recollected evidence amply supports the hypothesis that not only the sales of this rent-a-car company slumped down because of the economic crisis ushered Argentina to adopt restrictive measures to stop the pandemic, but also confirms hypothesis 3 which infers that the cars are now rented by new professional segments mainly oriented to essential industries as factories, farming industry and other sectors unaffected by COVID19. With this information in mind, we have consulted to the marketing Director, Juan Carlos who says us:

We were wrapped in a difficult position if you ask me the worse crisis the World has faced in its history or the worse I experienced. The first days were the saddest of my life. We do not know what to do. We were a company aimed for renting to foreigner tourists! Brazilians, Chileans, Americans who travelled to Bariloche, Iguazu and Calafate to enjoy our tourist destinations, now we were radically forced to swerve and attend other segments as medicines, scientists, police, farmers who are limited to use the public transport. We discovered that after all renting a car is the most secure form of traveling! (Juan Carlos, male 49 years old)

Table 2. Number of created bookings

Months	Created Bookings
ene-19	8661
feb-19	4800
mar-19	8299
abr-19	7474
may-19	8036
jun-19	8960
jul-19	10074
ago-19	10156
sep-19	10862
oct-19	12008
nov-19	10901
dic-19	11187
ene-20	10151
feb-20	8599
mar-20	4917
abr-20	333
may-20	332
jun-20	398
jul-20	504

Source: self elaboration, 2020

The above mentioned excerpt reveals that renting a car still remains as a valid option for drivers who need to move inside Argentina, though it enthralls as a substitute transport mean by the lack of airplanes and trains. This suggests that the effects in locations, far from being homogenous, vary on their nature and segmentation.

We have drawn to different groups: industrial and tourist locations. The former are grouped with locations where exist a great industrial manpower and infrastructure (i.e. Buenos Aires, Cordoba, Neuquen, Mar del Plata and Rosario). The latter, rather, is formed by locations based exclusively in leisure and tourist activities (i.e. Bariloche, Calafate, NOA, Northwestern Argentina & Iguazu and Mendoza).

Figure 2. Bookings

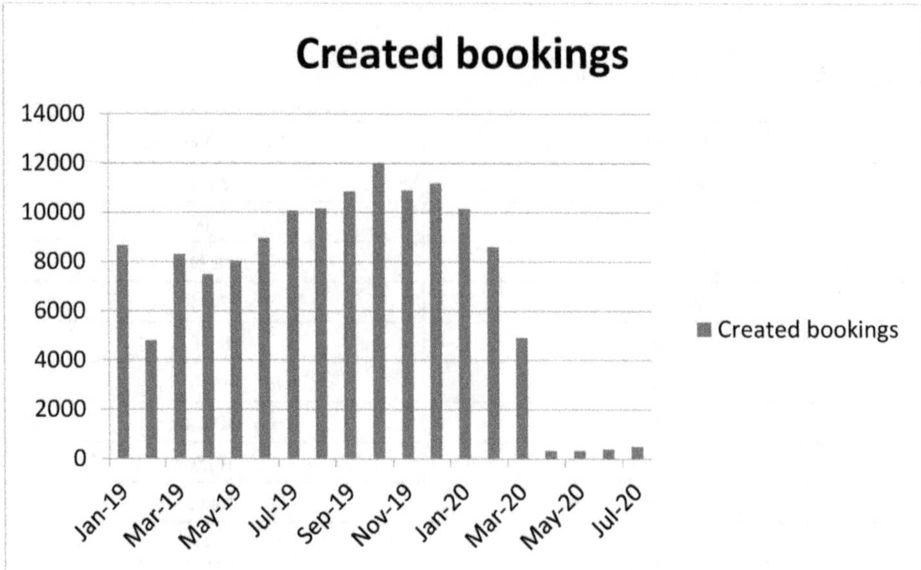

Sales per Locations

Industrial Destinations

Per table 4, Buenos Aires city billed ARS 9.873.919 in April of 2019, but this amount declined to ARS 3.387.136 in April of 2020. The variation was 60%. In May and June, the variation goes from 55% and 61% respectively. Neuquen has not the same luck. This industrial location has sales for ARS 3.170.416 in April 2019 but it reduced strongly to April 2020 in ARS 240.293, the variation percentage is for 92% or near circa. Cordoba appears to offer a similar scenario. The tendency repeats for the months of May and June. Mar del Plata is another third example which evolves in its own dynamic. In April of 2019, Mar del Plata registers marks for rental services for ARS 703.398 which changes to ARS 520.201 in April 2020. The variation oscillates between only 26% and 28% in the months of April and May, but if we compare July 2019 to July 2020, the variation is % 81. Let´s explain readers that Mar del Plata is a hybridized destination which offers a product of sun & beach in the summer period, combining the low season with a fishing industry. To some

Table 3. Sales in industrial destinations

Months	Buenos Aires	Neuquen	Mar del Plata	Rosario	Cordoba
ene-19	$ 16.407.212,00	$ 1.959.542,00	$ 2.141.632,00	$ 1.384.666,00	$ 3.654.305,00
feb-19	$ 17.488.564,00	$ 5.128.987,00	$ 2.084.238,51	$ 2.231.075,00	$ 4.832.411,00
mar-19	$ 13.851.681,00	$ 3.573.132,00	$ 1.697.116,00	$ 2.734.960,00	$ 3.126.214,00
abr-19	$ 9.873.919,00	$ 3.170.416,00	$ 703.398,00	$ 793.294,00	$ 2.458.165,00
may-19	$ 10.645.311,00	$ 2.742.421,00	$ 554.998,00	$ 842.742,00	$ 1.959.542,00
jun-19	$ 9.127.561,00	$ 2.752.426,00	$ 538.918,00	$ 893.052,00	$ 1.590.277,00
jul-19	$ 11.687.592,00	$ 2.747.791,00	$ 524.104,00	$ 831.586,00	$ 2.280.718,00
ago-19	$ 9.361.832,00	$ 4.792.920,00	$ 684.662,00	$ 1.109.579,00	$ 2.093.109,00
sep-19	$ 11.703.226,00	$ 4.293.294,00	$ 535.901,00	$ 2.362.072,00	$ 2.651.467,00
oct-19	$ 10.232.142,00	$ 3.247.217,00	$ 2.595.859,00	$ 1.183.781,00	$ 2.595.859,00
nov-19	$ 13.446.148,00	$ 4.453.225,00	$ 862.631,00	$ 1.012.234,00	$ 2.844.833,00
dic-19	$ 16.032.321,00	$ 3.153.993,00	$ 973.857,00	$ 1.223.158,00	$ 3.605.012,00
ene-20	$ 24.248.628,00	$ 4.786.935,00	$ 2.235.770,00	$ 4.786.635,00	$ 6.116.465,00
feb-20	$ 17.488.564,00	$ 5.128.987,00	$ 2.084.228,00	$ 2.231.075,00	$ 4.832.411,00
mar-20	$ 13.854.487,00	$ 3.573.132,00	$ 1.697.116,00	$ 2.734.960,00	$ 3.126.214,00
abr-20	$ 3.387.136,00	$ 240.293,00	$ 520.201,00	$ 2.229.804,00	$ 1.305.821,00
may-20	$ 4.777.740,00	$ 259.904,00	$ 207.452,00	$ 712.471,00	$ 392.700,00
jun-20	$ 3.564.951,00	$ 120.261,00	$ 537.558,00	$ 694.125,00	$ 186.969,00
jul-20	$ 4.829.467,00	$ 131.026,00	$ 94.615,00	$ 238.695,00	$ 87.710,00

Source: self elaboration, 2020

extent, July meets with the Winter Recess where thousands of city-dwellers coming from different states of the country go to Mar del Plata for tourism and relax. Having said this, the economic consequences of the lockdown in the industrial destinations is far from being lineal or homogeneous. Locations as Buenos Aires city experience a nuanced decline while other locations –as the example of Cordoba shows- do not reach the same criterion.

Table 4. Sales in tourist destinations

Months	Bariloche	Calafate	Mendoza	NOA e Iguazu
ene-19	$ 29.799.384,00	$ 6.282.780,00	$ 4.424.763,00	$ 9.452.008,00
feb-19	$ 43.355.202,00	$ 13.102.181,00	$ 8.738.333,00	$ 14.310.851,00
mar-19	$ 22.226.856,00	$ 9.626.625,00	$ 6.641.901,00	$ 10.171.384,00
abr-19	$ 11.521.252,00	$ 4.663.143,00	$ 1.063.272,00	$ 12.733.894,00
may-19	$ 5.701.413,00	$ 2.553.804,00	$ 4.518.902,00	$ 13.716.809,00
jun-19	$ 7.620.628,00	$ 1.220.948,00	$ 3.508.111,00	$ 10.085.040,00
jul-19	$ 21.099.285,00	$ 1.395.359,00	$ 5.221.935,00	$ 15.868.679,00
ago-19	$ 22.487.297,00	$ 1.416.449,00	$ 5.614.311,00	$ 19.382.045,00
sep-19	$ 16.878.021,00	$ 2.359.275,00	$ 5.686.458,00	$ 16.878.021,00
oct-19	$ 12.969.476,00	$ 5.955.109,00	$ 6.902.404,00	$ 18.542.965,00
nov-19	$ 16.987.705,00	$ 8.519.779,00	$ 6.802.795,00	$ 19.250.664,00
dic-19	$ 23.570.766,00	$ 10.291.812,00	$ 6.632.215,00	$ 12.655.631,00
ene-20	$ 43.004.083,82	$ 12.564.524,00	$ 9.776.888,00	$ 16.115.603,00
feb-20	$ 43.357.149,00	$ 13.145.075,00	$ 8.738.333,00	$ 14.310.981,00
mar-20	$ 22.231.335,00	$ 9.651.286,00	$ 6.641.901,00	$ 10.176.398,00
abr-20	$ 350.897,00	$ -	$ 182.206,00	$ 1.400.990,00
may-20	$ 693.832,00	$ -	$ 35.939,00	$ 2.261.313,00
jun-20	$ 47.972,00	$ -	$ 266.334,00	$ 869.172,00
jul-20	$ 77.097,00	$ -	$ 106.709,00	$ 687.216,00

Source: self elaboration, 2020

Tourist Destinations

Per our inferences previously outlined, tourist locations are more sensitive to the lockdown tan industrial ones. In fact, it is evident when we glimpse the evolution of sales in Bariloche and Calafate, even this latter reduced the sales to zero, but this does not happen with NOA & Iguazu. The volume of sales in Bariloche for April

2019 was ARS 11.521.252, but going down in April 2020 to ARS 77.097, which represents a variation of 96%. Even in June, this variation goes to 99%. Similar results are obtained in the successive months. Contrariwise, NOA has sales for ARS 12.733.384 in April 2019 and ARS 1.400.900. Needless to say, the variation is for 88%. Mendoza, which is probably affected by the Chileans who cross annually the border, profited almost ARS 1.063.272 on April 19 and ARS 182.206 on April 20 consolidating a variation of 82% in the decline. With the benefits of hindsight, though tourist destinations are clearly more affected than industrial location no less true is that further investigation is needed in this direction since the results are not homogenous.

Table 5. One way fees

Months	Drop Off Sales
ene-19	$ 1.228.986,00
feb-19	$ 3.193.261,00
mar-19	$ 2.242.677,00
abr-19	$ 942.694,00
may-19	$ 950.089,00
jun-19	$ 549.508,00
jul-19	$ 1.009.384,00
ago-19	$ 1.259.363,00
sep-19	$ 1.154.542,00
oct-19	$ 1.840.775,00
nov-19	$ 3.188.153,00
dic-19	$ 2.371.178,00
ene-20	$ 3.714.028,00
feb-20	$ 3.193.121,00
mar-20	$ 2.242.677,00
abr-20	$ 1.323.600,00
may-20	$ 712.662,00
jun-20	$ 766.315,00
jul-20	$ 477.494,00

Source: self elaboration, 2020

One of our consulted key informants, CEO Martin (male 62 years old) replied *when the pandemic, we asked the government for a bail out, not credit, the government accessed to pay partly the wages of our staff, our intention is not to fire people, we are a family and in this epic challenge we will remain all together, we are not afraid of the situation, we passed for similar situations in the past, for example in 2001 when the domestic market here in Argentina collapsed. Months after this disaster, we received the flux of tourists coming from neighboring countries. We reconverted our mission and vision, we became a tourist company, and recycled as a rent-a-car leader in the tourist market. I feel now we have to forget tourism for a long time, and focus on domestic drivers!.*

One Way Fees

For the sake of clarity, one-way fees or drop off fees are charges held to the renters when the car is not delivered in the same location where it has been picked up. In normal conditions, one-way fees are encouraged in Argentina all but discouraged outside the country. This contrasts radically with the European market where cars can be picked up and returned in different countries of the European Union. The one-way fee denotes the demand of renters who often rent a car as the main form of transport when strikes in others subsectors as civil airlines or long-distance buses come into fruition. In April 2019 the company earned ARS 972.694 in one way fee. In April of 2020, the sales in the same concept are rising to ARS 1.323.000. This reflects that hypothesis 4 is confirmed. Renters take advantage of renting a car to replace other less secure forms of transport. In May 2019 the sales were for ARS 950.089 while in 2020 for ARS 712.662. June 19 shows sales for ARS 549.508 which increases to ARS 766.315 for June 20. These numbers support the idea that rent-a-car company situates as an alternative mode to boost local tourism in the post-Covid 19 contexts.

To validate this, we interviewed mechanic Chief Staff (Marcelo, males, 42 years old) who claimed:

This situation has good and bad things. People are worried, shocked, they do not know what to do to return to their homes. Thousands of Argentineans have been stranded abroad, or in other states. The lockdown prohibits the public circulation, a fundamental right as the free movement has been cancelled, suspended by the government to avoid the sickness dispersion. In this mayhem, we received a lot of consult of young and elderly clients who needed a car to come back to home. The problem was that they drove thousands of mileage and sometimes the cars were not delivered in good conditions. In the months of May and June the situation normalized. Only those professional workers allowed by the government were renting cars. No foreign tourists anymore!

Figure 3. Drop off sales

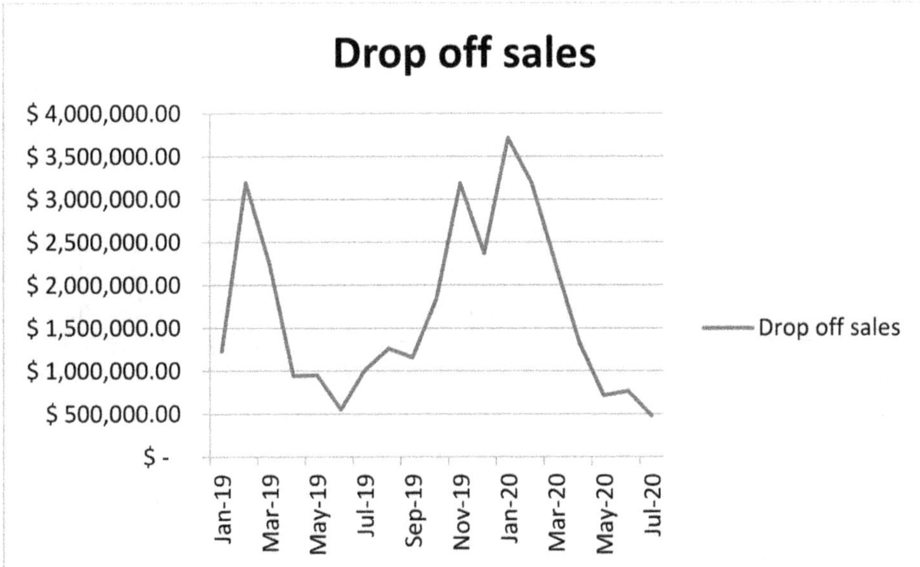

Renters

The last table is self-explanatory, the number of foreign renters has resoundingly fallen. Three months were grouped to better understand the dilemma in this case. In January 2019, the percentage of foreigner renters was 51%, while in January 2020 it was 52%. In April 2019, which is part of low season, 41% of renters come from abroad while this percentage slumps to 32% in April 2020. Not surprisingly, in May 2019 we have 65% of foreigner renters while it was for 16% only in May 2020. The results evince that international demands is blocked and the rent-a-car company suffered serious financial losses for the lockdown associated to the pandemic.

Table 6. Segmentation of renters.

Months	Argentinians	% VAR	Foreigners	% VAR2	Total
jan-19	3370	49%	3558	51%	6928
jan-20	3795	48%	4172	52%	7967
abr-19	3630	59%	2623	41%	6253
abr-20	122	68%	57	32%	179
may-19	1902	35%	3636	65%	5538
may-20	110	84%	22	16%	132

Source: self-elaboration, 2020

Managerial Perspectives

As stated, the outcomes of this research are not statistical representative of all rent-a-car companies. For that reason, they cannot be extrapolated to other universes or samples. Nevertheless of this, they infer interesting point of connection for expanding future policies in the tourism industry as well as managerial perspectives. At a first look, rent-a-car market offered a fertile ground to substitute mass transport means as bus or airplanes which are often fluid disseminators of COVID19. In this vein, the rent-a-car companies were the less affected subsector of the industry in comparison to cruise, hotels, civil airplanes, and buses. Therefore, the rent-a-car subsector though historically overlooked for applied research in the fields of tourism, offers nowadays a promising option to placate the devastating effects of COVID19 in the industry. Secondly, renting a car situates as one of the safer ways of moving, because the driver and his family are part of the same support bubble. Unlike other transport means there is no interaction with other passengers. Governments have emphasized on the needs of interacting with the same support bubble which is a group of people with whom one have personal daily contact. The support bubble prevents the close contact with other bubble households. Last but not least, the information gathered in this chapter shows not only rent-a-car has a considerable resiliency before the global pandemic it takes part of a broader system of connection which never stopped. Henceforth, policy makers and scholars who are interested in mitigating the effects of COVID19 in the tourism and hospitality industries should pay considerable attention to this neglected sector.

CONCLUSION

Metaphorically speaking, COVID19 not only interrogated furtherly the human omnipotence but also put the world on its knees. Its rapid dissemination in the modern means of transport escaped to all the sanitarian measures orchestrated in Asia. Only in months, COVID19 was wreaking havoc in Italy, Spain, and the US. Some additional months were needed for the Virus to arrive in Brazil, Ecuador, Chile and Argentina. The pandemic, in view of the lack of an effective vaccine, had a significant impact on the tourism and hospitality industries, resulting from travel-ban and the closure of borders. Academicians from the four corners of the world theorized on the effects and opportunities of the industry before this global hazard. Some voices claimed on the needs of adopting more sustainable forms of consumption while others remarked the urgency to create new methodological forms of research in a world without tourists. Still further, these studies lacked a robust empirical basis to understand the real effects of COVID in small-size companies.

Though debatable to some extent, the present chapter provided with valuable information and evidence on the effects of the pandemic in a rent-a-car company, a subsector which was relegated in the agenda of academicians. The outcomes are not heterogeneous but confirm that not only the studied company was seriously affected by the lockdown in Argentina, but also that tourist locations are more prone to experience a rapid downturn in respect to industrial ones. What is more important, in the absence of firm commercial aviation industry, rent-a-car subsector presents as a promising option of economic recovery to boost local tourism. It is important to caution that these results cannot be extrapolated to other universe and should be limited only to the sample in question. This seems to be a much deeper issue which needs further investigation.

REFERENCES

Amujo, O. C., & Otubanjo, O. (2012). Leveraging rebranding of 'unattractive' nation brands to stimulate post-disaster tourism. *Tourist Studies*, *12*(1), 87–105. doi:10.1177/1468797612444196

Barker, M., Page, S. J., & Meyer, D. (2003). Urban visitor perceptions of safety during a special event. *Journal of Travel Research*, *41*(4), 355–361. doi:10.1177/0047287503041004004

Bianchi, R. (2006). Tourism and the globalisation of fear: Analysing the politics of risk and (in) security in global travel. *Tourism and Hospitality Research*, *7*(1), 64–74. doi:10.1057/palgrave.thr.6050028

Blake, A., & Sinclair, M. T. (2003). Tourism crisis management: US response to September 11. *Annals of Tourism Research*, *30*(4), 813–832. doi:10.1016/S0160-7383(03)00056-2

Brouder, P., Teoh, S., Salazar, N. B., Mostafanezhad, M., Pung, J. M., Lapointe, D., Higgins Desbiolles, F., Haywood, M., Hall, C. M., & Clausen, H. B. (2020). Reflections and discussions: Tourism matters in the new normal post COVID-19. *Tourism Geographies*, *22*(3), 1–12. doi:10.1080/14616688.2020.1770325

Carr, A. (2020). COVID-19, indigenous peoples and tourism: A view from New Zealand. *Tourism Geographies*, 1–12.

Crossley, É. (2020). Ecological grief generates desire for environmental healing in tourism after COVID-19. *Tourism Geographies*, 1–10.

Floyd, M. F., & Pennington-Gray, L. (2004). Profiling risk perceptions of tourists. *Annals of Tourism Research, 31*(4), 1051–1054. doi:10.1016/j.annals.2004.03.011

Franklin, A. (2007). The problem with tourism theory. In The critical turn in tourism studies (pp. 153-170). Routledge. doi:10.1016/B978-0-08-045098-8.50014-3

Fuchs, G., & Reichel, A. (2006). Tourist destination risk perception: The case of Israel. *Journal of Hospitality & Leisure Marketing, 14*(2), 83–108. doi:10.1300/J150v14n02_06

Gallego, I., & Font, X. (2020). Changes in air passenger demand as a result of the COVID-19 crisis: Using Big Data to inform tourism policy. *Journal of Sustainable Tourism*, 1–20. doi:10.1080/09669582.2020.1773476

Galvani, A., Lew, A. A., & Perez, M. S. (2020). COVID-19 is expanding global consciousness and the sustainability of travel and tourism. *Tourism Geographies*, 1–10.

George, R. (2010). Visitor perceptions of crime-safety and attitudes towards risk: The case of Table Mountain National Park, Cape Town. *Tourism Management, 31*(6), 806–815. doi:10.1016/j.tourman.2009.08.011

Gössling, S., Scott, D., & Hall, C. M. (2020). Pandemics, tourism and global change: A rapid assessment of COVID-19. *Journal of Sustainable Tourism*, 1–20.

Gretzel, U., Fuchs, M., Baggio, R., Hoepken, W., Law, R., Neidhardt, J., ... Xiang, Z. (2020). e-Tourism beyond COVID-19: A call for transformative research. *Information Technology & Tourism*, 1.

Hall, C. M., Williams, A. M., & Lew, A. A. (2004). *Tourism: Conceptualizations, institutions, and issues. In A companion to tourism*. Blackwell Publishing.

Hannam, K., Sheller, M., & Urry, J. (2006). Mobilities, immobilities and moorings. *Mobilities, 1*(1), 1–22. doi:10.1080/17450100500489189

Higgins-Desbiolles, F. (2020). Socialising tourism for social and ecological justice after COVID-19. *Tourism Geographies*, 1–14.

Jonas, A., Mansfeld, Y., Paz, S., & Potasman, I. (2011). Determinants of health risk perception among low-risk-taking tourists traveling to developing countries. *Journal of Travel Research, 50*(1), 87–99. doi:10.1177/0047287509355323

Korstanje, M. E. (2011). Rent-a-car industry: A case study in Argentina. *Tourismos, 6*(1), 1–15.

Korstanje, M. E. (2018). *The mobilities paradox: a critical analysis*. Edward Elgar Publishing. doi:10.4337/9781788113311

Laws, E., Prideaux, B., & Chon, K. S. (Eds.). (2007). *Crisis management in tourism*. Cabi. doi:10.1079/9781845930479.0000

Lepp, A., & Gibson, H. (2008). Sensation seeking and tourism: Tourist role, perception of risk and destination choice. *Tourism Management, 29*(4), 740–750. doi:10.1016/j.tourman.2007.08.002

Mair, J., Ritchie, B. W., & Walters, G. (2016). Towards a research agenda for post-disaster and post-crisis recovery strategies for tourist destinations: A narrative review. *Current Issues in Tourism, 19*(1), 1–26. doi:10.1080/13683500.2014.932758

Matyas, C., Srinivasan, S., Cahyanto, I., Thapa, B., Pennington-Gray, L., & Villegas, J. (2011). Risk perception and evacuation decisions of Florida tourists under hurricane threats: A stated preference analysis. *Natural Hazards, 59*(2), 871–890. doi:10.100711069-011-9801-0

McKercher, B., & Chon, K. (2004). The over-reaction to SARS and the collapse of Asian tourism. *Annals of Tourism Research, 31*(3), 716–719. doi:10.1016/j.annals.2003.11.002 PMID:32572281

Miles, L., & Shipway, R. (2020). Exploring the COVID-19 pandemic as a catalyst for stimulating future research agendas for managing crises and disasters at international sport events. *Event Management, 24*(4), 537–552. doi:10.3727/152599519X15506259856688

Monterrubio, J. C. (2010). Short-term economic impacts of influenza A (H1N1) and government reaction on the Mexican tourism industry: An analysis of the media. *International Journal of Tourism Policy, 3*(1), 1–15. doi:10.1504/IJTP.2010.031599

Novelli, M., Burgess, L. G., Jones, A., & Ritchie, B. W. (2018). 'No Ebola… still doomed'–The Ebola-induced tourism crisis. *Annals of Tourism Research, 70*, 76–87. doi:10.1016/j.annals.2018.03.006 PMID:32572288

Packer, J. (2008). *Mobility without mayhem: Safety, cars, and citizenship*. Duke University Press. doi:10.2307/j.ctv123x6xj

Park, K., & Reisinger, Y. (2010). Differences in the perceived influence of natural disasters and travel risk on international travel. *Tourism Geographies, 12*(1), 1–24. doi:10.1080/14616680903493621

Prayag, G. (2020). Time for Reset? COVID-19 and Tourism Resilience. *Tourism Review International, 24*(2-3), 179–184. doi:10.3727/154427220X15926147793595

Reisinger, Y., & Mavondo, F. (2005). Travel anxiety and intentions to travel internationally: Implications of travel risk perception. *Journal of Travel Research*, *43*(3), 212–225. doi:10.1177/0047287504272017

Reisinger, Y., & Mavondo, F. (2006). Cultural differences in travel risk perception. *Journal of Travel & Tourism Marketing*, *20*(1), 13–31. doi:10.1300/J073v20n01_02

Ritchie, B. W. (2004). Chaos, crises and disasters: A strategic approach to crisis management in the tourism industry. *Tourism Management*, *25*(6), 669–683. doi:10.1016/j.tourman.2003.09.004

Robinson, L., & Jarvie, J. K. (2008). Post-disaster community tourism recovery: The tsunami and Arugam Bay, Sri Lanka. *Disasters*, *32*(4), 631–645. doi:10.1111/j.1467-7717.2008.01058.x PMID:18479472

Romagosa, F. (2020). The COVID-19 crisis: Opportunities for sustainable and proximity tourism. *Tourism Geographies*, *22*(3), 1–5. doi:10.1080/14616688.2020.1763447

Santana, G. (2004). Crisis management and tourism: Beyond the rhetoric. *Journal of Travel & Tourism Marketing*, *15*(4), 299–321. doi:10.1300/J073v15n04_05

Seabra, C., Abrantes, J. L., & Kastenholz, E. (2014). The influence of terrorism risk perception on purchase involvement and safety concern of international travellers. *Journal of Marketing Management*, *30*(9-10), 874–903. doi:10.1080/0267257X.2014.934904

Séraphin, H., Butcher, J., & Korstanje, M. (2017). Challenging the negative images of Haiti at a pre-visit stage using visual online learning materials. *Journal of Policy Research in Tourism, Leisure & Events*, *9*(2), 169–181. doi:10.1080/19407963.2016.1261146

Seraphin, H., Gowreesunkar, V., Roselé-Chim, P., Duplan, Y. J. J., & Korstanje, M. (2018). Tourism planning and innovation: The Caribbean under the spotlight. *Journal of Destination Marketing & Management*, *9*, 384–388. doi:10.1016/j.jdmm.2018.03.004

Séraphin, H., Korstanje, M., & Gowreesunkar, V. G. (Eds.). (2020). *Post-Disaster and Post-Conflict Tourism: Toward a New Management Approach.* CRC Press. doi:10.1201/9780429276224

Sheller, M., & Urry, J. (2004). *Tourism mobilities: Places to play, places in play.* Routledge. doi:10.4324/9780203340332

Sigala, M. (2020). Tourism and COVID-19: Impacts and implications for advancing and resetting industry and research. *Journal of Business Research, 117*, 312–321. Advance online publication. doi:10.1016/j.jbusres.2020.06.015 PMID:32546875

Solarin, S. A. (2015). September 11 attacks, H1N1 influenza, global financial crisis and tourist arrivals in Sarawak. *Anatolia, 26*(2), 298–300. doi:10.1080/13032917.2014.991933

Tarlow, P. (2014). *Tourism security: strategies for effectively managing travel risk and safety*. Elsevier.

Urry, J. (2006). Inhabiting the car. *The Sociological Review, 54*(1, suppl), 17–31. doi:10.1111/j.1467-954X.2006.00635.x

Verma, S., & Gustafsson, A. (2020). Investigating the emerging COVID-19 research trends in the field of business and management: A bibliometric analysis approach. *Journal of Business Research, 118*, 253–261. Advance online publication. doi:10.1016/j.jbusres.2020.06.057 PMID:32834211

Wen, J., Wang, W., Kozak, M., Liu, X., & Hou, H. (2020). Many brains are better than one: The importance of interdisciplinary studies on COVID-19 in and beyond tourism. *Tourism Recreation Research*, 1–4. doi:10.1080/02508281.2020.1761120

Wolff, K., & Larsen, S. (2014). Can terrorism make us feel safer? Risk perceptions and worries before and after the July 22nd attacks. *Annals of Tourism Research, 44*, 200–209. doi:10.1016/j.annals.2013.10.003

Yüksel, A., & Yüksel, F. (2007). Shopping risk perceptions: Effects on tourists' emotions, satisfaction and expressed loyalty intentions. *Tourism Management, 28*(3), 703–713. doi:10.1016/j.tourman.2006.04.025

Zenker, S., & Kock, F. (2020). The coronavirus pandemic–A critical discussion of a tourism research agenda. *Tourism Management, 81*, 104164. doi:10.1016/j.tourman.2020.104164 PMID:32518437

ADDITIONAL READING

Ashworth, G. J., & Tunbridge, J. E. (2000). *The tourist-historic city*. Routledge. doi:10.4324/9780080519470

Bell, D. (2012). Hospitality is society. *Hospitality & Society, 1*(2), 137–152. doi:10.1386/hosp.1.2.137_1

Comaroff, J. L., & Comaroff, J. (2009). *Ethnicity, Inc*. University of Chicago Press. doi:10.7208/chicago/9780226114736.001.0001

George, S. K. (2009). Hospitality as openness to the other: Levinas, Derrida and the Indian hospitality ethos. *Journal of Human Values*, *15*(1), 29–47. doi:10.1177/097168580901500103

Gössling, S., Scott, D., & Hall, C. M. (2020). Pandemics, tourism and global change: A rapid assessment of COVID-19. *Journal of Sustainable Tourism*, 1–20.

Higgins-Desbiolles, F. (2020). Socialising tourism for social and ecological justice after COVID-19. *Tourism Geographies*, *22*(3), 1–14. doi:10.1080/14616688.2020.1757748

Hottola, P. (2014). Somewhat empty meeting grounds: Travelers in South India. *Annals of Tourism Research*, *44*, 270–282. doi:10.1016/j.annals.2013.10.007

Innerarity, D. (2017). *Ethics of hospitality*. Taylor & Francis. doi:10.4324/9781315618241

Judd, D. R. S., & Fainstein, S. S. (Eds.). (1999). *The tourist city*. Yale University Press.

Knox, A. (2008). Gender desegregation and equal employment opportunities in Australian luxury hotels: Are we there yet? *Asia Pacific Journal of Human Resources*, *46*(2), 153–172. doi:10.1177/1038411108091756.

Knox, A. (2011). Upstairs, downstairs: An analysis of low paid work in Australian hotels. *Labour & Industry*, *21*(3), 573–594. doi:10.1080/10301763.2011.10669421

Korstanje, M. (2012). Reconsidering cultural tourism: An anthropologist's perspective. *Journal of Heritage Tourism*, *7*(2), 179–184. doi:10.1080/1743873X.2011.639883

Korstanje, M. E. (2017). *Terrorism, Tourism and the End of Hospitality in the 'West*. Springer.

Korstanje, M. E. (2018). *The mobilities paradox: a critical analysis*. Edward Elgar Publishing. doi:10.4337/9781788113311

Lashley, C. (2015). Hospitality and Hospitableness. *Research in Hospitality Management*, *5*(1), 1–7. doi:10.1080/22243534.2015.11828322

Lashley, C. (2019). Towards a Theoretical understanding of hospitality. In B. Rowson & C. Lashley (Eds.), *Experiencing hospitality* (pp. 1–24). Nova Science Publishers.

Lugosi, P. (2014). Hospitality and organizations: Enchantment, entrenchment and reconfiguration. *Hospitality & Society*, *4*(1), 75–92. doi:10.1386/hosp.4.1.75_7

Lynch, P., Molz, G. J., Mcinthosh, A., Lugosi, P., & Lashley, C. (2011). Theorizing hospitality. *Hospitality & Society, 1*(1), 3–24. doi:10.1386/hosp.1.1.3_2

MacCannell, D. (1976). *The tourist: A new theory of the leisure class*. University of California Press.

MacCannell, D. (1992). *Empty meeting grounds: The tourist papers*. Psychology Press.

Martin, R. J. (1998). *Professional management of housekeeping operations*. John Wiley and Sons.

Morrison, A., & O'Gorman, K. (2008). Hospitality studies and hospitality management: A symbiotic relationship. *International Journal of Hospitality Management, 27*(2), 214–221. doi:10.1016/j.ijhm.2007.07.028

Nichols, J. (2011). *Hospitality among the Romans. The Oxford Handbook of Social Relations in the Roman World*. Oxford University Press.

O'Gorman, K. (2007a). Dimensions of hospitality: exploring ancient and classical origins. In *Hospitality: A social lens* (pp. 31–46). Routledge. doi:10.1016/B978-0-08-045093-3.50005-9

O'Gorman, K. D. (2005). Modern Hospitality: Lessons from the past. *Journal of Hospitality and Tourism Management, 12*(2), 141–151.

O'Gorman, K. D. (2006). Jacques Derrida's philosophy of hospitality. *Hospitality Review, 8*(4), 50–57.

O'Gorman, K. D. (2007b). Discovering commercial hospitality in ancient Rome. *Hospitality Review, 9*(2), 44–52.

O'Gorman, K. D. (2007c). The hospitality phenomenon: Philosophical enlightenment? *International Journal of Culture, Tourism and Hospitality Research, 1*(3), 189–202. doi:10.1108/17506180710817729

Pagden, A. (1995). *Lords of all the World: Ideologies of Empire in Spain, Britain and France c. 1500-c. 1800*. Yale University Press.

Palmer, C. A. (1994). Tourism and colonialism: The experience of the Bahamas. *Annals of Tourism Research, 21*(4), 792–811. doi:10.1016/0160-7383(94)90084-1

Poulston, J. (2015). Expressive labour and the gift of hospitality. *Hospitality & Society, 5*(2-3), 145–165. doi:10.1386/hosp.5.2-3.145_1

Ritzer, G. (2019). Inhospitable hospitality? In B. Rowson & C. Lashley (Eds.), *Experiencing hospitality* (pp. 73–90). Nova Science Publishers.

Romagosa, F. (2020). The COVID-19 crisis: Opportunities for sustainable and proximity tourism. *Tourism Geographies*, 1–5.

Urry, J. (2002). *The tourist gaze*. Sage.

Virilio, P. (1989). *War and cinema. The Logistics of Perception*. Patrick Camille.

Wen, J., Wang, W., Kozak, M., Liu, X., & Hou, H. (2020). Many brains are better than one: The importance of interdisciplinary studies on COVID-19 in and beyond tourism. *Tourism Recreation Research*, 1–4. doi:10.1080/02508281.2020.1761120

KEY TERMS AND DEFINITIONS

COVID-19: New virus outbreak originated in Wuhan, China, which started a new global pandemic that affected most of the industrial economies of the world in 2020.

Economic Downturn: This means a term associated to stock and market crisis or an economic cycle constriction which generates an economic decline.

Fear: It signals to a basic emotion based on the needs of avoiding or confronting an external real or imagined threat.

Rent-a-Car Market: Market that hires and offers car rentals to tourists and travelers.

Tourism: It is considered a sector industry oriented to promote travel experiences and consumption. Today tourism is considered one of the largest industries worldwide.

Chapter 8
A Terrorism-Inflicted Causality in Kashmir:
Perspectives From Travel Agencies and Tour Operators

Anubha
National Institute of Tourism and Hospitality Management (NITHM), India

Mahender Reddy Gavinolla
https://orcid.org/0000-0002-3476-3132
National Institute of Tourism and Hospitality Management (NITHM), India

ABSTRACT

Terrorist incidents at tourist destinations create several negative impacts. Many studies have focused on the intent of the tourists to visit a place affected by terrorism. However, there are no studies gauging the perceptions of travel agencies and tour operators. In this regard, this particular study intends to understand the detrimental effects of terrorism on various travel agencies and tour operators operating in Kashmir in India, which also form a part of MSMEs. Case study and semi-structured interviews have been used. The findings clearly illustrate that terrorism does have detrimental effects on a tourist destination and, in turn, deleterious effects on the travel agencies and tour operators operating in the region. It is also evident that there are political instabilities and lack of promotion and marketing of the destination. It is hoped that the political stability and proper promotion will lead to positive destination image and better management of the destination.

DOI: 10.4018/978-1-7998-6996-2.ch008

INTRODUCTION

Tourism as an industry is extremely vulnerable to turbulent situations like terrorist attacks and political violence and instability (Araña & León, 2008). Terrorism risk is more pronounced as compared to other catastrophic risks as it is driven by human intent and intelligence, which is a factor that remains absent in case of naturally occurring unexpected disasters (Major,2002). According to Pain (2014), the key intent behind a terrorist attack is to incite fear globally. Terrorism has enlarged its horizon and has grown into a global concern by affecting the tourism industry, tourists and the host communities (Mansfeld & Pizam, 2006). Terrorism plays a significant role in affecting the tourism industry negatively thus travels related decisions (Sönmez, 1998)

Terrorist attacks can impede business activity and lower customer confidence, which in turn affects the demand, investment and consumption, leading to substantially hampering overall economic growth. There are four major economic repercussions of terrorist attacks: firstly, being the annihilation of human and physical capital. Secondly, increased mistrust and suspicion, thirdly, the diversion of funds from productive sectors to defense to enhance the security and fourth being negative impacts on highly vulnerable industries like tourism (Abadie and Gardeazabal, 2008). In this regard, several studies conducted in the past have elucidated and reinforced that terrorism impacts tourist arrivals and in turn there is a decline in tourism revenue in case of a destination affected by terrorism. For instance, effect of terrorism on demand for tourism (Buigut and Amendah, 2016), impact on Mediterranean regions (Drakos and Kutan, 2003), transnational terrorism and tourism (Enders and Sandler, 1991), economic impact (Sloboda, 200; Pizam and Fleischer, 2002; Enders et al., 1992) crisis and recovery of tourism from terrorist activities (Mansfeld, 1999), political violence (Neumayer, 2004) political instability and impact on tourism development (Saha and Yap, 2014; Thompson, 2011).

Further, there are quite a few studies focused on terrorism and tourism and its implications on Indian tourism industry. For Instance, a study focused on impact of terrorist activities on inbound tourism and foreign exchange earnings (Parida et al. 2015), home country terrorism impact on Indian medical tourism (Jha, 2020), tourism, migration and terrorism (Taylor, 2012), terror incidents and its impact on international tourist arrivals in India (Bhattacharya & Basu, 2010), bilateral cooperation and peace building in terror inflicted areas (Swain, 2009), terrorism as a thread and disorder to the development of Indian tourism (Dhariwal, 2005). Tourism is considered as an ambassador for peace creation and regional development (Becken & Carmignani, 2016). However, a study conducted on Kashmir, a major and popular destination of India hindered the smooth functioning and growth of tourism due to the spread of terrorist activities in the region over the last two decades (Chauhan

& Khanna, 2008). Although there are several studies conducted on the impact of terrorism on tourism in general, there are several aspects related to tourism and terrorism are still under researched (Arana & Leon, 2008). However, there are no studies gauging the perceptions of travel agencies and tour operators.

Kashmir is one such tourist destination with an enormous potential to attract tourists. However, terrorism has plagued this touristic destination and remains a challenge, like any other developed country even the developing economies are grappling with this issue and the problem is deep rooted in Jammu and Kashmir. Ever since India and Pakistan came into existence, Jammu and Kashmir has been a contested territory even became the root cause of the four wars between two countries (Narain, 2016). The conflict between two countries is well researched and documented in several studies (Ganguly, 2002; Lyon, 2008; Paul, 2005; Wolpert, 2011). For instance, role of Pakistan in the growing militancy in the region with the logistic support and militant training (Evans, 2000; Puri, 1993; Ray, 1997; Subramanian, 1990; Swami, 2003a, p. 15). However, researchers have chronicled the rise of ethno-religious sentiments in the oppressed Muslim class, the unexpected dismissal of local leaders and electoral malfeasances being the trigger point for the rise of militancy in Kashmir Behera, 2000, p. 164; Evans, 2000; Mohanty, 2018). The recent change in the status of Jammu and Kashmir was in the first week of August 2019, the government of India abrogated Article 370 and 35 A, stripping Jammu& Kashmir of special status, limiting land ownership and permanent residency of the natives and bifurcating the state into two new union territories: 'Jammu and Kashmir' and 'Ladakh'. By abrogating the Article 370 and 35 A and stripping the state of its special status thus removing the law that ascertained land ownership and permanent residence to natives (Venkatesan, 2020). Dutta (2019) reported that a vast population in Kashmir believed that this could result in drastically altering the demographic framework of Jammu and Kashmir. And this demographic change could trigger a fresh wave of militancy (Bhat, 2019). In the case of Kashmir, the central government scrapped off Article 370, which provided special autonomous status to the state and was taken as a step to control militancy and terrorist attacks in the state (Financial Express. (2019). The home minster of Government of India emphasized that eliminating terrorism would pave a path of development in Jammu and Kashmir, including tourism (Financial Express. (2019). In case of Kashmir, a massive drop of almost 60% is recorded after the state gained a new status of Union territory with only 4, 99,584 tourists visiting in 2019, which is the lowest in the past 10 years (Firdous Hassan, 2020). Significance of tourism as an economic driver in the state cannot be denied with an economic survey conducted by the J & K government in 2017, it was observed that tourism is a key sector and contributed 6.98% of the state GDP. (Directorate of Economics & Statistics J&K Government, 2020).

In this regard, this particular study intends to understand the detrimental effects of terrorism on various tourism service providers such as travel agencies and tour operators operating in Kashmir, which forms a part of Micro, Small and Medium Scale Enterprises (MSMSEs) also. The case study methodology is adopted to investigate the real-time problems faced by the tourism service providers, by interviewing the proprietors of the travel agencies and tour operators. Qualitative research technique such as interview method, with open-ended questions is conducted to examine the views and opinions of the respondents to have a better perspective on the ordeals in providing the tourism services in the region vested by terrorism activities. In order to do so, the study examines the difficulties faced and the steps that are taken to mitigate the effect of terrorism on their business that are MSME in size and nature.

In doing so, the study is divided into five parts. The first part of the chapter provides theoretical insights on terrorism and its implications by introducing the research gap and study area. Second part elucidates the review of the existing studies. Third part deals with the methodological approach of the study, followed by the next part with an insight on the study area, results of the study with discussions and implications. The last section focuses on concluding remarks with limitations and future scope of the research.

LITERATURE REVIEW

Terrorism, as a word traces its origins to the Latin word terrere, which means 'to frieghten' or 'to scare' and was adapted to describe terrorist attacks (Weimann, 2008) in developing a ubiquitous understanding. A much more comprehensive definition is "the systematic threat or use of violence, often across national borders, to attain a political goal or communicate a political message through fear, coercion, or intimidation of non-combatant persons or the general public" (Czinkota et al., 2004). A terrorist attack is defined as the physical or psychological attack of any armed group or gang on the civil society (Ruiz Estrada & Park, 2008). Hence, terrorism not only damages and destructs the physical assets, but also disrupts life of civilians psychologically. It has been investigated and reported by (Frey et al.,2007) that terrorist attacks have detrimental effects of financial markets and have economic effects, the study presents the compounded effect of terrorist attack, which according to the researchers majorly affects eight different activities, such as tourism, investments as well as FDI (foreign direct investment), savings and consumption, urban economy, national income, foreign trade, urban economy, national income, and stock markets. Based on a study conducted by (Major, 2002) it was reported that terrorism risk is more pronounced as compared to other catastrophic risks as it is driven by human intent and intelligence, which is a factor that remains absent in case of naturally occurring unexpected disasters. Tourism as an industry is extremely vulnerable to terrorist attacks and political violence and instability (Araña & León, 2008). The primary requisite of a destination to attract tourism is to be free from any activities in relation to terrorism (Araña & León, 2008). Large numbers of studies in terms of real and perceived risk, terrorism, tourist choice of destination, impacts of terrorism on international tourists' arrivals, tourism demand, crime, political violence, and instability have been conducted to understand and measure the consequences of terrorism (Korstanje & Tarlow, 2012; Perpiña et al. 2019; Weber, 1998).

Tourism plays an important role as an economic driver in case of tourist destinations by providing employment opportunities directly or indirectly (Pablo-Romero & Molina, 2013; Gavinolla et al. 2021). Act of terrorism is directly related to tourism demand, which is reestablished in various tourist destinations by various tourism economists (Fleischer and Buccola 2002; Llorca-Vivero 2008; Araña and León 2008; Feridun and Shahbaz 2010). Single terrorist attack can disrupt the tourist inflows and takes time to recover with the time taken for the incident to be forgotten (Avraham 2013; Fleischer and Buccola 2002). For the countries facing a reoccurring act of terrorism, can face dissolution or annihilation of tourism industry (Enders and Sandler 1991; Pizam and Fleischer 2002; Raza and Jawaid 2013). There has been substantial work done on economic impacts of tourism in general (Abadie and Gardeazabal 2008; Barth et al. 2006; Collier and Hoeffler 2004). Brady and Widdows (1988), further these studies, have established a fact that an economy driven by tourism can be negatively impacted by terrorism activities. There are several empirical studies to support the claim that there is a decreased demand for tourism in the destinations affected by terrorism (Wall 1996; Wahab 1996; Pizam and Smith, 2000).

Tourism industry worldwide is shaken by many natural disasters like earthquakes, floods (Becken, 2013; Cochrane, 2010) and man-made or influenced disasters such as wars and terrorist attacks (Richter & Waugh Jr, 1986; Sönmez, 1998). In the past, many studies have reported that a region jolted repeatedly by terrorist activity is bound to have a significant decline in tourist intent to travel to the affected region (Pizam and Mansfeld 1996; Sönmez and Graefe 1998; Sönmez 1998; McKercher and Hui 2004; Krakover 2005; Araña and León 2008). India too as a country face repeated jolts of terrorist attacks and political instability, one such region is northern state of Jammu and Kashmir, which has suffered with infiltration of militants and also cross-border terrorism (Bose, 2009). In case of India, Parida et al., (2015), also examined the linkage between terrorism, economic expansion on inbound tourism and resultant impact in terms of foreign exchange earnings from tourism earnings was summated and analyzed from year 1980 to 2001.It was established in the study that terror activities have detrimental effects on foreign tourist arrival and foreign exchange earnings (Parida et al.2015). In their study conducted by Parida et al. (2015)

also suggested that in a developing country like India, the development of tourism as a sector can generate large-scale employment and can as a result help in poverty reduction, however terror activities hinder these developmental activities. Also, in a study conducted by (Raja & Ankathi, 2020), the multiplier effects of terrorism on tourism is reviewed and elucidates the role of terrorism and its subsequent effects on the hospitality industry, first and foremost being rise in unemployment in the service sector, which eventually effects the revenue generated from tourism, and finally impacts the nations' economy. Further in their study, featuring 11 different countries around the world affected by terrorism, explicated, that India is a country with high potential tourism generation markets; terrorism has played a devastating role on tourism.

The above-mentioned studies conducted in various contexts revealed the implications of terrorist activities on tourism in various destinations of the world. Further, these studies have mainly focused on economic impacts on destination, impact on economic activities, terrorist activities as a hindrance for the development of tourism, and also with psychological and behavioral aspects of tourists and their reluctance to visit a terrorist affected destination. However, there are no studies examining the travel agents and tourism operators' perspectives, particularly in the context of MSME's. It is this regard the study elucidates mainly on the perspectives of the travel services providers in Kashmir, India.

KASHMIR AS A DESTINATION

Kashmir as a region is laced with various attractions which can appeal to a wide range of tourists, ranging from historical, cultural, archeological, scenic, climatic, and culinary and adventure resources (Chahal & Devi, 2015). It is because of these attributes Kashmir is regarded as "paradise on earth" and is identified as a prime tourist destination. Several studies conducted in the past also have established that, "Natural attractions" like the scenic beauty of the place, climate, unique flora and fauna in a destination plays an important role in attracting tourists to a destination (Gearing et al. 1974; Var et al. 1977; Ritchie and Zins, 1978; Ferrario, 1979; Elwin, 1989; Meinung, 1989; Tang and Rochananond, 1990; Hu and Ritchie, 1993; Weber, 1997; Vaughan and Edwards, 1999; Morachat, 2003).

However, revisiting the past of Kashmir is crucial to understand the current situation faced by the region. The northern state of Jammu and Kashmir is a vivid combination of three religiously and culturally distinct regions, Hindu majority region being Jammu, Muslim majority region being Kashmir and a Bussdhist majority region of Ladakh. However, In August 2019, the Indian government passed the Jammu and Kashmir Reorganization Act which has changed the status of statehood of Jammu and Kashmir and after the amendment the state is reorganized into two Union territories -Ladakh and Jammu and Kashmir. In August 2019, the Indian government passed the Jammu and Kashmir Reorganization Act which reorganized the state into two Union Territories – Ladakh, and Jammu and Kashmir. Back in the time of partition of British India into India and Pakistan, Kashmir was a princely state, with a Muslim dominated populace was ruled by a Hindu ruler Maharaja Hari Singh. The political climate so changed that the ruler joined hands with India by signing an "instrument of accession" following the fear of Pakistan backed tribesmen invading the region from northwest. In the turn of events a war broke in the region after the Indian military forces took over the region overthrowing the invaders. the region has seen a ground of dispute between the countries, resulting in a war in 1965 and then again, the Kargil war in 1999 A ceasefire was brokered by the intervention of the UN, with Pakistan acquiring one third of the area and the rest being secured as an Indian territory. However, the region has seen a ground of dispute between the countries, resulting in a war in 1965 and then again, the Kargil war in 1999.

Kashmir is selected as a study area, since then the region of Kashmir is known for its natural and cultural beauty not only in India but also across the globe, has faced several crises ranging from political instability, wars, to natural disasters like earthquakes and floods and, Covid-19 pandemic, and most importantly the terrorist inflicted activities impacted the destination negatively. Considering these characteristics, the study outcome will provide several important insights and suggestions that can be wider implications at global level.

RESEARCH METHODOLOGY

The primary objective of the study is to understand the perspectives of travel agencies and tour operators, operating in a terrorist affected zone, Jammu and Kashmir, and to understand the challenges faced. Travel and leisure sector are highly vulnerable to terrorist threats and actions. (Sullivan-Taylor, B., & Wilson, D. C. (2009). It has been also observed by Czinkota et al. (2005) that terrorism has directly affected the fortunes of the travel and aviation industry and was also established by a study conducted by Mayntz (2004), that organizations are a soft target for terrorists induced threats. As travel companies operating in the region can be classified under medium and small-scale industries (MSMEs). The primary purpose of the study is to explore the causal linkages between terrorism, travel and leisure industry with focus on travel agents and tour operators and the various issues and challenges faced by the proprietors of the travel agency and tour operators and the various steps being taken to mitigate the effects of terrorism on their industry. Adapting from the definition devised by Morrison (1998) to support the study conducted: "SME travel agents are those, which are independently owned, managed in a personalized manner by the owner-manager and not contracted out to an agent, represent the primary source of income for the owner-manager, and employ less than 50 employees".

A desk-based research was undertaken and various sources of information such as newspaper articles, websites information, research articles were explored in order to draw cognizance of the situation (Gowreesunkar & Gavinolla, 2020; Reddy et al. 2021). Substantiating the study with the above given definition by Morrison (1998), qualitative research methodology was adopted to conduct exploratory semi structured interviews with the proprietors of the travel agencies and tour operators were held (Livina & Reddy, 2017; Gowreesunkar & Gavinolla, 2020; Reddy et al. 2021). Case study method is commonly applied when the situation is lesser known to research and requires analysis supported and substantiated with individual instances (Yin et al. 2009; Veal, 2011). On these lines several studies used case study method to study lesser-known phenomenon with individual examples (Gowreesunkar et al., 2009; Bassil et al., 2019; Adeloye & Carr, 2019). This empirical study consists of several phases, starting with the first phase with the authors reviewing relevant literature to understand the phenomenon and finding the gaps, followed by designing the questionnaire. Second stage of the study consists of data collection through telephonic interviews with the respondents. Respondents were approached with the assistance of a native of the state, who also happens to be a close friend of one of the researchers, which resulted in opportunities of formal and informal interactions with the respondents participating in the study. Telephonic interviews were conducted rather than face to face interviews, due to the various restrictions and the current unfavorable conditions due to COVID-19 pandemic.

Respondents were selected voluntarily by explaining the importance of the study. A total of 12 proprietors were interviewed operating in the state of Jammu and Kashmir, by using the semi-structured questioning framework. Semi structured interviews are considered as an important method to obtain the data in the case study method (Yin et al. 2009). The duration of the survey lasted from a minimum of 15 minutes to a maximum time period of 25 minutes. The interviews were held in the first week of November 2020. Questionnaire consists of three parts with 10 questions covering the broader aspects such as, description of the company, opinion on impacts of terrorism on tourism in Kashmir, and followed by questions on recovery and resilience strategies adopted/adapted. The resulting data was consolidated and content was analyzed for recurrent themes, whereby contributing to generation of important challenges and issues faced and the also helped in comprehending the various steps being taken by the organizations to mitigate the effect of terrorism induced effects on the travel business. Furthermore, we have incorporated direct quotes from the respondents to support the study.

FINDINGS

Study findings are derived from the semi-structured interviews held with proprietors of travel and tourism service providers such as travel agencies and tour operators. Results of the interviews with tourism proprietors of travel agencies and tour operators shed focus on the current scenario, viewpoint on the impact and implications of terrorist activities in the Kashmir region. Thus, further provides insights on necessary strategies to be implementing in order to eradicate or lessen these negative impacts on tourism businesses in Kashmir. Total 16 proprietors were approached for the data collection, in which 4 respondents denied to provide information and opinion on the mentioned questions, thus the final sample size for the study was 12 responses (n=12) were obtained. To better derive the result, the authors first tried to understand the description of the company.

Description of the Company

To understand the description of the company authors asked the questions on numbers of employees or the workforce employed working in the company, how old is the company and the various tourist activities or services that are provided to the tourists in Kashmir. Out of 12 companies 6 companies mentioned that they are 6 to 8 years old with 5 employees. Another 5 companies mentioned that they are minimum of 12 to 14 years with 20 employees. One company mentioned that there are 30 employees working and one of the oldest companies in the region with 25 years history. It is widely understood that the older the company, the higher the number of employees.

For the question on what kind tourism services does the company offer to the tourist? All the respondents mentioned common services like tour packages, hotel accommodation booking, and transport services, however tour packages vary from company to company. While all the companies provide general tour packages to major attractions of Kashmir, only four companies are specialized in trekking and adventure sports tourism. However, two companies specialize in providing temple tours, family tours and honeymoon tours.

Opinion on Importance of Tourism in Kashmir

Further to understand the opinion on the importance of tourism in general on Kashmir as well as on their business. In response to the question 'do you think tourism is contributing positively to Kashmir?, all the respondents are of the opinion that tourism contributes positively to Kashmir, both to the government and also to the people of Kashmir. Further they have mentioned that tourism is the primary source of income for many people in Kashmir, a place with tourism driven economy, there is a scope for establishing travel related businesses in the region. Respondents agreed that there is an improvement in economic and employment benefits due to tourism in Kashmir over the years.

Opinions on Impacts of Terrorism on Terrorism in Kashmir

In this section, we asked what are the biggest risks facing the company due to terrorist activities in Kashmir? There were mixed responses over these questions, and one of the respondents said:

It goes without saying that when you hear terrorist activities are prevalent here, the majority of the first-time travelers are very skeptical about travel, leading to loss of business, cancellation of bookings, low foreign travellers and sale of good packages at rates reaping no profits.

Another response from the proprietor is in opinion that:

Convincing a person visiting the state is difficult, tourists visiting from abroad rely on media reporting of incidents and most of the time are cynical to make travel plans. However, there is one thing that I want to share that never ever has a tourist being harmed in any way in the state.

Other respondents mentioned that the company will fully take care the tourist with proper safety and security, and moreover they have insisted that they will conduct the tours where there is no risk and secure to visit the tourist as they are already known about the destinations like what is what. Another diverse opinion obtained; one should be aware that Kashmir have had no tragedies reported which involved visitors. However, it can be said that the Kashmir is safe for outsiders but unsafe for the residents.

Tourism in Kashmir consists of various attractions, thus having various stakeholders in providing the services like House Boats, Cab services, Restaurant, Accommodation Providers, DMOs and so on. These are the active stakeholders of tourism and also the vendors for the travel agencies and tour operators. In this regard, another important question we asked is that how the suppliers or partners do view the threat of terrorism?

We do work in collaboration with house boat owners, hotels, cab service providers and even the local restaurants who are serving authentic Kashmiri food. I must say things have been difficult for all of us, whether it is a terrorist attack or political amendments like removal of Article 370 and disconnection caused due to restricted phone services and internet services. Survival has become difficult now.

Another respondent says that, however it is affecting the vendors due to terrorist activities now, it is important to consider that Articles 370 had major impact on tourism and hoping that in long it will help the tourism.

We all have got majorly hit after the removal of Article 370, after the telephone connections were barred and internet was shut down. We had real trouble in getting bookings as we get our reservations either on telephone or internet. We had major cancellations and had to refund the advances paid, which was a major setback. Government should support us as after removal of Article 370 militancy was curbed, they should now promote the state as a safe and secure tourist destination so that we all will be confident and optimistic towards tourism.

Other respondents were in similar opinion that the terrorist activities lead to curfews which completely disables the business. Due to Article 370 removal Kashmir was shut down from August 4, 2018. Kashmir was refrained from any phone connectivity. People and businesses were completely disconnected, and a complete shutdown was the common situation. While other respondents mentioned terrorism is part of life now. People here have learnt to live with the obstacles caused. Series of events like the Pulwama attack, removal of article 370 and advisory issued by the government for the tourists to leave immediately created panic and concern, and with this pandemic now, they are trying all possible ways to attract tourists and minimize losses, but the cost is too high.

In connection with impacts of terrorist activities on tourism in Kashmir, further, we left the question open for the purpose of steering further discussion, for which each one of them had an opinion that difficult thing to persuade tourists wanting to come to Kashmir and overcoming their fear dilemma though there are several beautiful attractions in the region. Unfortunately, the media projects the image of the whole state being disturbed, though there are fewer incidents in some parts of the region. However, they have recovered from many incidents, but the challenges they face are continuous and the government should take steps to save tourism in Kashmir.

RECOVERY AND RESILIENCE STARTEGIES

While it is understood that the terrorist activities are impacting the tourism businesses in Kashmir, it is important to know how and what are the recovery and resilience strategies adopted by the company and other stakeholders, such as the government. In this background, we have asked a question on how the company operates generally to risks and crises. Most of the respondents mentioned that they are providing a safe and secure environment to the tourist with necessary information, security, guides. All the companies mentioned that they get proper permissions and clearness from all authorities to ensure smooth commute from one destination to another and they will organise the tours in the safer and secured places only. They brief all tourists with clear information and do's and don'ts to be followed in the region and activities they are supposed to participate while taking tour.

When the respondents were asked in particular, how do the companies reassure customers and suppliers about their level and extent of risk management? Responses include that the company will make sure that they will get proper clearances from authorized government bodies and provide the personalized guides and security, and arranging alternative options for the accommodation as well as alternative destinations in case of any issues arises before the travel dates and organizing the tours in the risk-free area. Enabling the option of changing dates and sometimes providing full refund, and ending up losing the customer forever is quite common. It was mentioned that they will make videos of the customers describing their services and experiences in the state as a tourist, and the same will be shared on social media platforms to convey a welcoming message to their future customers

Further, we have asked for the kind of support that they are expecting from the government. Along with the terrorist activities, the political and civil unrest due to Article 370 and continuous media attention, impacted has negatively impacted the tourist's intent to travel, and the companies were not prepared to serve the tourist in this situation. It is important to note the response came from one of the respondents that:

Government should support Kashmir as a tourist destination. We don't have resources as travel agents to present Kashmir as a tourist destination on the world stage by means of marketing and promotion. I think, if a celebrity endorses our state like Amitabh Bachchan does for Gujarat tourism, I am sure it will have positive effects. If the government endorses us on trains and flights, we will be known as a destination one must visit. I also think we should invite media personnel for a touristic experience such as a familiarization trip to promote Kashmir by the media so that we will do great as a tourist destination.

The region is under media watch all the time due to undue incidents which are very unfortunate. What they are expecting mainly is that the media and government should promote the region as a safer and secured destination with necessary measures. House boats are a major tourist attraction, but with a falling number of tourists due to terrorist activities and political unrest, the situation is appalling for the house boat owners. Government must support house boat owners by subsidizing timber required for repairs of a house boat.

To the question on how the resources are allocated towards combating terrorism and who makes such decisions? Responses mainly trolled towards promotion and marketing. All the respondents highlighted they have to divert/invest/commit a lot of funds on advertising and marketing to mitigate the effects of terrorist incidents. In fact, a major chunk of revenue is spent on marketing and the promotional activities showcasing the safety and security aspects.

Additionally, they have responded on resilience strategies to overcome the impacts of terrorism on tourism in Kashmir includes several things. They are expecting the local and national government to respond on these issues on a regular basis and most importantly they should take responsibility for marketing and promotion covering the aspects of safety and security of the destination so that their revenue is not siphoned on promotional activities of the destination. Promotional activities under the ambit of state or central government helps in generating confidence in the perspective tourists, as most of the tourists rely on government sources of information rather than a campaign run by a travel or tour company. It will also generate the revenue for the government. Renewal of licenses and insurance on a timely basis, support from the central government in terms of subsidies on timber for the house boats is expected from the government. Most importantly companies are expecting the Government should try to identify the root cause of the problem and diagnose it accordingly to recover from the disease of terrorism and political and civil unrest permanently, not just temporary remedy. Unless and until this happens all travel agencies and other tourism service providers just need to be happy with the limited business, they are able to do.

DISCUSSION AND IMPLICATIONS

Research in tourism with regards to terrorist inflicted destinations is evolving continuously. However, terrorism and its implications in reference to the tourist perceptions has been well studied in past. However, the empirical studies on perception on other stakeholders from the small and medium scale travel and tourism service providers are limited. This section contextualizes the study and validates the results about the implications of terrorist activities on tourism, more specifically, than from the tourism service providers.

Terrorist activities lead to the physical and psychological damage of the civil society (Ruiz Estrada & Park, 2008). This study has further proved that the terrorist inflicted activities are damaging the destination brand image and creates the fear and safety and security concerns at the destination. As it was discussed tourism is vulnerable to terrorist attacks and political instability (Araña & León, 2008), in the previous studies, it is also the similar situation in Kashmir that the terrorist activities creating the political unrest and restricted activities due to militant activities are some of them. This study confirms that due to terrorist activities the region got affected negatively leading to change in the perception of the tourist on safety and security. This has been a common phenomenon in several instances that the terrorist activities will have negative impact mainly on tourist behavior and psychology and also economic impact to the destination service providers, leading to loss of jobs, poverty.

It is widely observed that the political instability and terrorism together will have more damage to the tourism in the destination and the same was the situation in the Kashmir over the years.

In this regard a study suggests that the destinations or the countries that are heavily dependent on tourism and the policy makers should consider effective crisis recovery strategies while considering the political instability to reap the benefits of tourism (Saha & Yap, 2014). However, in the case of scrapping the article 370 is hoped to have political stability in the long run, though there are several short-term implications on tourism. Researchers have also made an attempt to understand the opinion of twitter users on revocation of Article 370, and, it's impact on terrorism and trade by performing a sentiment analysis on global basis and found that positive tweets came from India, while in comparison less positive tweets were from Pakistan and, people from Pakistan being more concerned abouts its impact on trade, and, on the other hand Twitter users in India were very positive about the revocation of Article 370 and its relevance in curbing terorism(Patil, R., Gada, N., & Gala, K., 2019).

Several studies revealed that the tourist decision making is affected by the destination image, safety, security and travel risk (Garg, 2015) and this has been the same case in Kashmir as several respondents mentioned these aspects. Kashmir is under media watch all the time due to undue incidents; however, the media as a powerful medium can advocate the readers or tourists in positive manners showcasing the positive aspects while highlighting the necessary actions to be taken care by the government for better resilience and crisis management. It was insisted that the media can play a significant role in rebuilding the destination image and encouraging the tourist to visit (Binoy, 2009). In this regard, as travel companies suggested it is important to consider that the media and government should promote the region as a safer and secured destination with necessary measures.

Tourist satisfaction plays a significant role in revisiting the destination and also providing the positive experience with friends and relatives and various social media platforms will increase the destination image. While the travel service providers in Kashmir considered these important strategies for the promotion it is equally important to promote the same by the local and national government and also celebrity driven promotional campaigns like other states of India following. This has been validated by several studies in the past (Jalilvand et al. 2013; Mair et al. 2016). This can be further hoped to have a better management of tourism in terms of marketing and promotion. There are several studies in the past that explained the rise of terrorist activities and militancy, these studies argued that militancy, secular and ethnic nationalism and border issues (Chowdary, 2012; Jacob, 2017). However, it is hoped that the political stability, particularly in connection with removal of article 370 may be expected in the near future. The unrest in the region have created long-term impact on the psychological wellbeing of the people of Kashmir and the only practical solution to this long unresolved problem is abrogation of Article 370 and equating the political rights of natives of Kashmir as the rest of the Indian citizens and, hence, bringing stability to the region and ensuring the physical and physiological wellbeing of Kashmiri's as other Indian citizens (Ghafur, 2020). The landmark decision will redraw geography and is looked upon as the stepping stone for curbing the prolonged militancy in the state (Golechha, 2020)

Based on the findings of the study, we offer several suggestions to overcome the issues faced by Kashmir tourism. While considering the fact that the terrorist inflicted activities affected negatively due to various reasons, it is important to maintain the political stability at the detonation so that it will further lead to normalcy at the destination. While ensuring political stability (as it is hoped due to the removal of article 370), the government should consider several recovery and resilient strategies for the better management and promotion of the destination. It is important note that the principles laid down by Faulkner (2001) are important to apply in these destinations. Thus, creating a destination disaster management plan is pivotal. It should have coordinated team approach and consider the range of tourism stakeholders providing various tourist services along with the government agencies and ensure that the team provides emergency services whenever required. It is essential to have clearly underlined protocols to ensure that the emergency activities are well managed. Further, there should be effective media communication and monitoring strategy. It should make sure that the contradictory and misleading information is not dispersed. Instead, media should play a crucial role in the emergency time to bring the situation in to normalcy and providing the reliable and necessary information on the status of tourism services at the destination (Faulkner, 2001). This can be possible by ensuring the local and national governments in terms of promotion and marketing by sharing the quality and positive experiences of the tourist. Government should support the local and small and medium scale service providers like house boats, hotels, restaurants and guides, affected due to the terrorist inflicted activities or political unrest.

CONCLUSION

The study of terror inflicted casualties in the context of Kashmir, particularly from the perspective of travel service providers that are small and medium scale companies in nature was conducted. Paper first reviewed the existing landscape of terrorist inflicted impacts and implications on various destinations in general and Kashmir in particular. The major reason for carrying out this study is that Kashmir is blessed with a significant number of cultural or natural attractions or tourism products, in which the state can gain the maximum benefits of tourism. However, due to terrorist inflicted activities it has been negatively affecting the stakeholders like service providers and tourists. This study confirms that the political and civil unrest and terror inflicted activities are affecting the destination leading to fear, negative attitude towards and concern over the safety and security at the destination from the perspective of travel agents and tour operators. This has further created several problems like loss of jobs, business, negative brand image, and decreased tourist number so on. Study suggests that creating political stability will have major positive implications on tourism. However, the removal of article 370 had major short term negative impacts on tourism, it is hoped that in long run it will lead to a political stability thus contributing to the better management of the destination from terror inflicted activities. The study also suggests media and both the local and national government can play a significant role in promoting the destination at the international arena as it has got the international repute attractions in Kashmir.

Like any other studies, this too has several limitations. First, the study is focused on the perspectives of travel agencies and tour operators of Kashmir, as small and medium scale organisations in nature. Further, there is a scope to study from the perspective of large scale organisations and also other stakeholders like restaurants, guides, house boaters and the local community. There are a significant number of studies suggesting that political stability will certainly contribute to lessen the impact of terrorist inflected causalities. In this regard, there has been a serious step taken by the government to maintain political stability by removing the article 370, so there is a need to focus the future studies on how far this will contribute to political stability, thus leading to better tackle with terror inflicted casualties in connection with tourism. Further there is a need for the future studies on how to build destination brand image, and also to what extent tourism contributes to the small and medium scale industries of tourism.

REFERENCES

Abadie, A., & Gardeazabal, J. (2008). Terrorism and the world economy. *European Economic Review*, *52*(1), 1–27. doi:10.1016/j.euroecorev.2007.08.005

Adeloye, D., & Carr, N. (2019). *17 Terrorism and tourism recovery cases*. Positive Tourism in Africa.

Ali, R. F. (2017). *History of armed struggles in Kashmir*. Jay Kay Books.

Arana, J. E., & León, C. J. (2008). The impact of terrorism on tourism demand. *Annals of Tourism Research*, *35*(2), 299–315. doi:10.1016/j.annals.2007.08.003

Avraham, E. (2013). Crisis communication, image restoration, and battling stereotypes of terror and wars: Media strategies for attracting tourism to Middle Eastern countries. *The American Behavioral Scientist*, *57*(9), 1350–1367. doi:10.1177/0002764213487733

Barth, J. R., Li, T., McCarthy, D., Phumiwasana, T., & Yago, G. (2006). Economic Impacts of Global Terrorism: From Munich to Bali. Capital Studies, Milken Institute.

Bassil, C., Saleh, A. S., & Anwar, S. (2019). Terrorism and tourism demand: A case study of Lebanon, Turkey and Israel. *Current Issues in Tourism*, *22*(1), 50–70. doi:10.1080/13683500.2017.1397609

Becken, S. (2013). Developing a framework for assessing resilience of tourism subsystems to climatic factors. *Annals of Tourism Research*, *43*, 506–528. doi:10.1016/j.annals.2013.06.002

Becken, S., & Carmignani, F. (2016). Does tourism lead to peace? *Annals of Tourism Research*, *61*, 63–79. doi:10.1016/j.annals.2016.09.002

Behera, N. C. (2000). State, Identity & Violence: Jammu, Kashmir, and Ladakh. Manohar Publishers.

Bhat, I. (2019). New Delhi's demographic designs in Kashmir. *Foreign Policy*. https://foreignpolicy.com/2019/08/16/new-delhis-demographic-designs-in-kashmir

Bhattacharya, M., & Basu, K. (2010). Impact of terror incidents on the foreign tourist arrivals in India: An econometric exploration. *International Journal of Tourism Policy*, *3*(3), 213–222. doi:10.1504/IJTP.2010.037993

Binoy, T. A. (2009). The Media Strategy for Destination Image Building in Crisis. *Atna Journal of Tourism Studies*, *4*(1), 138–150. doi:10.12727/ajts.4.11

Bose, S. (2009). *Kashmir: Roots of conflict, paths to peace*. Harvard University Press.

Bradyand, J., & Widdows, R. (1988). The impact of world events on travel to Europe during the summer of 1986. *Journal of Travel Research*, *26*(3), 8–10. doi:10.1177/004728758802600304

Buigut, S., & Amendah, D. D. (2016). Effect of terrorism on demand for tourism in Kenya. *Tourism Economics*, *22*(5), 928–938. doi:10.5367/te.2015.0467

Chahal, H., & Devi, A. (2015). Destination attributes and destination image relationship in volatile tourist destination: Role of perceived risk. *Metamorphosis*, *14*(2), 1–19. doi:10.1177/0972622520150203

Chauhan, V., & Khanna, S. (2008). Tourism: a tool for crafting peace process in Kashmir, J&K, India. Academic Press.

Cochrane, J. (2010). The sphere of tourism resilience. *Tourism Recreation Research*, *35*(2), 173–185. doi:10.1080/02508281.2010.11081632

Collier, P., & Hoeffler, A. (2004). Greed and grievance in civil war. *Oxford Economic Papers*, *56*(4), 563–595. doi:10.1093/oep/gpf064

Czinkota, M., Knight, G. A., & Liesch, P. W. (2004). Terrorism and international business: conceptual foundations. *Terrorism and the international business environment: The security-business nexus*, 43-57.

Dhariwal, R. (2005). Tourist arrivals in India: How important are domestic disorders? *Tourism Economics*, *11*(2), 185–205. doi:10.5367/0000000054183522

Directorate of Economics & Statistics J&K Government. (2020). *Economic Survey 2017*. Srinagar, Jammu: Directorate of Economics & Statistics Jammu & Kashmir Government.

Drakos, K., & Kutan, A. M. (2003). Regional effects of terrorism on tourism in three Mediterranean countries. *The Journal of Conflict Resolution*, *47*(5), 621–641. doi:10.1177/0022002703258198

Dutta, P. K. (2019). Kashmir: Curious case of demographic realities and perceptions. *India Today*.

Elwin, J. (1989). Site selection in tourism. In S. F. Witt & L. Moutinho (Eds.), *Tourism Marketing and Management Handbook* (pp. 403–406). Prentice Hall.

Enders, W., & Sandler, T. (1991). Causality between transnational terrorism and tourism: The case of Spain. *Studies in Conflict and Terrorism*, *14*(1), 49–58.

Enders, W., Sandler, T., & Parise, G. F. (1992). An econometric analysis of the impact of terrorism on tourism. *Kyklos*, *45*(4), 531–554. doi:10.1111/j.1467-6435.1992.tb02758.x

Estrada, M. A. R., & Park, D. (2008). Korean unification: How painful and how costly? *Journal of Policy Modeling*, *30*(1), 87–100. doi:10.1016/j.jpolmod.2007.07.001

Evans, A. (2000). The Kashmir insurgency: As bad as it gets. *Small Wars & Insurgencies*, *11*(1), 69–81. doi:10.1080/09592310008423261

Express, F. (2019). *Article 370 removal will result in end of terrorism; bring development in Jammu and Kashmir, reaffirms: Amit Shah*. Retrieved on November 11, 2020 from https://www.financialexpress.com/india-news/article-370-removal-will-result-in-end-of terrorism-bring-development-in-jammu-and-kashmir-reaffirms-amit-shah/1672413/

Faulkner, B. (2001). Towards a framework for tourism disaster management. *Tourism Management*, *22*(2), 135–147. doi:10.1016/S0261-5177(00)00048-0

Feridun, M., & Shahbaz, M. (2010). Fighting terrorism: Are military measures effective? Empirical evidence from Turkey. *Defence and Peace Economics*, *21*(2), 193–205. doi:10.1080/10242690903568884

Ferrario, F. F. (1979). The evaluation of tourist resources: An applied methodology. *Journal of Travel Research*, *17*(4), 24–30. doi:10.1177/004728757901700405

Fleischer, A., & Buccola, S. (2002). War, terror, and the tourism market in Israel. *Applied Economics*, *34*(11), 1335–1343. doi:10.1080/00036840110099252

Frey, B. S., Luechinger, S., & Stutzer, A. (2007). Calculating tragedy: Assessing the costs of terrorism. *Journal of Economic Surveys*, *21*(1), 1–24. doi:10.1111/j.1467-6419.2007.00505.x

Ganguly, Š. (2002). *Conflict unending: India-Pakistan tensions since 1947*. Columbia University Press.

Garg, A. (2015). Travel risks vs tourist decision making: A tourist perspective. *International Journal of Hospitality and Tourism Systems*, *8*(1), 1–9. doi:10.21863/ijhts/2015.8.1.004

Gavinolla, M. R., Kaushal, V., Livina, A., Swain, S. K., & Kumar, H. (2021). (in press). Sustainable consumption and production of wildlife tourism in Indian tiger reserves: A critical analysis. *Worldwide Hospitality and Tourism Themes*. Advance online publication. doi:10.1108/WHATT-08-2020-0091

Gearing, C. E., Swart, W. W., & Var, T. (1974). Establishing a measure of touristic attractiveness. *Journal of Travel Research*, *12*(4), 1–8. doi:10.1177/004728757401200401

Ghafur, A. (2020). India's concern for Kashmir. *Lancet*, *396*(10255), e47. doi:10.1016/S0140-6736(19)32076-8 PMID:31493894

Golechha, M. (2020). Integration of Kashmir for peace and prosperity. *Lancet*, *396*(10255), e48. doi:10.1016/S0140-6736(19)32077-X PMID:31493897

Gowreesunkar, G. V., & Gavinolla, M. R. (2020). Urbanism and Overtourism: Impacts and implications for the city of Hyderabad. In A. M. Morrison & J. A. Coca-Stefaniak (Eds.), *Routledge Handbook of Tourism Cities* (pp. 101–120). Routledge. doi:10.4324/9780429244605-7

Gowreesunkar, V., Cooper, C. P., & Durbarry, R. (2009). The role of internal marketing in sustainable destination management: Grand Bay, Mauritius, as a case study. *International Journal of Environmental, Cultural, Economic and Social Sustainability*, *5*(5), 21–24. doi:10.18848/1832-2077/CGP/v05i05/54684

Hassan, F. (2020). *Kashmir's tourism sector faces losses over Rs 9,000 cr; 1.4 lakh people lose jobs.* Retrieved on November 11, 2020 from https://www.forbesindia.com/article/special/kashmir039s-tourism-sector-faces-losses-over-rs-9000-cr-14-lakh-people-lose-jobs/57275/1

Hu, Y., & Ritchie, J. B. (1993). Measuring destination attractiveness: A contextual approach. *Journal of Travel Research*, *32*(2), 25–34. doi:10.1177/004728759303200204

Jalilvand, M. R., Ebrahimi, A., & Samiei, N. (2013). Electronic word of mouth effects on tourists' attitudes toward Islamic destinations and travel intention: An empirical study in Iran. *Procedia: Social and Behavioral Sciences*, *81*, 484–489. doi:10.1016/j.sbspro.2013.06.465

Jha, S. (2020). Impact of Home Country Terrorism on Medical Tourism in India January 2020. *International Journal of Asian Social Science*, *10*(8), 407–416. doi:10.18488/journal.1.2020.108.407.416

Kapur, S. P. (2006). The India-Pakistan Conflict: An Enduring Rivalry. *Canadian Journal of Political Science*, *39*(4), 966–967. doi:10.1017/S0008423906339960

Korstanje, M. E., & Tarlow, P. (2012). Being lost: Tourism, risk and vulnerability in the post-'9/11' entertainment industry. *Journal of Tourism and Cultural Change*, *10*(1), 22–33. doi:10.1080/14766825.2011.639455

Krakover, S. (2005). Estimating the effect of atrocious events on the flow of tourists to Israel. *Horror and human tragedy revisited: The management of sites of atrocities for tourism*, 183-194.

Livina, A., & Reddy, M. (2017). Nature Park as a Resource for Nature Based Tourism. In *Proceedings of the 11th International Scientific and Practical Conference* (Vol. 179, p. 183). 10.17770/etr2017vol1.2590

Llorca-Vivero, R. (2008). Terrorism and international tourism: New evidence. *Defence and Peace Economics*, *19*(2), 169–188. doi:10.1080/10242690701453917

Lyon, P. (2008). *Conflict between India and Pakistan: an encyclopedia*. Abc-Clio.

Mair, J., Ritchie, B. W., & Walters, G. (2016). Towards a research agenda for post-disaster and post-crisis recovery strategies for tourist destinations: A narrative review. *Current Issues in Tourism*, *19*(1), 1–26. doi:10.1080/13683500.2014.932758

Major, J. A. (2002). Advanced techniques for modeling terrorism risk. *The Journal of Risk Finance*, *4*(1), 15–24. doi:10.1108/eb022950

Mansfeld, Y. (1999). Cycles of war, terror, and peace: Determinants and management of crisis and recovery of the Israeli tourism industry. *Journal of Travel Research*, *38*(1), 30–36. doi:10.1177/004728759903800107

Mansfeld, Y., & Pizam, A. (2006). *Tourism, terrorism, and civil unrest issues. Tourism, Security and Safety*. Butterworth-Heinemann.

McKercher, B., & Hui, E. L. (2004). Terrorism, economic uncertainty and outbound travel from Hong Kong. *Journal of Travel & Tourism Marketing*, *15*(2-3), 99–115. doi:10.1300/J073v15n02_06

Meinung, A. (1989). Determinants of the attractiveness of a tourism region. *Determinants of the attractiveness of a tourism region*, 99-101.

Mohanty, S. (2018). The New Wave of Mobilisation in Kashmir: Religious or Political? *Social Change*, *48*(1), 59–71. doi:10.1177/0049085717743838

Morachat, C. (2003). *A Study of Destination Attractiveness through Tourists' Perspectives: A Focus On Chiang Mai, Thailand* (Unpublished doctoral thesis). Edith Cowan University.

Narain, A. (2016). Revival of violence in Kashmir: The threat to India's security. *Counter Terrorist Trends and Analyses*, *8*(7), 15–20.

Neumayer, E. (2004). The impact of political violence on tourism: Dynamic cross-national estimation. *The Journal of Conflict Resolution*, *48*(2), 259–281. doi:10.1177/0022002703262358

Pablo-Romero, M. D. P., & Molina, J. A. (2013). Tourism and economic growth: A review of empirical literature. *Tourism Management Perspectives*, *8*, 28–41. doi:10.1016/j.tmp.2013.05.006

Pain, R. (2014). Everyday terrorism: Connecting domestic violence and global terrorism. *Progress in Human Geography*, *38*(4), 531–550. doi:10.1177/0309132513512231

Parida, Y., Bhardwaj, P., & Chowdhury, J. R. (2015). Impact of terrorism on tourism in India. *Economic Bulletin*, *35*(4), 2543–2557.

Patil, R., Gada, N., & Gala, K. (2019, December). Twitter Data Visualization and Sentiment Analysis of Article 370. In *2019 International Conference on Advances in Computing, Communication and Control (ICAC3)* (pp. 1-4). IEEE. 10.1109/ICAC347590.2019.9036800

Perpiña, L., Camprubí, R., & Prats, L. (2019). Destination image versus risk perception. *Journal of Hospitality & Tourism Research (Washington, D.C.)*, *43*(1), 3–19. doi:10.1177/1096348017704497

Pizam, A., & Fleischer, A. (2002). Severity versus frequency of acts of terrorism: Which has a larger impact on tourism demand? *Journal of Travel Research*, *40*(3), 337–339. doi:10.1177/0047287502040003011

Pizam, A., & Mansfeld, Y. (1996). *Tourism, crime, and international security issues*. John Wiley & Son Ltd.

Pizam, A., & Smith, G. (2000). Tourism and terrorism: A quantitative analysis of major terrorist acts and their impact on tourism destinations. *Tourism Economics*, *6*(2), 123–138. doi:10.5367/000000000101297523

Puri, B. (1993). *Kashmir towards insurgency* (No. 4). UN.

Raja, P., & Raghu, A. (2020). *A critical review of terrorism effects and their impacts on tourist destination countries. In Materials Today: Proceedings*. Elsevier. doi:10.1016/j.matpr.2020.09.332

Ray, A. (1997). *Kashmir Diary: Psychology of Militancy*. Manas.

Raza, S. A., & Jawaid, S. T. (2013). Terrorism and tourism: A conjunction and ramification in Pakistan. *Economic Modelling*, *33*, 65–70. doi:10.1016/j.econmod.2013.03.008

Reddy, M., Kaushal, V., Livina, A., Swain, S., & Kumar, H. (2021). Sustainable Consumption and Production of Wildlife Tourism in Indian Tiger Reserves: A Critical Analysis. *Worldwide Hospitality and Tourism Themes.*

Richter, L. K., & Waugh, W. L. Jr. (1986). Terrorism and tourism as logical companions. *Tourism Management, 7*(4), 230–238. doi:10.1016/0261-5177(86)90033-6

Ritchie, J. B., & Zins, M. (1978). Culture as determinant of the attractiveness of a tourism region. *Annals of Tourism Research, 5*(2), 252–267. doi:10.1016/0160-7383(78)90223-2

Saha, S., & Yap, G. (2014). The moderation effects of political instability and terrorism on tourism development: A cross-country panel analysis. *Journal of Travel Research, 53*(4), 509–521. doi:10.1177/0047287513496472

Sloboda, B. W. (2003). Assessing the effects of terrorism on tourism by use of time series methods. *Tourism Economics, 9*(2), 179–190. doi:10.5367/000000003101298349

Sönmez, S. F. (1998). Tourism, terrorism, and political instability. *Annals of Tourism Research, 25*(2), 416–456. doi:10.1016/S0160-7383(97)00093-5

Sönmez, S. F., & Graefe, A. R. (1998). Influence of terrorism risk on foreign tourism decisions. *Annals of Tourism Research, 25*(1), 112–144. doi:10.1016/S0160-7383(97)00072-8

Subrahmanyam, K. (1990). Kashmir. *Strategic Analysis, 23*(11), 111–198.

Swain, A. (2009). The Indus II and Siachen Peace Park: Pushing the India–Pakistan Peace Process Forward. *The Round Table, 98*(404), 569–582. doi:10.1080/00358530903151854

Swami, P. (2003). Terrorism in Jammu and Kashmir in theory and practice. *India Review, 2*(3), 55–88. doi:10.1080/14736480412331307082

Tang, J. C., & Rochananond, N. (1990). Attractiveness as a tourist destination: A comparative study of Thailand and selected countries. *Socio-Economic Planning Sciences, 24*(3), 229–236. doi:10.1016/0038-0121(90)90005-R

Taylor, S. C., Torpy, D. J., & Das, D. K. (Eds.). (2012). *Policing global movement: Tourism, migration, human trafficking, and terrorism.* CRC Press. doi:10.1201/b12892

Thompson, A. (2011). Terrorism and tourism in developed versus developing countries. *Tourism Economics, 17*(3), 693–700. doi:10.5367/te.2011.0064

Vaughan, D. R., & Edwards, J. R. (1999). Experiential perceptions of two winter sun destinations: The Algarve and Cyprus. *Journal of Vacation Marketing*, *5*(4), 356–368. doi:10.1177/135676679900500405

Veal, A. J. (2011). *Research Methods for Leisure and Tourism: A Practical Guide* (4th ed.). Prentice Hall.

Venkatesan, V. (2020). *Abrogation of Articles 370 and 35A: Assault on the Constitution*. Frontline.

Wahab, S. (1996). Tourism and Terrorism: synthesis of the problem with emphasis on Egypt. In A. Pizam & Y. Mansfeld (Eds.), *Tourism, crime &International security Issues* (pp. 175–188). Wiley.

Wall, G. (1996). Terrorism and Tourism: An Overview and an Irish Example. In A. Pizam & Y. Mansfeld (Eds.), *Tourism, Crime and International Security Issues* (pp. 143–158). Wiley.

Weber, K. (1997). The assessment of tourist satisfaction using the expectancy disconfirmation theory: A study of the German travel market in Australia. *Pacific Tourism Review*, *1*(1), 35–45.

Weber, S. (1998). War, terrorism, and tourism. *Annals of Tourism Research*, *25*, 760–762.

Weimann, G. (2008). The psychology of mass-mediated terrorism. *The American Behavioral Scientist*, *52*(1), 69–86. doi:10.1177/0002764208321342

Wolpert, S. (2010). India and Pakistan: continued conflict or cooperation? Univ of California Press. doi:10.1525/9780520948006

Yin, Y., Zhang, X., Peng, D., & Li, X. (2009). Model validation and case study on internally cooled/heated dehumidifier/regenerator of liquid desiccant systems. *International Journal of Thermal Sciences*, *48*(8), 1664–1671. doi:10.1016/j.ijthermalsci.2008.12.017

ADDITIONAL READING

Ahmad, I. M., & Hussain, N. A. (2011). Impact of turmoil on tourism of Kashmir. *Journal of Economics and Sustainable Development*, *2*(7), 1–7.

Ahmed, A., & Verma, N. M. P. (2018). An Analysis of Growth and Performance of MSMEs in Jammu and Kashmir. *International Journal of Advance & Innovative Research, 5*(4), 93–99.

Behera, N. C. (2016). The Kashmir conflict: Multiple fault lines. *Journal of Asian Security and International Affairs, 3*(1), 41–63. doi:10.1177/2347797015626045

Mohan, A. (1992). The historical roots of the Kashmir conflict. *Studies in Conflict and Terrorism, 15*(4), 283–308. doi:10.1080/10576109208435908

Roy, A., Mishra, P., Bhatt, H., Chatterji, A. P., & Ali, T. (2011). *Kashmir: The case for freedom*. Verso Books.

Tavares, R. (2008). Resolving the Kashmir conflict: Pakistan, India, Kashmiris and religious militants. *Asian Journal of Political Science, 16*(3), 276–302. doi:10.1080/02185370802504316

KEY TERMS AND DEFINITIONS

Article 35A: This article in the Indian constitution empowered the J&K legislature to define state's "permanent residents" and empowering them with special rights and privileges.

Article 370: The Indian constitution granted special autonomy to the erstwhile state of Jammu and Kashmir, which allowed the state to have its own constitution and the Indian parliament needed the state's concurrence in applying laws in the state except for the matters pertaining defense, foreign affairs and communications.

Jammu and Kashmir: A union territory of India which was a state until October 31,2019, before the legislation was passed changing its status from a state to a union territory in August,2019.

Chapter 9
Different Strategies for Different Crises to Rescue Micro, Small, and Medium Travel Agencies

Almudena Otegui
https://orcid.org/0000-0003-4112-8580
University of Vigo, Spain

ABSTRACT

This chapter approaches crisis management for micro, small, and medium-sized travel agencies. In the first place, the author sets what a crisis is, which businesses should be considered as micro, small, and medium-sized, and which crises have hit travel agencies since the 1990s. Then, the author analyzes different strategies micro, small, and medium travel agencies have used for each of the above-mentioned crises, considers different strategies that could be implemented, and provides some relevant recommendations. Further, on the frame of the last crisis due to COVID-19, the author presents strategies that have already been adopted and others that could be implemented. Finally, the author highlights some differences between this last crisis and past crises of the tourism sector in general and micro, small, and medium travel agencies in particular.

INTRODUCTION

The literature on crisis management practices in travel agencies is sparse (del Alonso-Almeida et al., 2015; Shehawy & Ragab, 2017). A notable number of researches,

DOI: 10.4018/978-1-7998-6996-2.ch009

articles, and books have appeared on the subject of management crisis in tourism exclusively as a result of the economic crisis (Bodosca, 2015), but not specifically about micro, small and medium enterprises (SMEs) and even less about SME travel agencies, despite travel agency industry is dominated by small and micro enterprises (del Alonso-Almeida et al., 2015). For this reason, this chapter may be relevant and useful for both academics and professionals working in SMEs travel agencies.

Dealing with different crises is difficult for SME travel agencies, due to managers' limited control and uncertainty in this industry, which is highly susceptible to crises and where a small change in one part of the world can fast affect the other. Each crisis is different, so different approaches are necessary from managers (Bodosca, 2015), to leave behind the usual reactive strategy based only on costs cut, with a limited effect, and adopt a combination of different strategies, instead. This will allow managers to look at the crisis as an opportunity (Sigala, 2020), in a proactive way.

This chapter aims to help SME travel agencies' managers who have difficulties when it comes to overcoming the crisis and are not aware of the necessary change they must undertake in their management or, being aware, do not know how to undertake it. The author, conducting a literature review, analyzes different crises that travel agencies have faced, the strategies they used, some recurring strategies in different crises, other possible strategies, difficulties managers face to apply them, and relevant solutions. Finally, the author will present possible strategies that SME travel agencies can undertake when facing this new crisis produced by Covid-19, always bearing in mind the volatility and unpredictability that characterize the tourism sector.

BACKGROUND

Many definitions of crisis exist, in this chapter, the author refers to a tourism crisis as an event or any occurrence that can risk the normal operations and conduct of a tourism business or damage a tourism destination reputation of safety, attractiveness and comfort that interrupt the continuity of business by the reduction in tourist arrivals and expenditures (Israeli & Perl, 2011; Shehawy & Ragab, 2017).

Crises are characterized by four factors: Pressure of time, the extent and control of problems, concern about the scale or threat of such incidents, and the existence of limited response options (Fraiz Brea, & Rodríguez-Toubes, 2012). The events can be categorized also based on their severity and frequency (Tuclea et al., 2014), it is possible to classify crises as follows (Martens et al., 2016; Ritchie, 2004):

1. Immediate crises, where little or no warning occurs.
2. Emerging crises, which are slower in developing.

3. Sustained crises, which may last for weeks, months or even years.

Crisis management strategies have to be placed in their context (Fraiz Brea & Rodríguez-Toubes, 2012), and the enterprise's size is an important constraint when it comes to manage crisis strategies. According to Eurostat's (2017) data, 92,86% of the travel agencies in the European Union (28 countries) were in the category of microenterprise, 5,92% were in the category of small enterprise, and 1,22% were in the category of medium enterprise. The European Commission (2013) adopted a definition of micro, small, and medium-sized enterprises in the Commission Recommendation of 6 may 2003:

1. The category of micro, small, and medium-sized enterprises ("SMEs") is made up of enterprises which employ fewer than 250 persons and which have an annual turnover not exceeding EUR 50 million, and/or an annual balance sheet total not exceeding EUR 43 million.
2. Within the SME category, a small enterprise is defined as an enterprise which employs fewer than 50 persons and whose annual turnover and/or annual balance sheet total does not exceed EUR 10 million.
3. Within the SME category, a microenterprise is defined as an enterprise which employs fewer than 10 persons and whose annual turnover and/or annual balance sheet total does not exceed EUR 2 million. (p. 39)

This chapter focuses on SMEs travel agencies for whom dealing with different crises is difficult, due to managers' limited control and uncertainty in this industry. The list below details the principal and global crises that SME travel agencies have faced since the 90's:

1. Emergence and growth of online travel agencies (OTAs), being the first "Expedia" in 1996 (Expedia Group, n.d.), that has led to a major transformation of tourism industry, trigging a loss of market share to SME traditional travel agencies. This emergence and growth has jeopardized traditional SME travel agencies very much for some years (Canalis, 2020). It has been such a fast growth that in 2007 38% of travel agencies' sales was made by OTAs. In 2018, 74% of travels had one service which was booked online (Canalis, 2020). Although SME travel agencies where capable to adapt, at the beginning they had to compete with a strong competence coming from a digital space still unknown for them. Furthermore, OTAs were capable of offering a predatory pricing, pushing consumers to compare prices and booking the cheapest (Chaudary et al., 2020).

2. Directive 2006/123/EC of the European Parliament and of the Council of 12 December 2006 on services in the internal market (European Parliament, 2006), which is better known as the Bolkestein's Directive, was aimed to facilitate the freedom of establishment and the freedom to provide services, eliminating bureaucratic barriers and allowing, among other things, the sale of travel services without the need to have a travel agency license. This directive affected travel agencies in three ways. First, it was a "pull effect" for people without business, tourism formation or experience, that contributed to discredit travel agents' profession. Secondly, it produced defenselessness of customers in case of a breach of contract or bankruptcy of service suppliers. In these cases, customers would have to appeal to the country where the supplier was, due to the lack of establishment of the seller (Press, 2009). This lack of establishment allowed these new sellers to drive down prices and decrease their benefits, in order to avoid customers making reservations with a travel agency (Moneo, 2018). This choice was not possible for established travel agencies, who have to keep expenses from their activity (e.g., local sales, employees, mandatory especial insures for travel agencies, and guarantees required for some suppliers).
3. Economic and financial crisis of 2008 resulted in the closure of more than 5,000 travel agencies in Spain until 2013; in 2008, there were 13,000 travel agencies (Amadeus & Associació Corporativa d'Agències de Viatge Especialitzades [ACAVe], 2018). This means that about 40% percent of the travel agencies that were open in Spain in 2008 had to close because of the crisis. The majority of the travel agencies were poorly capitalized, and the demand contraction made them change their sales strategies or close definitively (BBC, 2009). Most of them had serious liquidity problems aggravated by the reluctance of banks that saw as a risk lending them money (Tuclea et al., 2014).
4. The current economic and health crisis due to Covid-19 does not yet allow to draw conclusions on the consequences for travel agencies, but the ACAVe has already announced that up to 60% of the country's travel agencies could be doomed to closure. The sector expects to invoice 70% less this year and the invoicing prior to the crisis will not recover until 2022 (Santana, 2020). The European Travel Agents and Tour Operators Association (2020) stated:

The travel bans and closing of borders imposed to help contain the Covid-19 spread has brought travel and tourism to a halt. Tourism hotspots are deserted, hotels and restaurants hauntingly empty, aircrafts parked on runways. The activity of a whole industry is suspended, with devastating impact for businesses, most of which are SMEs and family-run enterprises. (p. 1).

MAIN FOCUS OF THE CHAPTER

Issues, Controversies, and Problems

When a crisis comes and puts at risk the regular operations of a tourism business, managers could adopt reactive management strategies, which means a lack of planning and a diffident reaction after the crisis has happened, or they could look upon the crisis as an opportunity, adopting proactive management strategies (Fraiz Brea & Rodríguez-Toubes, 2012; Ritchie & Jiang, 2019; Tuclea et al., 2014) and being prepared before the crisis happened. An analysis of the strategies that SME travel agencies have implemented in each of the previous crises shows that most of the SME travel agencies adopt a reactive strategy in every crisis, almost at first, because they are not usually prepared for a crisis or because they do not have a strategic planning. This reactive strategy is usually limited to business costs cut, but this strategy of cost reduction has a limited effect, especially if it is a sustained crisis (Tuclea et al., 2014). The literature states that implementing general solutions from past experiences in future crisis is not the most suitable strategy to apply (Fraiz Brea & Rodríguez-Toubes, 2012; Ritchie, 2004). But, in the case of SME travel agencies, a reactive strategy based on costs cut seems to be a recurrent strategy in different crises.

There is a higher likelihood of financial constraints in small enterprises (Cowling et al., 2018) being the access to finance an important problem for SMEs (EIM Business & Policy Research, 2010); that is why the first strategy is cost reduction, so they do not lose too much capital. Amadeus and ACAVe's strategic study of the travel agency sector in Spain, published on 2018, that complements and updates the data from the first edition of the study published in 2009, reported that more than two-thirds of travel agencies had taken action to reduce costs, and the strategies to do it were reduction of sale points (23,50%), reduction of ordinary expenses (22,30%), reduction of employees (22,20%), and reduction in marketing and advertising (18,80%).

To combat the negative effects of the emergence of OTA's and of the Bolkestein's Directive, travel agencies used a strategy based on expertise and on the quality service, minimizing risk feelings and enforcing customer relationship management (CRM). CRM refers to strategic management of the relationship with the customers, formed by marketing plans, activities, and actions, with the objective of building loyalty and promoting recommendations from the customers to potential customers. In Spain, 53% of traditional travel agencies believe that the quality of their services is the principal added value they can offer to their customers, and 46,1% mention their competitive advantage in the travel advice they are capable to provided, opposite to companies on the Internet (Amadeus & ACAVe, 2018). Intangibility and heterogeneity are two characteristics of services that induce a risk feeling in consumers (Jun,

2020). European (Directive (EU) 2015/2302 of the European Parliament and of the Council of 25 November 2015) and national mandatory requirements as local sales, employees, mandatory special insurances for travel agencies, guarantees required for some suppliers, and the corresponding travel agency license, made it possible to SME travel agencies, along with CRM, to minimize this risk feeling.

Over time, online agencies greatly improved their Web pages, being able to offer more detailed and quality information and, therefore, as SMS travel agencies had been doing, offering security and minimizing risk feeking. This, added to the great offers and low-cost services (del Alonso-Almeida et al., 2015), attracted more and more customers. SME travel agencies were therefore push to adopt a new approach based on negotiate the tariffs with each customer who requests it (Tuclea et al., 2014), offering discounts or gratuities, which negatively affects profits and future business development. Travel agencies, as intermediaries, have low profit margins (Quintana et al., 2016), thus, it is necessary, when making profits, capitalized them, to ensure the survival in the long term. Nevertheless, with low profit margins (Quintana et al., 2016), to make benefits it is necessary to have a large volume of sales, and this was only possible for OTA's or large travel agencies.

The result was that, when the economic and financial crisis of 2008 started, the majority of travel agencies were poorly capitalized (BBC, 2009) and most of them had serious liquidity problems (Tuclea et al., 2014). In that moment, specialization based on a better expertise of a certain type of travel was the strategy of many SME travel agencies to gain a little more market (Sarrate, 2020). Specialization could be determined by the types of travel selling by the travel agency and/or by the target audience of these travels (Fakhrutdinova, 2018). The predominant specialization is that related with business segment (Amadeus & ACAVe, 2018), which allow to sell enough during a part of the year to capitalize for the rest of the year, when sales fell. However, this way of running the company carries a risk, namely having no financial reserves at the beginning of the next year's sale period, not to mention the almost nonexistent cash-flow by then.

That it was a risk has been shown with the last crisis of 2020, in which travels have been prohibited in practically the whole world, borders have been closed, and travel agencies have had to temporally close (Gebicki, 2021; Visontay, 2020). In addition, all bookings which have been already booked have been cancelled, many paid in full, with the consequent consumers seeking refunds for these cancelled bookings (Alderman et al., 2020). Travel agencies with lack of capitalization and/or with liquidity problems have been in great difficulties to carry out these refunds. Not knowing when the level of sales will be recovered (López, 2020) makes the situation even more difficult.

After the analysis of the strategies adopted by SME travel agencies in the different crises, it is possible to state that crises lived by SME travel agencies have been faced implementing similar strategies and adapting these strategies to the crisis and to the moment. Due to a limited scope of action, a strategic planning makes SME travel agencies better prepared to face a crisis. Management practices such as management support, customer focus and process management, appeared to be effective in small service firms (del Alonso-Almeida et al., 2015). It helps to adapt travel agencies when a crisis hits this business sector, and to implement a proactive strategy that foster a faster recovery. However, implementing a proactive strategy is not always simple (Fraiz Brea & Rodríguez-Toubes, 2012). The lack of time, resources, expertise, access to finance (EIM Business & Policy Research, 2010), and the unpredictable external environment act as barriers (Tuclea et al., 2014).

SOLUTIONS AND RECOMMENDATIONS

Capitalization, Specialization and Quality Relationships

Because difficulties of access to finance, one of the first steps for SME travel agencies must be capitalized. This will allow travel agencies to have a timeframe to prepare other strategies and adapt their strategic planning. Capitalization is difficult when most sales and benefits come only in a part of the year (example.g., benefits coming for the summer sales). Thus, it is important to bring different segments of customers that travel in different seasons in a year. Since many customers prefer to use the professional consultation provided by a travel agent (Santana H., 2020), specialization must remain as a strategy to carry out. Travel agencies must be experts in the types of travel they usually sell, but not limited only to one type. Having loyal customers (Quintana et al., 2016), who book more than one type of travel throughout the year, allows to have a better chance of recovering sales before the rest of the competitors. However, it is not about losing sight of specialization, indeed, selling many types of travel, without specializing in any, could finally take to lose the added value that SME travel agencies have (Quintana et al., 2016). This added value makes SME travel agencies capable of maintaining their market share compared to OTAs, other big travel agencies, and not formally established travel agencies.

How can SME travel agencies extend their customer segment, sell during most part of the year, and achieve customer engagement? The strategy begins by meeting all the requirements of the client, making them loyal (Quintana et al., 2016), and strengthening the relationship (Tuclea et al., 2014). SME travel agencies must build a positive word of mouth (Chaudary et al., 2020), taking and advantage of the customer-contact personnel to develop quality relationships (RQ) with the customers, because of the relevance of the role human capital plays in the process of achieving customer service quality and satisfaction in service companies (del Alonso-Almeida et al., 2015). RQ can be defined as the customers' satisfaction and trust relationship toward the information system quality and the perceived product value and service (Masri et al., 2020). A strong RQ indicates that the past contact has satisfied the customer and he/she wishes to repeat the purchase (Castellanos- Verdugo et al., 2014). CRM actions are necessary to have a strong RQ, so, in addition to the continued use of the service by this customer, word of mouth, customer loyalty, and customer reliability must be established. (Castellanos-Verdugo et al., 2014). For developing a CRM, it is important have a professionalized customer-contact personnel, with great education, expertise, and willingness (Dediu, 2018; Fakhrutdinova, 2018; Matović & Đurković, 2020). Also, it is important to implement some other strategies linked with the strategic planning, as part of a normal run of the business, that would help in a context of crisis due to positive effects that quality management practices have on the customers feeling of secure (del Alonso-Almeida et al., 2015).

Digitalization and Innovation

Nowadays, one of the most important management strategy is the digitalization strategy. A simple definition of digitalization is the process of converting analog information into its digital form (Kumar & Shekhar, 2020). However, its importance lies in being a source of information, it also serves as a source of trust because it is "the cover letter" of a business (Chaudary et al., 2020). Besides, it is very useful as a complement to the activity of assessing to improve satisfaction and to reinforce the commitment with the customers (Santana, H, 2020). A travel Web page is a source of benefits for customers because it allows them to make comparisons in pricing (Castellanos-Verdugo et al., 2014), have an easy tool of booking, and a fast source of communication with the travel agency before, during and after the travel. Furthermore, it is possible to assert that combining a travel Web page with a personnel relationship gives travel agencies a competitive advantage, compared to OTAs (Castellanos-Verdugo et al., 2014).

Digitalization is always important and necessary, for the above-mentioned reasons, but in a crisis, it is even most important because it allows to capture demand (Bennasar et al., 2020). Digitalization allows to know which customers are more willing to travel, their travel desires and preferences, and their contact references. Also allows to know which customers could give high return on investment and increase travel agencies 'incomes (Bennasar et al., 2020), knowing which complementary products and services could increase the customer experience and sales. It is necessary to increase sales of complementary products and services, increase the incomes, and help the businesses to reach the target of capitalization, allowing them to maintain a cash flow in the expectation of the end of the crisis and an increase in the demand.

However, digitalization is not a priority to SME (Smartpoint, 2019), about 80% of travel agencies in Spain have some presence in social media, being Facebook and WhatsApp the most used, followed by Twitter and Instragram. However, only 9% have a blog, only 14% have their own app, and a very small proportion of travel agencies use digital tools to sell their products and services (Amadeus & ACAVe, 2018). SME travel agencies barriers to establish digitalization strategies have been management's attitude, firm's financial situation, perceived risks, technological issues, and lack of reliability and trust (Chaudary et al., 2020; Smartpoint, 2019). How could SME travel agencies overcome these barriers? There are public programs which are created specifically for the digital transformation of SMEs. They usually have two approaches, one is technical support, advice, and formation; the other is economical support, grants, and financial lines (Smartpoint, 2019). These programs come from central, regional, and local governments, such as SME Growth Subsidy in Belgium; Digitalization Boot in Denmark; vouchers for digitalization in Sweden; or Silicon Vilstal in Germany (European Commission, 2019). There are also programs funded by private enterprises, as the case of Google, which has developed, along with the Industrial Organization School of Spain, a program of digital marketing for SME of the tourism industry in Spain (Hosteltur, 2020).

SME travel agencies usually do not have enough economical resources to develop a great technological transformation that goes along with the digitalization strategy. They could gain it by outsourcing these services to a technological supplier, by paying an acceptable cost for it and allowing a good positioning in the digital market through search engines and meta search engines (Amadeus & ACAVe, 2018).

The objective of a technological transformation is to extract valuable information to help in decision-making and business processes. It allows to detect trends and behaviors, predict future behaviors, improve decision-making, and transform information into knowledge, for example, learn about new market segments. That is why digitalization is an opportunity and a tool for developing innovation, which is another important strategy (Ebojoh & Verma, 2020). Innovation made by the people who deliver the service is very important (Areej Shabib, 2020), has almost become a precondition for the survival and growth of SMEs (Divisekera & Nguyen, 2018) due to the competitive nature of the tourism industry (del Alonso-Almeida et al., 2015), that makes service providers always keen to satisfy their customers (Areej Shabib, 2020).

The literature is scarce in-service innovation in tourism in general (Areej Shabib, 2020), but as in any industry, the result of an innovation can be a new or improved service, process, organizational or managerial method or marketing technique (Divisekera & Nguyen, 2018). Innovation does not mean only making different things or be always changing but considering a process for introducing new elements in the business (Zabalza & Zabalza, 2012). The aim of innovation is to enhance the quality of services, customer satisfaction, and thereafter the customer loyalty (Areej Shabib, 2020).

Digital and technological transformation allows to create data storage and collection systems on how supply information, among others, about preferences of customers, allowing to develop products and services according to these preferences (Amadeus & ACAVe, 2018). With these data, it is possible to innovate, creating new or improved products and services capable to satisfy customers, maintaining their commitment with the travel agency. However, the innovation in the tourism industry goes farther than the technological innovation, or in the quality of the product. During and after a crisis changes in lifestyle behaviors have been reported (Balanzá-Martínez et al., 2021; Dotti Sani, 2018; Scuri et al., 2018) that is where the importance of innovation lies. SME travel agencies must be able to grasp what customers will need or will expect. Knowing their travel behavior intentions (Chebli & Ben Said, 2020) after the impact of a crisis.

SMEs do not usually have a clear innovation strategy, and are constrained, as with the digitalization strategy, by several factors, such as time, resources, competence, regulations, prioritizing tasks, managing the human capital, perception of too high cost or insufficient capacity to negotiate with suppliers (Amadeus & ACAVe, 2018; Ebojoh & Verma, 2020). But innovation does not have to be seen as something out of reach for SME travel agencies, innovation is a source of competitive advantage for all companies from the tourism industry, including SMEs (Rusu, 2016). SMEs can overcome their barriers in two ways. The first way implies the support from cluster organizations, which includes creating the awareness of digitalization, emphasizing the need and importance of crafting an innovation strategy, training, mentoring, and funding (Ebojoh & Verma, 2020). The second way is changing the strategic management, which leaves the idea that an innovation strategy requires great changes. Managers must wonder: What could my travel agency offer that other travel agencies do not?, the key is trying to make something different from the rest of competitors (San Juan, 2020). Any added value of products and services, that competitors are not offering yet, could make customers be willing to pay a higher price and achieve loyalty, as well (San Juan, 2020). A travel agency can innovate by a gesture as simple as removing the printers (Bennasar et al., 2020). This idea arose after a manager realized that, during the lockdown due to the Covid-19 pandemic, no one in the enterprise had used printers to do their job. After getting back to work, this manager thinks that, if they had been able to work without printers for a few months, they could continue without them. Only with this simple gesture an enterprise is innovating, leading to save money and make the business more sustainable and environmentally friendly.

Travel agencies committed with innovation and digitalization are capable to have a wide knowledge of which are nowadays demand trends: customers are very aware about environmental aspects and sustainable tourism, are looking for personalization and exclusivity, have a great value of aspects such as wellness and personal health, and have a desire of experiences (Romero, 2019). Travel agents cannot lose sight of these demand trends and, with a combination of customer relationship, digitalization, and innovation strategies, it will be possible for SME travel agencies to offer exactly what customers require in each moment. Without sales, there is no income, therefore there are no profits. Selling consists in satisfying the clients' needs through the acquisition of the travel agency's products and services, being a beneficial operation for both. "It is not about just selling, it is about selling the experience" (Beigel, 2015)

Flexibility

Although it is not a strategy itself, flexibility in cancellation policies is another issue to consider; one important aspect for tourism business is the cancelation policy, especially in crisis moments. A recommendation is to not ask for more than 10% as a deposit out of the total cost of the travel (Bodosca, 2015). SME travel agencies will benefit in two ways: They will be able to compete with OTAs that from the beginning of their existence have had more flexible booking conditions than those of SME traditional travel agencies; and they will be able to boost demand because customers will be more disposed to make reservations if they know they can cancel them in case of impossibility or refusal to travel.

To implement all these strategies, travel agencies managers have to be solicitous about the staff and their predisposal to changes (Group, 2019). Training and knowledge could be an important weakness for SMEs (Smartpoint, 2019). Therefore, in addition to their intrinsic characteristics, customer-contact personnel should be trained in technologies (Chaudary et al., 2020) and oriented to develop and enhance the quality of the workforce (Divisekera & Nguyen, 2018), and should receive a correct communication about strategies and targets to achieve with these strategies.

FUTURE RESEARCH DIRECTIONS

Although it is early to issue final conclusions about the consequences of Covid-19 pandemic crisis, as it is still active at the time of writing this chapter, this crisis, as other crises, leads to a necessary change in the business management of SME travel agencies and a necessary adaptation when organizing and selling trips. Stakeholders in the tourism industry need to be resilient and consider how to overcome all the apocalyptic predictions (Chebli & Ben Said, 2020).

This crisis is different from the previous crises in two aspects, namely the sudden and unexpected impact (Platje et al., 2020), and the temporary impossibility of acting due to impositions of governments about travel restrictions (i.e., quarantines, confinements, closures of tourist businesses). Although the strategies the experts outline as necessary to overcome this crisis are similar to those of other crises, the temporary impossibility of acting leads to a necessary request for aid to governments by various stakeholders of the tourism sector. Thus, for example, the travel agencies members of the National Observatory of outbound tourism in Spain (ObservaTUR) demand: Campaigns and messages to show Spain as a secure market, direct aids through European funds, a reduced VAT, free RT-PCR Covid-19 tests for corporate travel, exchangeable vouchers for trips to consume through Spanish agencies, and direct aid to digital transformation (Nexotur, 2020a). The Transport and Tourism Committee of the European Parliament, after a meeting with representatives of the tourism sector, have asked to the European Commission: Direct and specific financial support, consistent and transparent criteria for assessing risks across the EU, coordination within the EU of travel restrictions, hygiene and health protocols, and a EU policy and strategy on sustainable tourism (Ramón, 2020a). Other demands have been a decrease or delay in payment of taxes to be able to stay capitalized, flexibility in employment to adapt staff cost, specific aids for self- employed, and a campaign to support travel agents and the advantages offered by the professional channel in terms of security, trust, and solvency (González, 2020a). Based on these particularities of the crisis (i.e., the sudden and unexpected impact that gives no chances of action to SME travel agencies and the necessary aids from governments due to travel restrictions that make almost impossible to continue a business activity for travel agencies) within the XX Forum of ACAVE, the economist Antón Costas suggested to apply a strategy of "3 Rs" (González, 2020b):

1. Resist: "As far as you can" and, at this point, the government can do much with the aid.
2. Recover: Pandemic economies scar people's behavior, reduce and alter consumption patterns, and, as people tend to try to save more, recovery is different.
3. Reinvent: As consumption patterns change, competitiveness also does, and it is necessary to seek profitability elsewhere; reinventing the businesses is essential.

At the resist phase, government aids would be extremely helpful, but the capacity to resist until the recovering moment arrives depends on how capitalized each travel agency is. For some months, travel agencies have not had any deposit, as the author pointed at the beginning of this chapter, due to the mandatory closing of their business, and it is going to be difficult for less capitalized travel agencies to resist. At the resist phase, it would be very important to prepare the recovery phase, start to apply strategies in order to be ready for a faster and better recovery, and some strategies of the past can be useful again. For the moment, it is possible to asset that a reactive strategy of cost cut has been already implemented (Turijobs, 2020), based on a recent study of Turijobs (Mirando hacia el futuro: La lucha del sector turístico frente al Covid-19) made with more than 4,700 professionals and 427 companies in the tourism and hospitality sector from seven European countries (Germany, Austria, Spain, France, Italy, Portugal, and Switzerland). The main measure of cost cut carried out has been to resort to employment regulation files, followed by cut general expenses, and stop selection processes.

In services contexts, customers perceived uncertainty because of factors such as the intangibility, complexity, or the lack of familiarity with the service (Castellanos-Verdugo et al., 2014). In this crisis, this uncertainty is even stronger, because what is at risk is the health of the customers. Now, more than ever, a consistent CRM based in the expertise and advice of quality personnel is essential. Once again, a same strategy is valid in a different crisis and, again, it must be adapted. Customers are seeking for something more than advice about their trips. Now, travel agents must be expert in geopolitics, to know the state of the borders and the requirements of the governments, as well as a health consultant (Lyonnet, 2020). SMEs must turn on their added value and strengthen the figure of the agent, the expert who advises and explains different security conditions, new changes, and cancellations policies, offering reliance to the customers (Klempert, 2020). Amadeus and ACAVe's (2018) recent study pointed the importance of customer-contact personnel due to Covid-19; travelers will prefer advice and advice from a professional. Indeed, 70% of those surveyed expect support from the agent to solve problems, while 66% affirmed they need the same or more advice now with the pandemic than before. The study reflects that agency workers would play a large role in trips abroad, since 50% believe that they need assistance for international travel. In addition, 80% would appreciate having contact with their agent during the trip (Nexotur, 2020). Travelport has surveyed 5,000 people who made at least one trip in the last year, the conclusions report that 33% now use the travel agent more than before, compared to 49% who do it the same as before and 18% who do it less. The reasons why those want to turn to the agent now more than ever are security information, ability to manage changes, trust, and capacity to solve complex situations (Vargas, 2020).

Digitalization and innovation are two repeated strategies, now focusing digitalization as source of information, as it will help to know what customers demand, and innovation as a tool to offer these demands. Digitalization could allow to know which customers are more willing to travel, which are their travel desires, preferences, and contact references, as well as which complementary products and services could increase the customer experience and sales. It was expected that some customer segments would begin to travel before than others once borders and travel restrictions would open. For example, it was expected young people would start to travel before older people or have a larger demand of leisure trips than business trips, due to the capacity of the businesses to move their meetings from in-person to distance meetings, thanks to digital tools. However, surprisingly, demand of business trip fell less than leisure ones (Nexotur, 2020b).

During the summer of 2020, there was a change in demand. International tourism was replaced by internal tourism, bookings on workdays fell and rose in the weekends, and the advance time with which a reservation is made decreased, an average that drops from two to one week. Rural tourism, ecotourism, camping tourism, and nautical tourism with boat rental had a higher demand. Customers showed preference on less crowded places, leaving the traditional circuit (Hinojosa, 2020a), hotel reservations in large cities such as New York and Boston, which in 2019 were at the top of the most visited, were replaced by others, such as Dallas or Atlanta, where it is easier to keep social distance or that are not designed for large events (Vargas, 2020). If travel agencies have been not capable to anticipate these data, for which digitalization tools and innovation were totally necessary, and have not rushed to change their sales segment and destinations, they have not sold "anything" since March 2020 (Bennasar et al., 2020). This trend will continue, if they are not able to adapt to new requirements and ways of travel that will come once this crisis finish.

Flexibility and adaptation have become useful strategies, with a critical reduction in advance purchase, nonrefundable rates in Asia have virtually disappeared (Hinojosa, 2020b). It would help to calm the feelings of uncertainty that the clients could feel. It is for this reason that having travel insurance which includes the option of cancelling due to voluntary withdrawal and has guarantees with Covid-19 coverage in terms of health care at the destination, could help to create security to the clients and is a source of sales, too. Furthermore, travel insurance usually has a good profit margin to travel agencies. As a result, selling implies satisfying the needs of clients through the acquisition of the travel agency's products and or services, being a beneficial operation for both.

Adaptation is necessary because travel demand returns as soon as travel restrictions are lifted, but it also falls dramatically as soon as problems arise, an example of this is the Caribbean. There are therefore very radical and rapid changes in the behavior of demand (Hinojosa, 2020b), and travel agencies have to be prepare for these changes. Another recent adaptation of travel agencies, which could be defined as innovation because adaptation and innovation are very close, is that travel agencies are offering a quick test of Covid-19 to stimulate the demand (Ramón, 2020b). In terms of cost, it would be not affordable for SME travel agencies, but a solution that some have found is to offer it in association with their suppliers. Suppliers pay for the test and travel agencies buy the travel services from this supplier.

More coordination and association between SME travel agencies would allow to offer a quick test of Covid-19 to clients, even to the smallest travel agencies, if there is a union between SME travel agencies to buy a great quantity of test at lower prices than an only travel agency could obtain individually. Recently, some associations have been created, specific for SME travel agencies, such as the Travel Agency Rights Platform in Spain (PDAV), created during the pandemic, or the Association of SME Travel Agents and Agencies of Argentina (AVAECE), which was created in 2019 with the aim of meeting the needs of SME travel agencies that were not being met in other associations to which travel agencies of all sizes belong (González, 2019; Roa, 2020).

Vaccine tourism has been the most recent strategy to increase sales, creating with it a new type of tourism. There are trips to Russia, Dubai or Serbia that are being already sold by travel agencies, and includes, in addition of the vaccine, flights, transfers, guide, visa and a medical invitation letter (Dartford, 2021; Menor, 2021; Permuy, 2021; Punzón, 2021).

In the future, once this crisis finishes, it will be desirable to conduct research about which strategies were finally appropriate and effective in this crisis and the path traveled by the travel agency sector to adapt to new ways of traveling and interacting with clients in case it would be possible finally to confirm a tendency of changes in way of life, consumption, attitudes, desires or traveling.

CONCLUSION

An industry as tourism which is highly susceptible to crises, where a small change in one part of the world can fast affect the other, needs managers capable of dealing with crises and of implementing strategies to overcome these crises. Besides limited control and uncertainty, SME travel agencies have many difficulties in implementing useful strategies; thus, most of the times, they restrict their initiative to a reactive strategy of cost cut, until a better time comes and the demand emerges. However, it is not only possible, but absolutely necessary, for SME travel agencies to implement proactive strategies, such as specialization, CRM, QR, digitalization, innovation or flexibility. These strategies would be useful in every crisis, as long as they are implemented adapting them to the crisis and the context. Due to SME travel agencies' difficulties of financing, capitalization is particularly important to be able to resist a crisis while SME travel agencies implement these strategies. In addition, to implement these strategies SME travel agencies need a work force predisposed to changes, well-trained, and properly guide by managers about strategies to implement, ways to implement them, and expected objectives in implementing these strategies.

Finally, a better coordination and union between SME travel agencies would be desirable to face this Covid-19 crisis and the crises to come.

REFERENCES

Alderman, R., De Franceschi, A., Giancaspro, M., Howells, G., Lei, C., Lete, J., Micklitz, H.-W., Miscenic, E., Naude, T., Pichonnaz, P., Poillot, E., Ramsay, I., Terryn, E., Twigg-Flesner, C., & Wilhelmsson, T.COVID-19- Consumer Law Research Group. (2020). Consumer Law and Policy Relating to Change of Circumstances Due to the COVID-19 Pandemic. *Journal of Consumer Policy*, *43*(3), 437–450. doi:10.100710603-020-09463-z

Amadeus & Associació Corporativa d'Agències de Viatge Especialitzades. (2018). *Segundo estudio estratégico de las agencias de viajes en España*. Author.

Areej Shabib, A. (2020). *Service innovation in tour guiding*. https://2020.mtcon.org/wp-content/uploads/2020/09/MTCON-20-English.pdf

Balanzá-Martínez, V., Kapczinski, F., de Azevedo Cardoso, T., Atienza-Carbonell, B., Rosa, A. R., Mota, J. C., & De Boni, R. B. (2021). The assessment of lifestyle changes during the COVID-19 pandemic using a multidimensional scale. *Revista de Psiquiatría y Salud Mental*, *14*(1), 16–26. doi:10.1016/j.rpsm.2020.07.003 PMID:32962948

BBC. (2009). https://www.bbc.com/mundo/lg/economia/2009/07/090728_1226_agencias_viajes_pea

Beigel, S. (2015, July 11). *Inspirational ideas: "We don't just sell products; we sell the experience!"* EIP-AGRI - European Commission. https://ec.europa.eu/eip/agriculture/en/news/inspirational-ideas-%E2%80%9Cwe-don%E2%80%99t-just-sell-products-we-sell-experience%E2%80%9D-0

Bennasar, T., Carbajo, M., Gallego, A., & Tárraga, J. C. (2020). *XV edición de la Jornada sobre Gestión de Riesgos en el sector turístico*. Academic Press.

Bodosca, S. (2015). Sustainable decision making under the crisis for tourism businesses. *Ecoforum*, *4*(1), 127–131.

Canalis, X. (2020, January 23). *La resiliencia de las agencias frente a internet, en una infografía*. https://www.hosteltur.com/126352_la-resiliencia-de-las-agencias-frente-a-internet-en-una-infografia.html

Castellanos-Verdugo, M., Oviedo, M. A., & Vega-Vázquez, M. (2014). CRM and RQ as key factors in retail setting services in an economic crisis context. The case of travel agencies. Springer-Verlag.

Chaudary, M., Sharma, A., & Sharma, S. (2020). Are small travel agencies ready for digital marketing? Views of travel agency managers. *Tourism Management*, (79), 1–10.

Chebli, A., & Ben Said, F. (2020). The Impact of Covid-19 on Tourist Consumption Behaviour: A Perspective Article. *Journal of Tourism Management Research*, *7*(2), 196–207. doi:10.18488/journal.31.2020.72.196.207

Cowling, M., Liu, W., & Zhang, N. (2018). Did firm age, experience, and access to finance count? SME performance after the global financial crisis. *Journal of Evolutionary Economics*, *28*(1), 77–100. doi:10.100700191-017-0502-z

Dartford, K. (2021, abril 16). *What is vaccine tourism and is it ethical*? Euronews. https://www.euronews.com/travel/2021/04/16/what-is-vaccine-tourism-and-which-european-travel-companies-are-offering-trips

Dediu, L. (2018). Travel agency staff: The need for professional training in the information technologies context of use. *Cactus Tourism Journal*, *17*(1), 9.

Del Alonso-Almeida, M. M., Bagur-Femenías, L., & Llach, J. (2015). The adoption of quality management practices and their impact on business performance in small service companies: The case of Spanish travel agencies. *Service Business*, *9*(1), 57–75. doi:10.100711628-013-0218-6

Directive (EU) 2015/2302 of the European Parliament and of the Council of 25 November 2015 on package travel and linked travel arrangements, amending Regulation (EC) No 2006/2004 and Directive 2011/83/EU of the European Parliament and of the Council and repealing Council Directive 90/314/EEC, 32015L2302, CONSIL, EP, OJ L 326 (2015). https://data.europa.eu/eli/dir/2015/2302/oj/eng

Divisekera, S., & Nguyen, V. K. (2018). Determinants of innovation in tourism evidence from Australia. *Tourism Management*, *67*, 157–167. doi:10.1016/j.tourman.2018.01.010

Dolnicar, S., & Laesser, C. (2007). Travel Agency Marketing Strategy: Insights from Switzerland. *Journal of Travel Research*, *46*(2), 133–146. doi:10.1177/0047287507299573

Dotti Sani, G. M. (2018). The economic crisis and changes in work–family arrangements in six European countries. *Journal of European Social Policy*, *28*(2), 177–193. doi:10.1177/0958928717700566

Ebojoh, S., & Verma, N. (2020). *The influence of digitalization on the innovation strategy of small and medium-sized enterprises*. University of Gävle. http://urn.kb.se/resolve?urn=urn:nbn:se:hig:diva-31530

EIM Business & Policy Research. (2010). *European SMEs under pressure. Annual report on EU SMEs 2009*. https://ec.europa.eu/docsroom/documents/15767/attachments/1/translations/en/renditions/native

European Commission. (2013). *Commission recommendation concerning the definition of micro, small, and medium-sized enterprises*. https://eur-lex.europa.eu/LexUriServ/LexUriServ.do?uri=OJ:L:2003:124:0036:0041:en:PDF#:~:text=The%20category%20of%20micro%2C%20small,not%20exceeding%20EUR%2043%20million

European Commission. (2019). *Digitalisation support to SMEs*. https://ec.europa.eu/easme/sites/easmesite/files/dop_p2p_digital.pdf

Eurostat. (2017). *Services by employment size class*. https://appsso.eurostat.ec.europa.eu/nui/show.do?dataset=sbs_sc_1b_se_r2&lang=en

Expedia Group. (n.d.). *History of the online travel industry pioneer*. https://www.expediagroup.com/about/history/

Fakhrutdinova, L. R. (2018). Establishment of an effective enterprise in the touristic market. *Revista San Gregorio. Extra*, *25*, 131–137.

Fraiz Brea, J. A., & Rodríguez-Toubes, D. (2012). Desarrollo de una política de gestión de crisis para desastres en el turismo. *Tourism & Management Studies*, (8), 10–18.

Gebicki, M. (2021, January 18). *The death of the travel agent: Will 2021 be the year it happens?* Traveller. https://www.traveller.com.au/the-death-of-the-travel-agent-will-2021-be-the-year-it-happens-h1tf3a

González, T. (2019, September 12). *Crean Asociación de Agentes y Agencias de Viajes Pymes de Argentina*. Hosteltur: Toda la información de turismo. https://www.hosteltur.com/lat/131228_crean-asociacion-de-agentes-y-agencias-de-viajes-pymes-de-argentina.html

González, T. (2020a, September 10). *Las agencias con ventas al 10% y ni el 25% de la plantilla reincorporada*. https://www.hosteltur.com/139139_las-agencias-con-ventas-al-10-y-ni-el-25-de-la-plantilla-reincorporada.html

González, T. (2020b, October 22). *La estrategia de las tres R para afrontar la recesión pandémica.* https://www.hosteltur.com/140026_la-estrategia-de-las-tres-r-para-afrontar-la-recesion-pandemica.html?code=home-page%7B2020-10-22%7D&utm_source=newsletter- es&utm_medium=email&utm_campaign=demanda-latente-explosiva-las-pernoctaciones- subieron-un-544-en-el

Group, I. D. M. (2019, September 16). *Cómo afrontan las pymes la resistencia al cambio en su digitalización.* IT Reseller; IT Digital Media Group. https://www.itreseller.es/pyme/2019/09/como-afrontan-las-pymes-la-resistencia-al-cambio-en-su-digitalizacion

Hinojosa, V. (2020a, September 11). *¿En qué está trabajando la intermediación para asegurar su futuro?* https://www.hosteltur.com/139179_en-que-esta-trabajando-la-intermediacion-para-asegurar-su-futuro.html

Hinojosa, V. (2020b, September 15). *Flexibilidad y adaptación, la vía que marca Asia para recuperar el turismo.* https://www.hosteltur.com/139220_flexibilidad-y-adaptacion-mantra- para-la-recuperacion-del-turismo-en-asia.html?code=home-page%7B2020-09- 15%7D&utm_source=newsletter-es&utm_medium=email&utm_campaign=flexibilidad-y- adaptacion-la-via-que-marca-asia-para-recu

Hosteltur. (2020, September 22). *Turismo lanza un programa de digitalización para pymes de la mano de Google | Economía.* Hosteltur: Toda la información de turismo. https://www.hosteltur.com/139380_turismo-lanza-un-programa-de-digitalizacion-para- pymes-de-la-mano-de-google.html

Israeli, A., & Perl, Y. (2011). Crisis management in the travel agency sector: A case of study. *Journal of Vacation Marketing*, (17), 115–125.

Jun, S.-H. (2020). The Effects of Perceived Risk, Brand Credibility and Past Experience on Purchase Intention in the Airbnb Context. *Sustainability*, *12*(12), 5212. doi:10.3390u12125212

Klempert, R. (2020, September 29). *Agencias y pospandemia: Así será el nuevo modelo de distribución.* https://www.hosteltur.com/lat/139521_agencias-y-pospandemia-asi-sera-el- nuevo-modelo-de-distribucion.html

Kumar, S., & Shekhar. (2020). Digitalization: A Strategic Approach for Development of Tourism Industry in India. *Paradigm*, *24*(1), 93–108. doi:10.1177/0971890720914111

López, P. (2020, June 30). *La prohibición de la llegada de cruceros pone en jaque a las agencias de viajes.* https://tur43.es/transportes/la-prohibicion-de-la-llegada-de-cruceros-pone-en-jaque-a-las-agencias-de-viajes.html

Lyonnet, J. (2020, September 9). *Las tendencias que observa Amadeus para el reinicio del turismo.* https://www.hosteltur.com/lat/139122_las-tendencias-que-observa-amadeus-para-el-reinicio-del-turismo.html

Martens, H. M., Feldesz, K., & Merten, P. (2016). Crisis Management in Tourism – A Literature Based Approach on the Proactive Prediction of a Crisis and the Implementation of Prevention Measures. *Athens Journal of Tourism, 3*(2), 89–101. doi:10.30958/ajt.3-2-1

Masri, N. W., You, J.-J., Ruangkanjanases, A., Chen, S.-C., & Pan, C.-I. (2020). Assessing the Effects of Information System Quality and Relationship Quality on Continuance Intention in E-Tourism. *International Journal of Environmental Research and Public Health, 17*(1), 174. doi:10.3390/ijerph17010174 PMID:31881762

Matović, I. M., & Đurković, J. V. (2020). Contemporary trends in human resource management in travel agencies. *Tourism International Scientific Conference Vrnjačka Banja-TISC, 5*(1), 248–263.

Menor, D. (2021, abril 19). *Viaje a Serbia con vacuna incluida por 500 euros.* La Voz de Galicia. https://www.lavozdegalicia.es/noticia/sociedad/2021/04/19/viaje-serbia-vacuna-incluida-500-euros/00031618826922629838512.htm

Moneo, M. (2018, July 7). *Las mayoristas sacan partido del intrusismo en las agencias.* https://www.preferente.com/noticias-de-agencias-de-viajes/las-mayoristas-sacan-partido-del- intrusismo-en-las-agencias-277847.html

Nexotur. (2020a, September 16). *Las agencias piden una campaña para potenciar las ventas en España.* Nexotur. https://www.nexotur.com/noticia/109183/nexotur/las-agencias-piden- una-campana-para-potenciar-las-ventas-en-espana.html

Nexotur. (2020b, September 14). *Coronavirus: Las agencias, en estado crítico.* Nexotur. https://www.nexotur.com/noticia/109136/nexotur/coronavirus:-las-agencias-en-estado-critico.html

Nexotur. (2020c, September 24). *El papel de los agentes de viajes será mucho más importante, según Amadeus.* Nexotur. https://www.nexotur.com/noticia/109296/nexotur/el- papel-de-los-agentes-de-viajes-sera-mucho-mas-importante-segun-amadeus.html

Permuy, I. (2021, abril 9). *Turismo de vacunas: A qué países se puede viajar para obtener antes la inoculación.* abc. https://www.abc.es/sociedad/abci-turismo-vacuna-paises-nsv-202104091208_noticia.html

Platje, J., Harvey, J., & Rayman-Bacchus, L. (2020). COVID-19 – reflections on the surprise of both an expected and unexpected event. *The Central European Review of Economics and Management*, *4*(1), 149–162. doi:10.29015/cerem.874

Press, E. (2009, February 2). *Economía/Empresas - Las agencias de viajes presentan alegaciones al anteproyecto de ley sobre la directiva Bolkestein*. Europa Press. https://www.europapress.es/economia/noticia-economia-empresas-agencias-viajes-presentan- alegaciones-anteproyecto-ley-directiva-bolkestein-20090202151820.html

Punzón, C. (2021, abril 17). *Agencias de Galicia y el resto de España venden viajes a Moscú para vacunarse con Sputnik V.* La Voz de Galicia. https://www.lavozdegalicia.es/noticia/sociedad/2021/04/17/agencias-galicia-resto-espana-venden-viajes-moscu-vacunarse-sputnik-v/0003_202104G17P3991.htm

Quintana, T. A., Gil, S. M., & Peral, P. P. (2016). How could traditional travel agencies improve their competitiveness and survive? A qualitative study in Spain. *Tourism Management Perspectives*, *20*, 98–108. doi:10.1016/j.tmp.2016.07.011

Ramón, D. (2020a, September 27). *El turismo está al borde del colapso: eurodiputados piden a la UE que actúe*. https://www.hosteltur.com/139476_el-turismo-esta-al-borde-del-colapso-eurodiputados-piden-a-la-ue-que-actue.html

Ramón, D. (2020b, October 1). *Agencias ofrecen pruebas rápidas de COVID-19 para impulsar los viajes*. https://www.hosteltur.com/139576_agencias-de-viaje-ofrecen-pruebas-rapidas-de-covid-19-a-sus-clientes.html

Ritchie, B. (2004). Chaos, crises, and disasters: A strategic approach to crisis management in the tourism industry. *Tourism Management*, *25*(6), 669–683. doi:10.1016/j.tourman.2003.09.004

Ritchie, B. W., & Jiang, Y. (2019). A review of research on tourism risk, crisis and disaster management: Launching the annals of tourism research curated collection on tourism risk, crisis and disaster management. *Annals of Tourism Research*, *79*, 102812. doi:10.1016/j.annals.2019.102812

Roa, M. (2020, April 29). *Nace PDAV, la asociación que da voz a las agencias pequeñas*. https://www.agenttravel.es/noticia-037724_Nace-PDAV-la-asociacion-que-da-voz-a-las- agencias-pequenas-.html

Romero, S. (2019). *Hostelería y turismo*. 7 Editores.

Rusu, B. (2016). The impact of innovations on the business model: Exploratory analysis of a small travel agency. *Procedia: Social and Behavioral Sciences*, *221*, 166–175. doi:10.1016/j.sbspro.2016.05.103

San Juan, J. C. (2020, March 6). *La dificultad de la propiedad hotelera. Mi experiencia en Casual Hoteles*. https://ahoraturismo.com/2020/03/06/del-bi-al-bi-08-abr/

Santana, H. (2020). Good practices for e-business in the travel agencies. Izvestia Journal of the Union of Scientists - Varna. *Economic Sciences Series*, *9*(1), 71–79.

Santana, M. (2020, June 23). *Un 60% de agencias de viajes podrían cerrar si no se prorrogan los ERTE*. https://www.hosteltur.com/137610_un-60-de-agencias-de-viajes-podrian-cerrar-si-no-se-prorrogan-los-erte.html

Sarrate, M. (2020). *La realidad de las empresas turísticas ante la COVID-19. XX Foro ACAVe*. El renacer del Turismo.

Scuri, S., Tesauro, M., & Petrelli, F. (2018). Implications of modified food choices and food- related lifestyles following the economic crisis in the Marche Region of Italy. *Annali Di Igiene Medicina Preventiva e Di Comunnità*, *2*, 173–179. doi:10.7416/ai.2018.2208 PMID:29465154

Shehawy, Y., & Ragab, A. (2017). Crisis Management Actions in the Egyptian Travel Agencies: A Managerial Perspective. *Journal of the Faculty of Tourism and Hotels- University of Sadat City*, *1*(1), 111–131. doi:10.21608/mfth.2017.26071

Sigala, M. (2020). Tourism and COVID-19: Impacts and implications for advancing and resetting industry and research. *Journal of Business Research*, *117*, 312–321. doi:10.1016/j.jbusres.2020.06.015 PMID:32546875

Smartpoint. (2019). *Informe eApyme*. Asociación eApyme. https://www.eapyme.com/app/download/17019151/Informe+eAPyme+2019.pdf

The European Travel Agents and Tour Operators Association. (2020). *Covid-19: Tourism Ministers and the Commission need to step up efforts for* [Press release]. Group of National Travel Agents' and Tour Operators' Associations within the EU. https://www.ectaa.org/Uploads/press-releases/PR-20200424-Tourism-Ministers-and- Commission-need-to-step-up-efforts.pdf

Tuclea, C., Vasile, C., Schiopu, F., & Marin, M. (2014). Facets of economic and financial crisis impact on strategic planning of travel agencies. *Amfiteatru Economic Journal*, *16*(8), 1222–1237.

Turijobs. (2020). *Mirando hacia el futuro: La lucha del sector turístico frente al Covid-19*. L'Hospitalet de Llobregat.

Vargas, A. (2020, September 30). *Oportunidades para el agente de viajes en un mundo transformado.* https://www.hosteltur.com/139520_oportunidades-para-el-agente-de-viajes-en- un-mundo- transformado.html?utm_content=buffer90ef0&utm_medium=social&utm_source=facebook& utm_campaign=noticia&fbclid=IwAR3DfaoVSO4bpVQAD6dvmE1TDB8_npqu56YRGmV n9scmmw-NauA6OtPlSZo

Visontay, E. (2020, July 18). Death of the travel agent: Covid-19 an existential threat for an industry without "any product to sell." *The Guardian*. http://www.theguardian.com/australia- news/2020/jul/19/death-of-the-travel-agent-covid-19-an-existential-threat-for-an-industry-without-any-product-to-sell

Zabalza, M., & Zabalza, A. (2012). *Innovación y cambio en las instituciones educativas*. Homo Sapiens.

ADDITIONAL READING

Albert Piñole, I. (2015). *Gestión de viajes, servicios y productos turísticos*. Centro de Estudios Ramon Areces SA.

Alves, V. (2020, April 9). A indústria do turismo sobreviverá à crise do Covid-19? https://www.uninter.com/noticias/a-industria-do-turismo-sobrevivera-a-crise-do-covid-19

Amadeus & Associació Corporativa d'Agències de Viatge Especialitzades. (2009). *Primer estudio estratégico de las agencias de viajes española.*

Fraiz Brea, J. A., & Rodríguez-Toubes, D. (2011). Proactive crisis management in tourism: A study experience in Galicia. *PASOS: Revista de Turismo y Patrimonio Cultural*, *9*(2), 279–289. doi:10.25145/j.pasos.2011.09.025

Kim, Y., Kim, S., & Yoo, J. (2011). Travel agency employees' career commitment and turnover intention during the recent global economic crisis. *Service Industries Journal*, *32*(8), 1247–1269. doi:10.1080/02642069.2010.545393

Michael, C. (1994). *Tourism and politics*. Wiley.

Pareskevas, A., & Altinay, L. (2013). Signal detection as the first line of defence in tourism crisis management. *Tourism Management*, *34*, 158–171. doi:10.1016/j.tourman.2012.04.007 PMID:32287739

Pinheiro, S., & Eniele, K. (2017). Gestão do turismo em tempos de crises e vulnerabilidades. *Revista de Turismo Contemporâneo*, *5*(1), 98–117.

KEY TERMS AND DEFINITIONS

Customer Relationship Management (CRM): Strategic management of the relationship with the customers, formed by marketing plans, activities, and actions, with the objective of building loyalty and promoting recommendations from the customers to potential customers.

Digital Transformation: Integration of technologies to improve or reinvent an organization's business model. It involves the entire organization by considering human capital, moving out old technology, and modifying the company culture.

Digitalization: Conversion process of adopting and using digital technologies to change a business to provide new opportunities.

Innovation: Development of new products, services, processes or organizational methods of business, marketing, or the improvement of the existing ones.

Quality of the Relationship (RQ): Evaluation of customers' satisfaction with the product or service provided, their trust in the supplier, and their commitment with the firm.

Small and Medium Enterprises: The category of micro, small, and medium-sized enterprises (SMEs) is made up of enterprises which employ fewer than 250 persons and which have an annual turnover not exceeding EUR 50 million and/or an annual balance sheet total not exceeding EUR 43 million.

Strategic Management: Initiatives taken by managers to enhance the performance of firms. They involve planning, monitoring, analysis, and assessment of the determined strategies in the strategic planning.

Strategic Planning: An activity which determines the objectives and considers the internal and external environment, available techniques and resources to design, implement, analyze, and adapt the strategies by which an organization needs to achieve their goals.

Word of Mouth: The communicative action by which customers pass information of a firm and its products or services to other people. The objective of the firm has to be that this information is positive, in order to attract new customers, increase sales, and gain more profits.

Chapter 10
Travel Agencies:
A Sector Hit by the COVID-19 Crisis

Noelia Araújo Vila
https://orcid.org/0000-0002-3395-8536
University of Vigo, Spain

Diego R. Toubes
https://orcid.org/0000-0001-7017-6659
University of Vigo, Spain

Lucília Cardoso
Centre for Tourism Research, Development, and Innovation (CITUR), Portugal

ABSTRACT

The whole world is going through an unprecedented health crisis, which has affected all sectors of global economy. In actual context, the tourism industry is particularly affected by COVID-19. In this context, the present chapter focuses on one specific subsector: travel agencies. These companies have not only stopped booking new trips, but also had to cancel bookings made prior to 2020. Therefore, many agencies did not manage to survive, and those that did are struggling. Considering this scenario, the present study aims to characterise the state of the art on this phenomenon – the effects of COVID-19 on travel agencies. To this end, all studies containing the keywords "covid" and "travel agency" published on the Scopus database were retrieved and subjected to a content analysis. The findings reveal a greater use of new technologies and online platforms. Moreover, studies point to a focus on sustainable and inclusive tourism, as well as on so far neglected market segments, as alternatives for recovering this sector.

DOI: 10.4018/978-1-7998-6996-2.ch010

INTRODUCTION

The Covid-19 crisis is unlike any other crisis in recent history, as previous pandemics were limited to specific regions, and have lasted in relatively shorter periods. Consequently, their death toll was much smaller compared to that of Covid-19 (Miao et al., 2021). Although this scenario affects every sector of the global economy, the tourism industry has been particularly harmed. Due to the intangible nature of the tourism product, destinations are especially vulnerable to perceptions of risk by consumers. There is a large body of academic literature examining how tourism reacts to and recovers from crises. Disruptive events that affect tourism are caused by many factors. Amongst those, geopolitical factors, such as political instability, military conflicts, terrorism, receive particular attention (Ivanov et al., 2020). However, tourism is often also severely affected by social, ecological, and economic crises.

Since March 13, 2020, the world has been experiencing an unprecedented health crisis, which has led to total lockdown periods in many countries. For over a year, mobility within and between countries and cities has been limited. This scenario has caused unprecedented damage to the travel and tourism industry, as it is essentially dependent on human mobility (Gössling, Scott & Hall, 2020). Low-income countries have been particularly affected, as they often lack transparency regarding public health measures, and have poor public health infrastructure (Karabulut, Bilgin, Demir & Doker, 2020). The world is currently facing the third wave of this pandemic, with expectations of entering a fourth wave soon. The hospitality and catering sectors have temporarily closed completely in many countries. In other countries, such as Spain, these sectors have been partially closed.

One of the key subsectors of the tourism industry is that of travel agencies, which have also been severely affected by the health crisis. As travel agencies work with bookings in advance, they have sustained considerable losses, since they had to cancel reservations made before the beginning of the crisis, and on top of that, are severely limited in terms of making new reservations. Moreover, they must play the role of intermediaries between their customers and providers regarding cancellations. The period is of extreme uncertainty, as mobility restrictions depend on the regulations established by national and local governments, which are frequently changing in response to the evolution in the number of Covid cases. Therefore, the demand for travel reservations has significantly decreased, and many tourists had to cancel reservations due to changes in the mobility restrictions either in their own country or in the destination. Many studies have examined the devastating impacts of Covid-19 on the tourism industry, which is to be expected since tourism is known to decline in response to health risks (Yang, Zhang and Chen, 2020). In this context, the present chapter focuses on the particularly affected subsector of travel agencies.

The goal of the present study is to characterise the state of the art on effects of Covid-19 on travel agencies, highlighting the main problems the sector is going through, and the strategies it must follow to survive in this disruptive scenario. To this end, a review of the literature on travel agencies was first carried out. In particular, two headings are presented: Adaptation of the travel agency sector to the digital world and the context of "smart" industry and The Impact of the Covid-19 pandemic on travel agencies. The methodology of the study is described below. All academic literature published containing the keywords "covid" and "travel agency", was collected from the Scopus database, resulting in a total of 56 articles. These articles were subjected to a deductive content analysis. The following section shows the results, and finally the main conclusions are presented. The findings show that the travel agency sector is constantly reinventing itself to adapt to new conditions. During the present crisis, the analysed studies reveal new trends, such as a focus on sustainable and inclusive tourism, as well as on the far neglected market segments.

BACKGROUND

Adaptation of the Travel Agency Sector to the Digital World and the Context of "Smart" Industry

Digitalization has greatly changed the tourism industry, turning it into a "smart" industry, that is, an innovative and technologically advanced market that is fully integrated into the Industry 4.0 paradigm. However, what lies behind the term "smart tourism" is often not clear. In this context, Gretzel et al. (2015) and Femenia-Serra et al. (2019) emphasise how the term "smart" represents the current universal buzzword, used to describe technological, economic, and social sign of progress led by sensor-based technologies, big data, digital connectivity, and information exchange. In this context, the meaning of "smart tourism" is often closely related to the use of intelligent systems and devices. These include new technologies such as the Internet of Things (IoT), wireless or RFID communication, and near-field communication (NFC). It also includes assistive technology devices, which are also becoming "smart" devices, as they gain versatility and connectivity features (Pencarelli, 2020). Naturally, all these devices, mobile (i.e., smartphones) or not (smart TVs), are used by tourism destination and product managers to add value to tourists' experience in the context of smart tourism.

Essentially, the concepts of tourism 4.0 and intelligent tourism are related to the development of new digital technologies, which represent a platform that integrates these two concepts (Pencarelli, 2020). However, smart tourism also builds on a sustainable and human-oriented development of scientific and technological innovation. Therefore, the effectiveness of digital tourism depends on the existence and interoperability of qualified human and social capital. This means that heterogeneous organisations must have the ability to interact with people, share mutually beneficial goals, and use common communication codes to exchange information and knowledge.

Another relevant concept in the context of smart tourism is electronic word of mouth (eWOM), as its role in consumers' decision-making process is increasingly relevant. In many industries, tourists are increasingly relying on the experiences of other consumers, rather than on the official information provided by companies. Companies are increasingly aware of the need to focus on the digital part of their business to achieve higher revenue and profitability (Ailawady, 2015). Retailers in general (GR) are using social media as a tool to strengthen customer relationships (Rapp et al., 2013). In the tourism industry, this information source becomes even more relevant due to the intangible nature of tourism products. In sum, tourists buy the promise of an experience, and to increase the chances of having a pleasant experience, they pay a lot of attention to the experiences of other people who have visited a certain destination, stayed in a certain hotel, and used the services of a certain tourism agency.

In sum, the digitalization of the tourism market and the increasing relevance of online reviews have led to dynamic changes in the travel industry. The latter have increased sharply in two main areas: online aggregation service providers, such as Kayak, and online review platforms, such as TripAdvisor (Bigné et al., 2016). Moreover, Online Travel Agencies (OTAs) meet a wide range of travel needs, allowing customers to explore, plan, and book flights, hotels, transportation, and other travel services. In addition to traditional distribution channels, these service providers also offer their internet travel resources (Law, Law, & Wai, 2002; Kim & Lee 2004). Therefore, offering services online increases the competitiveness of OTAs (Kourtesopoulou et al., 2019) in comparison to the offline agency model. This represents a threat for traditional travel agencies, which had to adapt their business models to survive.

Familiarity with Internet use is the most important factor impacting online bookings (Varma et al., 2000). Digital consumers are very familiar with the online environment. They can search for information on tourist services, compare options (i.e., different destinations, hotels, or attractions), and of course, book online. By doing so, they save time, as well as costs, as they can access the most advantageous offers (Conyette, 2015; Park et al., 2007). Therefore, their purchasing behaviour is different from that of the non-tech-savvy consumer.

With the development of e-commerce, the O2O (online to offline) model is becoming increasingly popular. O2O refers to the use of mobile and online technologies to promote local sales or offline exchanges (Fitzgerald, 2012). In other words, online shopping and marketing drive offline consumption. Therefore,

OTAs attract tourists and business travellers by featuring information on hotels, price comparisons, discounts, and reviews, on their websites.

Currently, OTAs and hotels simultaneously collaborate and compete in a multi-channel environment. Hotels offer valuable packages on their website to convince returning guests to book directly with them (Chang et al., 2019). In sum, these two channels cooperate to attract new customers but compete for regular customers' reservations.

From the perspective of competing for regular customers, the transaction process can be described as an O2O2O model (online to offline to online). The first O (online) refers to the attraction of new customers by OTAs; The second O (offline) refers to the experience of onsite services by customers; and the third O (online) refers to the repeat reservations either through the OTA's or through the hotel's website (Chang et al., 2019). Therefore, the digitalization of this market has significantly changed the relationship between hotels and travel agencies. Traditional travel agencies have long been important intermediaries in the tourism industry (Hui & Wan, 2005; Lai, 2014). OTAs, on the other hand, are not only intermediaries between customers and hotels, but also business partners, and to a great extent, competitors (Kim & Lee 2004). In this context, studies like Kristianto (et al., 2018) have concluded that the relationship between the hotel and the OTAs, even if it is not hostile, is in conflictions.

The Impact of the Covid-19 Pandemic on Travel Agencies

As addressed in the previous section, travel agencies have had to reinvent themselves to adapt to the online world and the consequent changes in consumption patterns. However, at a time when agencies can take advantage of the internet at all stages of the purchase, from searching for information to post-purchase comments (online reputation), comes an unprecedented global pandemic that severely hits this sector.

Following the spread of Covid-19 worldwide, many countries have been implementing short-term travel restrictions. For example, the United States first cancelled travels to and from China, and subsequently imposed a travel ban to the United Kingdom and Ireland (ABC News, 2020). Since the end of 2019, these restrictions have been implemented throughout the whole world. Professionals have been looking back to previous disasters, such as the SARS outbreak in 2003 (Mao et al., 2010) and the tsunami in Arugam Bay, Sri Lanka, in 2004 (Robinson & Jarvie, 2008), in search of information on similar scenarios. However, the tourism industry has entered uncharted territory: compared to SARS-Cov, Covid-19 has brought much more dire consequences to the global market. As the UNWTO (2020) has noted, the current situation is also much worse than the consequences of the 2009 global economic crisis. In 2003, the number of international tourists decrease by 4%, and SARS caused a 0.4% drop. In comparison, in the first five months of 2020, international tourist arrivals decreased by 56%, and $320 billion was lost in tourism exports, more than three times what was lost during the 2009 global economic crisis. In sum, the crisis has severely affected the tourism industry in every country in the world, from Italy, where tourism accounts for 6% of GDP, to Palau, where the activity generates almost 90% of exports. The United Nations (2020) recognise that this crisis has been a major shock to developed economies, and a real tragedy for vulnerable and developing countries.

Naturally, this scenario has also caused considerable losses to travel agencies. In Spain, for example, by the end of October 2020, 20% of travel agencies had closed or were about to close. Only 40% of agencies have opened offices, and many were only partially open. Almost all travel agencies have had to claim to "Temporary Employment Regulatory Records (ERTE)", and more than 20% of agencies have had to furlough employees. Employability in the industry has fallen by 14%, and during the summer of 2020, 95% of agencies have claimed that their sales had fallen sharply or simply stagnated. An indicator of the recession of this sector is employment: 87.6% of companies made a total or partial ERTE in 2020, and 21.6% have had to furlough one or more workers. The most concerning segment is leisure travel. In a research carried out by Hosteltur (2020), 93.3% of agents claimed that leisure tourists play the greatest role in their companies' revenue. However, business travel has also been greatly affected, as 40% of travel agencies have also claimed that the cancellation of events, congresses, and conferences has had a significant impact on their businesses.

The current pandemic has made all travel agencies, no matter their size, go into survival mode. However, the right path to survival in this uncertain scenario is far from clear. Compensatory consumption of domestic tourism has increased, and so have accommodation costs. However, agencies are still struggling to deal with huge marketing and personnel costs in the absence of significant revenue. Therefore, the industry has a lot to learn and can only learn and take one step at a time (Wu, 2020). Some studies provide useful information that travel agencies can leverage to adapt their offer. Previous research has shown that perceived risks have a negative impact on visitors' perceptions of destinations (Khan et al., 2017; Loureiro & Jesus, 2019). These risks are mainly related to safety and protection, including health-related issues. The pandemic can also affect travel plans in general. As certain parts of the world begin to recover from this crisis, people's travel planning needs to be carefully organised to reduce potential health risks and minimise stress in tourists. For example, tourists should be required to purchase a travel insurance when booking a trip. Relevant policies may even require travel insurance to be included in travel bookings as an automatic charge (Wen et al., 2020). In addition, after the COVID-19 outbreak, tourists are more likely to use facilities and visit destinations equipped with high-quality medical services. Destinations should strive to protect tourists from public health problems during their journey. Accordingly, travel agencies must offer safe services (e.g., transport and accommodation) to their customers, to which they have to carefully evaluate their providers.

Even before the current crisis, travel agencies were having problems in capturing a significant market share. They normally attract only 22% of travellers, according to data from the Tourism Barometer (SmartTravel, 2020) – see Figure 1. Moreover, their participation in the market is much more prominent among older travellers.

Figure 1. Travel shopping channels
Source: own elaboration from SmartTravel (2020)

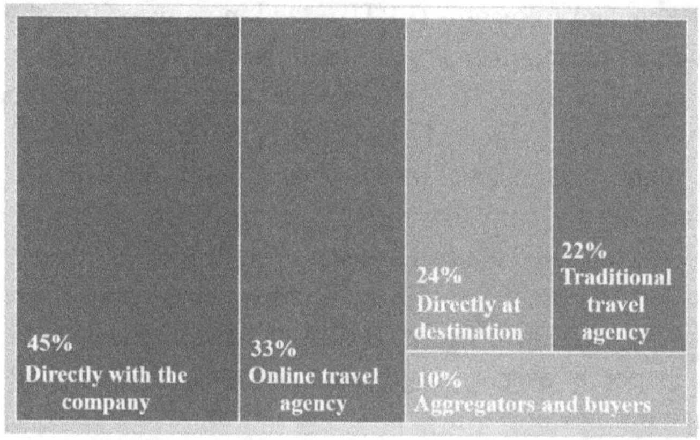

The situation is critical, and to mitigate it, in addition to enhancing safety, travel agencies can adopt the following strategies:

- Attract young travellers, a segment that traditionally does not feel the need to use agencies' services. After this health crisis, older people will eventually become the most vulnerable group (although not the only one), and young people will be the first to start booking trips again. NEXTGEN's travel barometer shows that younger people travel two to three times more than other age groups. Therefore, they must become the key target of travel agencies in their quest to recovery. The BRAINTRUST NEXTGEN travel barometer data also provides abundant information on youngsters' travel behaviour. For example, it corroborates that younger generations are the ones who make the least use of travel agencies. However, this is mostly because agencies have not yet tried to convey their value to this segment. One way to do that is to cater for new concepts of travel that are popular within this segment, such as bleisure (43% have extended a business trip to include leisure activities) or fly&drive (which is relevant for 71% of this segment).

- Travel agencies can also offer travel consulting as a differentiating value proposition. While a travel agency cannot compete with other industry players on factors such as price or convenience, they can provide travellers with a unique experience by advising and customising their trip. After many hours of training and many years of travel experience, travel agents are extremely skilled to guide customers and suggest options based on their needs. Although the information on destinations and tourist services are abundant on online forums or travel websites, travellers are unlikely to find this kind of tailored advice elsewhere. Moreover, factors such as proximity, safety, and quality of service provided by the agent can further add to the value of face-to-face channels.
- Finally, travel agencies need to learn from the experience and adapt to the new, post Covid-19 reality. To this end, they must be disciplined and rigorous, rather than conducting "business as usual". Personnel need to be constantly trained, not only on technical but also on relational (or soft) skills, which are essential to provide a strong value proposition.

METHODOLOGY

The present study's objective is to characterise the state of the art on effects of Covid-19 on travel agencies, highlighting the main problems the sector is going through, and the strategies it must follow to survive in this disruptive scenario. To this end, a bibliographic approach was adopted. The first step consisted of retrieving all the academic literature on the topic published on the Scopus Database. First, a search for documents containing the words "covid" and "travel agency", in titles, abstracts, and keywords, was carried out. This procedure took place on February 15, 2021 and resulted in a total of 56 documents: 46 research articles, five research notes, four reviews, and one letter. Moreover, 37 documents were published in 2020, while 19 were published in 2021.

The retrieved documents were subjected to deductive content analysis (applied to titles, abstracts and keywords). Moreover, a correlation analysis was further carried out within the addressed research topics. Words were adopted as the registry unit, and frequency was the primary enumeration rule for coding. Finally, an intraclass correlation coefficient analysis was performed using Pearson's statistic. This coefficient is used as an indicator of the probability of establishing a linear equation between two variables. That is, for each change of unit in one of the variables, a corresponded change (correlative) is expected in the other variable, regardless of the variables' magnitude and measurement scale (Zhou et al., 2016). The indicator is used to quantify the agreement between different measurements of a numerical variable. It can be applied even in cases where there are more than two observations per subject. It is an indicator of the reliability of a single measure, which is determined by the following expression:

$$\rho = \frac{Var\left(\pi\right)}{Var\left(\pi\right) + Var\left(\varepsilon\right)}$$

Finally, the intraclass correlation coefficient values (CCI) can range from 0 to 1. The maximum possible match is verified when it is= 1. On the other hand, when CCI = 0, the observed match is equal to that expected to occur by chance.

RESULTS

Amongst the retrieved works, as shown in Figure 2, Medicine is the most frequent field of research (30%), as although the words "travel agency" were used to filter the research, the results include articles on the pandemic and travel from a health point of view. The second most popular field, as expected, is Social sciences (18.9%).

Figure 2. Research fields of works
Source: Scopus (2021)

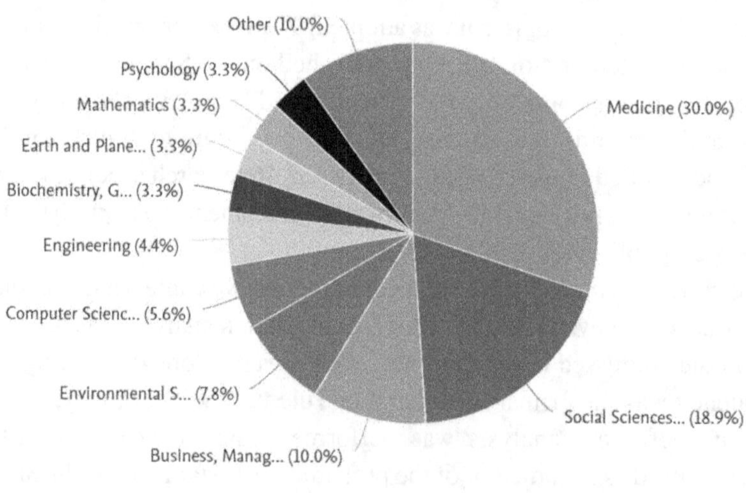

As shown in Figure 3, most authors have published only one of the retrieved works. The only exceptions were Heymann (London School of Hygiene & Tropical Medicine, London, United Kingdom) and Marston (Centers for Disease Control and Prevention, Atlanta, United States), who have published two works each.

Figure 3. Some authors of the works
Source: Scopus (2021)

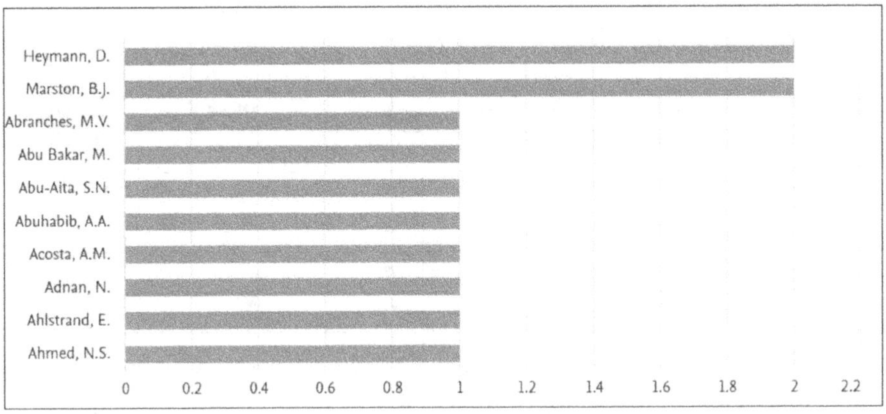

Like in most research topics, the United States leads the rank of publications with 15 works, followed by China, with 10, and the United Kingdom, with 8. Moreover, Australia, Malaysia, Sweden, India, Japan, and Spain have three works each (Figure 4).

Figure 4. Country of origin of works
Source: own elaboration

Regarding the content analysis results, titles were the first analysed elements. Frequency analysis shows that, after removing articles, prepositions, pronouns, and words with frequency lower than 11, the words "covid" and "19" are the most frequent, as expected. They are followed by "tourism", "pandemic", and "coronavirus", which are present in more than nine works. Other words with relatively high frequency are "health", "analysis", "crisis", "response" and "travel". Word frequency on titles is summarised in Table 1.

Table 1. Word frequency on titles (n=477)

Rank	Word	Repetitions	Rank	Word	Repetitions
1º	19	47	60º	Room	2
4º	covid	46	61º	patient	2
10º	tourism	11	62º	Stage	2
12º	pandemic	11	63º	Nigeria	2
13º	coronavirus	9	64º	evidence	2
20º	outbreak	5	65º	Zealand	2
21º	Health	5	66º	Hotel	2
22º	analysis	5	67º	mobile	2
23º	Crisis	5	68º	France	2
24º	response	5	69º	spread	2
25º	Travel	5	70º	lessons	2
28º	industry	4	71º	critical	2
29º	Data	4	75º	normal	2
30º	New	4	76º	Start	2
31º	Cov	4	77º	situation	2
33º	impact	4	78º	human	2
34º	SARS	4	80º	Limit	2
36º	transmission	4	81º	technology	2
37º	Novel	3	82º	2020	2
38º	Cases	3	83º	Safety	2
39º	China	3	84º	control	2
41º	Taizhou	3	85º	person	2
42º	patients	3	86º	2019	2
43º	sectors	3	87º	modelling	2
44º	pandemic	3	88º	Effect	2
45º	Public	3	89º	agencies	2

continues on following page

Table 1. Continued

Rank	Word	Repetitions	Rank	Word	Repetitions
46º	Model	3	90º	Swedish	2
47º	management	3	91º	patterns	2
49º	impacts	3	92º	context	2
50º	Social	3	95º	Road	2
51º	Risk	3	98º	characteristics	2
52º	perception	2	100º	measures	2
53º	cardiac	2	102º	epidemiology	2
55º	employee	2	103º	aviation	2
56º	respiratory	2	105º	Early	2
57º	Based	2	106º	sector	2
59º	anxiety	2			

The analysis of abstracts shows a total of 2,925 different words. After removing articles, prepositions, pronouns, and words that appear less than 11 times, once again, "covid" and "19" are the most frequent, followed by "travel", "2020", "cases", "pandemic", "health", "disease", "risk", "study", "agencies", "social", "patients", "model", and "public", all with frequency higher than 30. The word frequency analysis on abstracts is summarised in Table 2.

Table 2. Analysis of the studied abstracts (n=2,925)

Rank	Word	Repetitions	Rank	Word	Repetitions
7º	Covid	145	93º	China	17
8º	19	144	95º	impact	16
18º	tourism	73	96º	future	16
19º	Travel	71	97º	economic	16
20º	2020	69	100º	information	16
21º	Cases	67	101º	Rate	15
28º	pandemic	49	102º	2021	15
29º	Health	46	103º	control	15
31º	disease	40	105º	recovery	15
33º	Risk	38	107º	management	15
34º	Study	37	109º	hotels	14
36º	agencies	35	110º	Years	14

continues on following page

Table 2. Continued

Rank	Word	Repetitions	Rank	Word	Repetitions
37º	Social	34	111º	related	14
38º	patients	34	112º	policy	14
40º	Model	33	113º	case	14
41º	Public	31	114º	effects	14
43º	Data	30	116º	recommendations	13
44º	government	30	120º	factors	13
47º	coronavirus	28	122º	sectors	13
51º	Care	26	123º	paper	13
52º	2019	26	125º	contacts	12
53º	transmission	26	126º	rights	12
56º	Results	25	127º	potential	12
59º	analysis	23	128º	after	12
61º	countries	23	129º	human	12
62º	measures	23	130º	reserved	12
63º	research	22	131º	possible	12
65º	industry	22	132º	approach	12
66º	Crisis	22	134º	group	12
67º	agency	21	137º	authors	12
70º	World	21	141º	policies	11
71º	Spread	21	142º	age	11
74º	Global	20	144º	days	11
75º	outbreak	20	145º	employee	11
80º	reported	19	146º	community	11
81º	SARS	19	147º	virus	11
82º	limited	19	149º	identified	11
87º	findings	17	150º	online	11
88º	affected	17	151º	face	11
89º	people	17	152º	country	11
92º	sector	17	153º	distancing	11

Regarding keywords, the analysis shows 265 different words. Once again "Covid" and "19" were the most common words, followed by "pandemic", "tourism", "travel", "social", "health" and "coronavirus". The word frequency analysis on abstracts is summarised in Table 3.

Table 3. Analysis of the studied keywords (n=265)

Rank	Word	Repetitions	Rank	Word	Repetitions
1º	covid	29	33º	healthcare	2
2º	19	28	34º	control	2
3º	pandemic	12	35º	safety	2
4º	tourism	12	36º	prediction	2
5º	travel	7	37º	least	2
6º	social	6	38º	destination	2
7º	health	6	39º	prison	2
8º	of	6	40º	Sweden	2
9º	coronavirus	5	41º	SARS	2
10º	model	4	42º	scenarios	2
11º	analysis	4	43º	virus	2
12º	crisis	4	44º	recovery	2
13º	management	4	45º	stigma	2
14º	information	3	46º	distancing	2
15º	medicine	3	47º	technology	2
16º	theory	3	48º	classification	2
17º	transport	3	49º	remote	2
18º	risk	3	50º	Chinese	2
19º	sustainability	3	51º	communication	2
20º	regression	3	52º	labour	2
21º	policy	3	53º	learning	2
22º	system	3	54º	modeling	2
23º	virtual	3	55º	industry	2
24º	data	3	56º	disaster	2
25º	public	3	57º	structural	2
26º	rate	2	58º	resilience	2
27º	disease	2	59º	carbon	2
28º	immunity	2	60º	price	2
29º	covid19	2	61º	low	2
30º	event	2	62º	emission	2
31º	perception	2	63º	reduction	2
32º	trust	2			

The most common words on the works' titles, abstracts, and keywords are graphically represented. The 15 most common words were: "covid" and "19" (with the same frequency we can consider that they are a single word), "tourism", "pandemic", "coronavirus", "outbreak", "health", "analysis", "crisis", "travel", "disease", "risk", "response", "study", "model and "management". Analogous to previous studies (Porte, Saur-Amaral & Pinho, 2018), possible correlations, or thematic associations, were verified (Figure 5).

The Pearson statistic, points to relatively high correlations, especially between "tourism", "pandemic", "travel", "disease", "risk", "study", "health", and "model". Moreover, works employing the word "study" are more likely to use "tourism", "pandemic", "health", "travel", "disease", "risk" and "model". The correlation analysis between the topics addressed by the studies is summarised in Table 4.

Figure 5. Most common terms in the analysed works' titles, abstracts, and keywords

To dig deeper into the data, a categorical content analysis was carried out on the studies' abstracts, focusing on their contributions to the tourism industry and particularly to the travel agencies sector (Table 5). Results show that the pandemic has severely affected the tourism industry, especially through travel cancellation. The analysis also suggests that, during the pandemic, the use of new technologies and online platforms by travel agencies has increased substantially. As there has been a

Table 4. Correlations between the most common topics investigated in the effects of Covid on travel agencies

	C19	TO	PA	CO	OU	HE	AN	CR	TR	DI	RI	RE	ST	MO	MA
C19	**1,000**	0,987	0,986	1,000	0,995	0,986	-0,944	0,995	0,985	0,990	0,990	-0,370	0,990	0,985	0,976
TO	0,987	**1,000**	0,984	0,984	0,967	1,000	-0,985	0,998	1,000	1,000	1,000	-0,512	1,000	1,000	0,998
PA	0,986	1,000	**1,000**	0,983	0,965	1,000	-0,986	0,997	1,000	1,000	1,000	-0,520	1,000	1,000	0,999
CO	1,000	0,984	0,983	**1,000**	0,997	0,983	-0,938	0,994	0,982	0,987	0,987	-0,352	0,987	0,981	0,972
OU	0,995	0,967	0,983	0,997	**1,000**	0,965	-0,908	0,981	0,964	0,971	0,971	-0,277	0,971	0,963	0,972
HE	0,986	1,000	1,000	0,983	0,965	**1,000**	-0,986	0,997	1,000	1,000	1,000	-0,518	1,000	1,000	0,999
AN	-0,944	-0,985	-0,986	-0,938	-0,908	-0,986	**1,000**	-0,971	-0,987	-0,982	-0,982	0,655	-0,982	-0,987	-0,993
CR	0,995	0,998	0,997	0,994	0,981	0,997	-0,971	**1,000**	0,997	0,999	0,999	-0,457	0,999	0,997	0,992
TR	0,985	1,000	1,000	0,982	0,964	1,000	-0,987	0,997	**1,000**	1,000	1,000	-0,523	1,000	1,000	0,999
DI	0,990	1,000	1,000	0,987	0,971	1,000	-0,982	0,999	1,000	**1,000**	1,000	-0,500	1,000	1,000	0,997
RI	0,990	1,000	1,000	0,987	0,971	1,000	-0,982	0,999	1,000	1,000	**1,000**	-0,500	1,000	1,000	0,997
RE	-0,370	-0,370	-0,520	-0,352	-0,277	-0,518	0,655	-0,457	-0,523	-0,500	-0,500	**1,000**	-0,500	1,000	-0,564
ST	0,990	1,000	1,000	0,987	0,971	1,000	-0,982	0,999	1,000	1,000	1,000	-0,500	**1,000**	1,000	0,997
MO	0,985	1,000	1,000	0,981	0,963	1,000	-0,987	0,997	1,000	1,000	1,000	1,000	1,000	**1,000**	0,999
MA	0,976	0,998	0,999	0,972	0,972	0,999	-0,993	0,992	0,999	0,997	0,997	-0,564	0,997	0,999	**1,000**

C19: Covid-19; TO: tourism; PA: pandemic; CO: coronavirus; OU: outbreak; HE: health; AN: analysis; CR: crisis; TR: travel; DI: disease; RI: risk; RE: response; ST: study; MO: model; MA: management growing positive attitude towards technology, this may be the path for recovering this sector. Tourism has collapsed globally, forcing travel agencies to reinvent themselves yet again. The studies show that this reinvention includes an orientation towards more sustainable, resilient, and inclusive forms of tourism. Experts are quite optimistic regarding the recovery of the tourism industry in general. Enhanced use of online tools and platforms can lead to a speedy recovery once travel restrictions are relaxed.

DISCUSSION

The current health crisis has significantly affected all sectors globally. The COVID-19 pandemic has been completely unexpected and unprecedented in scale and scope. This outbreak has had a rapid impact on the scientific and academic world, with many studies and research with COVID-19 as the main subject of study. A phenomenon of such global impact aroused the interest of practically all areas of research. For this reason, thousands of articles have been published in just one year.

In the case of tourism, which is the subject of this work, the impact of the pandemic has been enormous, with the tourism industry completely paralysed the longest period in history, something unprecedented. For this reason, researchers in this area have also devoted considerable efforts to analysing the impact of this virus on the sector, as well as assessing the possible change in tourist behaviour and predicting recovery periods. Research in this area has been worldwide, with the United States standing out with the largest number of publications, but many other countries, such as China, Australia, India, Japan, Spain, France and Brazil, have dedicated effort and resources to research on this issue. As for the most researched topics in the area of travel agencies, the tourism subsector on which this chapter focuses, key terms such as Covid-19 or pandemic, and others such as risk, disease, government, countries, China or impact, are frequently repeated. These terms highlight the geographical scope of this pandemic, since its appearance in China, and the great concern about the impact and risk it entails for the travel agencies.

Content analysis provides more detailed information, such as the importance of new technologies at a time when Internet consumption has reached an all-time high. Online channels are the current means to reach consumers, to whom to provide tourism products and services when geographical mobility will not be restricted and travel is allowed. This consumption pattern is expected to remain after the pandemic, thus increasing the market segment that uses online media compared to the pre-pandemic situation. However, there is a need to improve online channels in order to adapt quickly to the new situation. The convenience of rethinking tourism has emerged in the academic debate. A more sustainable, local, inclusive and resilient tourism can be the new way of consuming tourism in the wake of this

Table 5. Content analysis of jobs related to travel agencies

Article	Authors	Topics
Blockchain technology for winning consumer loyalty: Social norm analysis using structural equation modelling	Pérez-Sánchez, M.Á., Tian, Z., Barrientos-Báez, A., Gómez-Galán, J., Li, H. (2021)	Crisis after Covid-19. Digital technologies. Blockchain. Loyalty generation. OTA (Online Travel Agencies). Positive attitude towards technology and greater usability perceived by OTAs.
Theorising tourism in crisis: Writing and relating in place	Gibson, C. (2021)	Covid-19 Pandemic. Collapse of tourism. Opportunity to reconstitute more sustainable, resilient and inclusive forms of tourism.
Sustainability of Spanish tourism startups in the face of an economic crisis	García-Gallo, M.D., Jiménez-Naharro, F., Torres-García, M., Guadix-Martín, J., Giesecke, S.L. (2021)	The study carried out in this article identifies tourist startups, the branch of activity (hotels, restaurants, transport and travel agencies) with identification of intangibles in their balance sheet. Parameters such as intangible investment, economic sustainability and financial sustainability are key variables in crises situations such as the current one.
	Liu, Y., Shi, H., Li, Y., Amin, A. (2021)	Explore factors influencing the post-pandemic intentions of Chinese residents to take part in travel. The results showed that attitude, subjective norms, perceived behavioural control, and past travel behaviour has significant positive effects on post-pandemic exit on travel intentions.
Tourism crisis management: evidence from Covid-19	Zhong, L., Sun, S., Law, R., Li, X. (2020)	Covid-19 severely affected the global tourism industry. Hospitality and tourism experts, including travel agencies, have a positive attitude towards the recovery of the tourism industry, while tourism professionals perceive the long-term impacts.
Recovering hotel room sales during the COVID-19 pandemic: lessons from OTA information using the quantile regression approach	Guo, L., Liu, K., Song, Y., Yang, Z. (2021)	Relationship between online travel agencies (OTA) and information and sales of hotel rooms in the context of the pandemic. Impulse pathway for the recovery of the sale of rooms.
Perception of Safety Tourism in Colombia	Antošová, G., Lima Santos, L., Stradová, S. (2020)	Analysis of travel during the Covid-19 to authentic and unknown places as part of the offer of travel agencies.
Impact of COVID-19: research note on tourism and hospitality sectors in the epicentre of Wuhan and Hubei Province, China	Knight, D.W., Xiong, L., Lan, W., Gong, J. (2020)	The purpose of this document is to present the initial findings of a vulnerability assessment based on the perceptions of professionals working in four tourism and hospitality sectors in Wuhan and Hubei Province: cruise lines, hotels, travel agencies and tourist attractions. Immediate economic loss due to Covid-19, as well as concerns and uncertainties of recovery.
Coping with the New Normal Imposed by the COVID-19 Pandemic: Lessons for Technology Management and Governance	Herath, T., Herath, H.S.B. (2020)	The dependence of many facets of society on technology and online platforms has grown exponentially during the Covid-19 pandemic.

pandemic, a path to which travel agencies must adapt. And it's time to broaden the target audience's spectrum, not only to older segments or business groups, but also to reach out segments that may have been lost, such as the young population, most affected by the negative impact of the crisis on disposable income. The current pandemic has caused a temporary pause in which travel agencies can design how to redefine business. After the pandemic, individuals will travel again, perhaps with new patterns of consumption, but before that travel agencies must be prepared and adapt their offerings.

CONCLUSION

The world has suffered previous economic crises. In the Great Depression of 1929, for example, the U.S. stock market crack had catastrophic consequences to the whole world, especially to European countries. In the 2009 global economic crisis, another stock market crack in the United States, this time caused by the real estate bubble, also had severe global repercussions. Such crises have led to dire consequences to the global economy, such as rising unemployment and devaluation of stock exchanges. Now, the world faces a crisis brought about by a virus, or, from the economic perspective, by the measures taken to stop it from spreading. Economists and industry specialists have been trying to anticipate actions based on previous health crises. However, other crises in the modern era do not compare to what the world is experiencing now. Lots of works have addressed forecasts and possible scenarios, but the future is still highly uncertain, as many variables cannot be controlled or even predicted.

The effects of this economic crisis may be short-term or long-term. This depends first on how medical science advances towards ways to cure and prevent the spread of the disease. It also depends on the adaptability of economic sectors, in each country. Within the tourism industry, one of the most affected subsectors has been travel agencies. New bookings have been paralysed due to the uncertainty regarding future mobility restrictions, and many trips booked before 2020 have been cancelled. Many agencies did not manage to survive, and the ones that did had to furlough or even fire part of their employees. Even before the current crisis, this sector had to continuously adapt to new consumption patterns brought about by the digitalization of the travel market. Consumers increasingly carry out each stage of their purchase online. They search for information on tourist portals, price comparators, hotel websites, and social networks; book online through OTAs or aggregation services; and share their experiences on review platforms.

During the current pandemic, internet use has grown even further, which experts see as an opportunity to recover the sector. Consumers who previously booked their trips in a physical office have discovered online platforms and are expected to keep using them after the pandemic. OTAs have been pointed out as the main path to recovery. Therefore, adaptation capabilities and the use of new technologies will be key factors for the survival of travel agencies.

In addition to new technologies, certain trends must also be useful for these companies, including the commitment to sustainability, inclusive tourism, and alternative tourism, which consumers increasingly perceive as valuable. Additionally, travel agencies are advised to offer travel consulting services. After all, personalised service is one of the main reasons for consumers to remain loyal to physical travel agencies in the first place.

Finally, to recover market share, travel agencies should also seek to cater for a segment they had long been neglecting: younger travellers. Young adults are expected to be the first to return to pre-covid-19 travel patterns once restrictions are relaxed. Therefore, offering services that provide value to this segment will be essential for the survival of travel agencies. In sum, the Covid-19 crisis has been causing a great recession in the travel agencies sector. Now, it is time to consider redesigning certain strategies to recover and prepare for possible future adverse situations.

The novelty of the research topic, which began with the outbreak of the pandemic at the beginning of 2020, is the main limitation of this study. Although the literature found is abundant, there is still uncertainty about how the pandemic will evolve and what the effects will be on the sector analysed. Therefore, as a future line of research, it is proposed to carry out a new review once the pandemic is over. It will then be possible to verify whether the studies carried out to date are conclusive or whether the future situation shows changes with respect to current expectations.

REFERENCES

Ailawadi, K. L., Pauwels, K., & Steenkamp, J.-B. E. M. (2018). Private-label use and store loyalty. *Journal of Marketing*, *72*(6), 19–30. doi:10.1509/jmkg.72.6.019

Bigné, E., Caplliure, E. M., & Miquel, M. J. (2016). eWOM on travel agency selection: Specialized versus private label. *Psychology and Marketing*, *33*(12), 1046–1053. doi:10.1002/mar.20938

Chang, Y. W., Hsu, P. Y., & Lan, Y. C. (2019). Cooperation and competition between online travel agencies and hotels. *Tourism Management*, *71*, 187–196. doi:10.1016/j.tourman.2018.08.026

Conyette, M. (2015). *21 century travel using websites, mobile and wearable technology devices.* Athens: ATINER'S Conference Paper Series.

Femenia-Serra, F., Neuhofer, B., & Ivars-Baidal, J. A. (2019). Towards a conceptualisation of smart tourists and their role within the smart destination scenario. *Service Industries Journal*, *39*(2), 109–133. doi:10.1080/02642069.2018.1508458

Fitzgerald, M. (2012). *O2O: O2 for local business*. Retrieved from the website: http://www. onlineeconomy. org/tag/online-to-offline

Gössling, S., Scott, D., & Hall, C. M. (2020). Pandemics, tourism and global change: A rapid assessment of COVID-19. *Journal of Sustainable Tourism*, *29*(1), 1–20. do i:10.1080/09669582.2020.1758708

Gretzel, U., Sigala, M., Xiang, Z., & Koo, C. (2015). Smart tourism: Foundations and developments. *Electronic Markets*, *25*(3), 179–188. doi:10.100712525-015-0196-8

Hosteltur. (2020). *La COVID puede abocar al cierre del 20% de las agencias de viaje*. Edición España. Available on https://www.hosteltur.com/140017_la-covid-puede-abocar-al-cierre-del-20-de-las-agencias-de-viaje.html

Hui, T. K., & Wan, D. (2006). Factors affecting consumers' choice of a travel agency: The case of Singapore. *Journal of Travel & Tourism Marketing*, *19*(4), 1–12. doi:10.1300/J073v19n04_01

Ivanov, S. H., Webster, C., Stoilova, E., & Slobodskoy, D. (2020). Biosecurity, crisis management, automation technologies and economic performance of travel, tourism and hospitality companies–A conceptual framework. *Tourism Economics*.

Karabulut, G., Bilgin, M. H., Demir, E., & Doker, A. C. (2020). How pandemics affect tourism: International evidence. *Annals of Tourism Research*, *84*, 102991. doi:10.1016/j.annals.2020.102991 PMID:32834226

Khan, M. J., Chelliah, S., & Ahmed, S. (2017). Factors influencing destination image and visit intention among young women travellers: Role of travel motivation, perceived risks, and travel constraints. *Asia Pacific Journal of Tourism Research*, *22*(11), 1139–1155. doi:10.1080/10941665.2017.1374985

Kim, W. G., & Lee, H. Y. (2005). Comparison of web service quality between online travel agencies and online travel suppliers. *Journal of Travel & Tourism Marketing*, *17*(2-3), 105–116. doi:10.1300/J073v17n02_09

Kourtesopoulou, A., Theodorou, S. D., Kriemadis, A., & Papaioannou, A. (2019). The impact of online travel agencies web service quality on customer satisfaction and purchase intentions. In *Smart Tourism as a Driver for Culture and Sustainability* (pp. 343–356). Springer. doi:10.1007/978-3-030-03910-3_24

Kristianto, D. A., Syamsu, M. N., & Wisnumurti, A. (2018). Online Travel Agents Use for Sales and Promotion of Hotel in Kaliurang Yogyakarta. *Kepariwisataan: Jurnal Ilmiah*, *12*(01), 15–28.

Lai, I. K. W. (2014). The role of service quality, perceived value, and relationship quality in enhancing customer loyalty in the travel agency sector. *Journal of Travel & Tourism Marketing*, *31*(3), 417–442. doi:10.1080/10548408.2014.883346

Law, R., Law, A., & Wai, E. (2002). The impact of the internet on travel agencies in Hong Kong. *Journal of Travel & Tourism Marketing*, *11*(2-3), 105–126. doi:10.1300/J073v11n02_06

Loureiro, S. M. C., & Jesus, S. (2019). How perceived risk and animosity towards a destination may influence destination image and intention to revisit: The case of Rio de Janeiro. *Anatolia*, *30*(4), 497–512. doi:10.1080/13032917.2019.1632910

Mao, C. K., Ding, C. G., & Lee, H. Y. (2010). Post-SARS tourist arrival recovery patterns: An analysis based on a catastrophe theory. *Tourism Management*, *31*(6), 855–861. doi:10.1016/j.tourman.2009.09.003 PMID:32287733

Miao, L., Im, J., Fu, X., Kim, H., & Zhang, Y. E. (2021). Proximal and distal post-COVID travel behavior. *Annals of Tourism Research*, *88*, 103159. doi:10.1016/j.annals.2021.103159

News, A. B. C. (2020). *Coronavirus update: Morrison brings in travel restrictions, Australian cases pass 250, Donald trump's COVID-19 test is clear.* Available at: www.abc.net.au/news/2020-03-15/coronavirus-update-latest-news-us-travel-ban-extended-trump-test/12057094

Park, Y. A., Gretzel, U., & Sirakaya-Turk, E. (2007). Measuring web site quality for online travel agencies. *Journal of Travel & Tourism Marketing*, *23*(1), 15–30. doi:10.1300/J073v23n01_02

Pencarelli, T. (2020). The digital revolution in the travel and tourism industry. *Information Technology & Tourism*, *22*(3), 455–476. doi:10.100740558-019-00160-3

Porte, M., Saur-Amaral, I., & Pinho, C. (2018). Research in auditing: Main themes. *Revista Contabilidade & Finanças*, *29*(76), 41–59. doi:10.1590/1808-057x201804410

Rapp, A., Beitelspacher, L. S., Grewal, D., & Hughes, D. E. (2013). Understanding social media effects across seller, retailer, and consumer interactions. *Journal of the Academy of Marketing Science*, *41*(5), 547–566. doi:10.100711747-013-0326-9

Robinson, L., & Jarvie, J. K. (2008). Post-disaster community tourism recovery: The tsunami and Arugam Bay, Sri Lanka. *Disasters, 32*(4), 631–645. doi:10.1111/j.1467-7717.2008.01058.x PMID:18479472

SmartTravel. (2020). *Qué pueden hacer las agencias de viajes ante la crisis del coronavirus*. Available on https://www.smarttravel.news/pueden-las-agencias-viaje-ante-la-crisis-del-coronavirus/

Naciones Unidas. (2020). *Informe de políticas: La Covid-19 y la transformación del turismo*. Agosto 2020.

Varma, A., Sprott, D. E., Silverman, S. N., & Stem, D. E. (2000). Adoption of Internet shopping: The role of consumer innovativeness. *Industrial Management & Data Systems*.

Wearing, S., & Neil, J. (1999). *Ecoturismo: Impacto, Tendencias y posibilidades*. Editorial Síntesis.

Wen, J., Kozak, M., Yang, S., & Liu, F. (2020). COVID-19: potential effects on Chinese citizens' lifestyle and travel. *Tourism Review*.

Wu, P. J. (2020). *A Study on the Impact of COVID-19 on Travel Agencies in Taiwan and their Response*. Academic Press.

Yang, Y., Zhang, H., & Chen, X. (2020). Coronavirus pandemic and tourism: Dynamic stochastic general equilibrium modeling of infectious disease outbreak. *Annals of Tourism Research, 83*, 102913. doi:10.1016/j.annals.2020.102913 PMID:32292219

ADDITIONAL READING

Abu-Rayash, A., & Dincer, I. (2020). Analysis of mobility trends during the COVID-19 coronavirus pandemic: Exploring the impacts on global aviation and travel in selected cities. *Energy Research & Social Science, 68*, 101693. doi:10.1016/j.erss.2020.101693 PMID:32839706

Albattat, A. (2020). The Impact of Online Marketing in Travel Agency. In *The Emerald Handbook of ICT in Tourism and Hospitality*. Emerald Publishing Limited. doi:10.1108/978-1-83982-688-720201006

Brun, I., Rajaobelina, L., Ricard, L., & Amiot, T. (2020). Examining the influence of the social dimension of customer experience on trust towards travel agencies: The role of experiential predisposition in a multichannel context. *Tourism Management Perspectives, 34*, 100668. doi:10.1016/j.tmp.2020.100668 da Silva, G. L., & Mendes Filho, L. (2018). Perfil e frequência de uso das On-line Travel Agency (OTA) por consumidores na rede hoteleira de Natal/RN. *Revista Acadêmica Observatório de Inovação do Turismo*, *12*(1), 22–44.

Dudek, A., Jaremen, D. E., Michalska-Dudek, I., & Walesiak, M. (2019). Loyalty Model Proposal of Travel Agency Customers. *Sustainability*, *11*(13), 3702. doi:10.3390u11133702

Granados, J. C., Pérez, L. M., Pedraza-Rodríguez, J. A., & Gallarza, M. G. (2021). Revisiting the quality-value-satisfaction-loyalty chain for corporate customers in the travel agency sector. *European Journal of Tourism Research*, *27*, 2711–2711.

Jain, S., & Sharma, T. (2020). Social and travel lockdown impact considering coronavirus disease (COVID-19) on air quality in megacities of India: Present benefits, future challenges and way forward. *Aerosol and Air Quality Research*, *20*(6), 1222–1236. doi:10.4209/aaqr.2020.04.0171

Li, Z., Hua, C., Fu, X., & Liu, X. (2021). Beyond complainers: Reclassifying customers of travel agency regarding post-failure behavior and loyalty performance. *Asia Pacific Journal of Tourism Research*, *26*(3), 329–346. doi:10.1080/10941665.2020.1862886

Liew, V. K. S. (2020). The effect of novel coronavirus pandemic on tourism share prices. *Journal of Tourism Futures*.

Rajaobelina, L. (2018). The impact of customer experience on relationship quality with travel agencies in a multichannel environment. *Journal of Travel Research*, *57*(2), 206–217. doi:10.1177/0047287516688565

Ray, A., Bala, P. K., & Rana, N. P. (2021). Exploring the drivers of customers' brand attitudes of online travel agency services: A text-mining based approach. *Journal of Business Research*, *128*, 391–404. doi:10.1016/j.jbusres.2021.02.028

Sharma, A., Sharma, S., & Chaudhary, M. (2020). Are small travel agencies ready for digital marketing? Views of travel agency managers. *Tourism Management*, *79*, 104078. doi:10.1016/j.tourman.2020.104078

Škare, M., Soriano, D. R., & Porada-Rochoń, M. (2021). Impact of COVID-19 on the travel and tourism industry. *Technological Forecasting and Social Change*, *163*, 120469. doi:10.1016/j.techfore.2020.120469

Uğur, N. G., & Akbıyık, A. (2020). Impacts of COVID-19 on global tourism industry: A cross-regional comparison. *Tourism Management Perspectives*, *36*, 100744. doi:10.1016/j.tmp.2020.100744 PMID:32923356

Wells, C. R., Sah, P., Moghadas, S. M., Pandey, A., Shoukat, A., Wang, Y., Wang, Z., Meyers, L. A., Singer, B. H., & Galvani, A. P. (2020). Impact of international travel and border control measures on the global spread of the novel 2019 coronavirus outbreak. *Proceedings of the National Academy of Sciences of the United States of America, 117*(13), 7504–7509. doi:10.1073/pnas.2002616117 PMID:32170017

Xie, L., Guan, X., Cheng, Q., & Huan, T. C. T. (2020). Using customer knowledge for service innovation in travel agency industry. *Journal of Hospitality and Tourism Management, 45*, 113–123. doi:10.1016/j.jhtm.2020.08.001

Yeh, S. S. (2020). Tourism recovery strategy against COVID-19 pandemic. *Tourism Recreation Research*, 1–7. doi:10.1080/02508281.2020.1805933

Zhang, H., Song, H., Wen, L., & Liu, C. (2021). Forecasting tourism recovery amid COVID-19. *Annals of Tourism Research, 87*, 103149. doi:10.1016/j.annals.2021.103149

KEY TERMS AND DEFINITIONS

Alternative Tourism: Refers to those trips that are intended to carry out recreational activities in direct contact with nature and cultural expressions that envelop you with an attitude and commitment to know respect, enjoy and participate in the conservation of natural and cultural resources. Alternative tourism covers a wide range of activities, but more generally, it is defined as "forms of tourism consistent with natural, social and community values and that allow both hosts and visitors to enjoy positive and highly appreciable interaction and a shared experience" (Wearing & Neil, 1999, p. 3).

Blockchain: Single record, consensual, and distributed across multiple nodes in red. Usability: Quality of the website or computer program that are easy to use and facilitate the reading of texts, quickly download the information and present simple functions and menus, so the user finds their queries and comfortable their use.

Digital Industry: Introduction and application of digital technology to all factory production systems and processes, as in the relationship with customers.

Inclusive Tourism: It is tourism that allows equal opportunities, of all people, to develop all the actions that make up the tourist activity in a safe, comfortable, autonomous, and standardised way, and that also tries to seek economic profitability, seeking to benefit both the person with disabilities and the companies of the sector.

Resilience: Ability to adapt to an adversary of situation. Smart tourism: refers to the application of information and communication technologies, similar to smart cities, for the development of innovative tools and approaches to improve tourism.

Compilation of References

Abadie, A., & Gardeazabal, J. (2008). Terrorism and the world economy. *European Economic Review*, *52*(1), 1–27. doi:10.1016/j.euroecorev.2007.08.005

Abd-Elhamid, H. F., Fathy, I., & Zeleňáková, M. (2018). Flood prediction and mitigation in coastal tourism areas, a case study: Hurghada, Egypt. *Natural Hazards*, *93*(2), 559–576. doi:10.100711069-018-3316-x

Adeloye, D., & Carr, N. (2019). *17 Terrorism and tourism recovery cases*. Positive Tourism in Africa.

Agència Catalana de Turisme. (2020). *Informació de deguiment de la COVID-19 en la activitat turística de Catalunya*. http://act.gencat.cat/informacio-seguiment-activitat-turistica-covid-19/

Ailawadi, K. L., Pauwels, K., & Steenkamp, J.-B. E. M. (2018). Private-label use and store loyalty. *Journal of Marketing*, *72*(6), 19–30. doi:10.1509/jmkg.72.6.019

Akis, S., & Warner, J. (1994). A descriptive analysis of North Cyprus tourism. *Tourism Management*, *15*(5), 379–388. doi:10.1016/0261-5177(94)90092-2

Alderman, R., De Franceschi, A., Giancaspro, M., Howells, G., Lei, C., Lete, J., Micklitz, H.-W., Miscenic, E., Naude, T., Pichonnaz, P., Poillot, E., Ramsay, I., Terryn, E., Twigg-Flesner, C., & Wilhelmsson, T.COVID-19- Consumer Law Research Group. (2020). Consumer Law and Policy Relating to Change of Circumstances Due to the COVID-19 Pandemic. *Journal of Consumer Policy*, *43*(3), 437–450. doi:10.100710603-020-09463-z

Alegre, I., & Berbegal-Mirabent, J. (2016). Social innovation success factors: Hospitality and tourism social enterprises. *International Journal of Contemporary Hospitality Management*, *28*(6), 1155–1176. doi:10.1108/IJCHM-05-2014-0231

Ali, R. F. (2017). *History of armed struggles in Kashmir*. Jay Kay Books.

Aliyev, K., & Ahmadova, N. (2020). Testing tourism-led economic growth and economic-driven tourism growth hypotheses. *Toκrism*, *68*(1), 43–57. doi:10.37741/t.68.1.4

Amadeus & Associació Corporativa d'Agències de Viatge Especialitzades. (2018). *Segundo estudio estratégico de las agencias de viajes en España*. Author.

Ambrož, R., & Omerzel, D. G. (2018). Business model innovation in tourism: How to survive in highly competitive environments. *Academica Turistica-Tourism and Innovation Journal, 10*(2), 175–183. doi:10.26493/2335-4194.10.175-183

Amit, R., & Zott, C. (2012). *Creating value through business model innovation.* Academic Press.

Amujo, O. C., & Otubanjo, O. (2012). Leveraging rebranding of 'unattractive' nation brands to stimulate post-disaster tourism. *Tourist Studies, 12*(1), 87–105. doi:10.1177/1468797612444196

Andersen, P. N., & Andersen, J. B. (2012). *Green business model innovation in the tourism and experience economy.* Nordic Innovation.

Anestis, F., Stathis, P., & Tzung-Cheng, H. (2021). The good, the bad, and the ugly on COVID-19 tourism recovery. *Annals of Tourism Research, 87*, 103117. doi:10.1016/j.annals.2020.103117 PMID:33518847

Angelidou, M. (2015). Smart cities: A conjuncture of four forces. *Cities (London, England), 47*, 95–106. doi:10.1016/j.cities.2015.05.004

An, S., Choi, Y., & Lee, Ch. (2021). Virtual travel experience and destination marketing: Effects of sense and information quality of flow and visit intention. *Journal of Destination Marketing & Management, 19*, 1–10. doi:10.1016/j.jdmm.2020.100492

An, T. B., & Thi, D. B. V. (2012). Research on factors affecting competitiveness of four-star hotels in Thua Thien Hue. *Journal of Economic Development*, (261), 21–29.

Aramendia, E., & Ollo, A. (2013). ICT impact on tourism industry. *International Journal of Management Cases, 15*(2), 87–98.

Arana, J. E., & León, C. J. (2008). The impact of terrorism on tourism demand. *Annals of Tourism Research, 35*(2), 299–315. doi:10.1016/j.annals.2007.08.003

Arbussà Reixach, A., Bikfalvi, A., & Marquès i Gou, P. (2017). Strategic agility-driven business model renewal: The case of an SME. *Management Decision, 55*(2), 271–293. doi:10.1108/MD-05-2016-0355

Arcese, G., Valeri, M., Poponi, S. & Elmo, G. C. (2020). Innovative drivers for family business models in tourism. *Journal of Family Business Management.* doi:10.1108/JFBM-05-2020-0043

Areej Shabib, A. (2020). *Service innovation in tour guiding*. https://2020.mtcon.org/wp-content/uploads/2020/09/MTCON-20-English.pdf

Aven, T., & Renn, O. (2018). Improving government policy on Risk: Eight key principles. *Reliability Engineering & System Safety, 176*, 230–241. doi:10.1016/j.ress.2018.04.018

Avraham, E. (2013). Crisis communication, image restoration, and battling stereotypes of terror and wars: Media strategies for attracting tourism to Middle Eastern countries. *The American Behavioral Scientist, 57*(9), 1350–1367. doi:10.1177/0002764213487733

Bain and Ernst & Young. (2020). *Impacto de COVID-19 en hostelería en España*. https://www.ey.com/Publication/vwLUAssets/ey-bain-company-impacto-de-covid-19-en-hosteleria-en-espana/$FILE/ey-bain-company-impacto-de-covid-19-en-hosteleria-en-espana.pdf

Baker, D. M. (2020). Tourism and terrorism: Terrorists' threats to commercial aviation safety and security. *Tourism, Terrorism and Security*, 163-181. doi:10.1108/978-1-83867-905-720201012

Balanzá-Martínez, V., Kapczinski, F., de Azevedo Cardoso, T., Atienza-Carbonell, B., Rosa, A. R., Mota, J. C., & De Boni, R. B. (2021). The assessment of lifestyle changes during the COVID-19 pandemic using a multidimensional scale. *Revista de Psiquiatría y Salud Mental*, *14*(1), 16–26. doi:10.1016/j.rpsm.2020.07.003 PMID:32962948

Ballarín, E. (2020). *Tendencias de la hostelería post coronavirus*. Live Horeca Live. Retrieved from https://www.smarttravel.news/10-tendencias-explican-sera-la-hosteleria-post-coronavirus/

Banco de España. (2020). *Artículos analíticos. Boletín Económico 4/2020. El impacto de la crisis del COVID-19 sobre la situación financiera de las empresas no financieras en 2020: Evidencia basada en la central de balances. Banco de España*. Retrieved from https://www.bde.es/bde/es/secciones/informes/boletines/relac/Boletin_Economic/Informes_trimest/

Banerjee, S., & Chua, A. (2020). How alluring is the online profile of tour guides? *Annals of Tourism Research*, *81*, 1–11. doi:10.1016/j.annals.2020.102887

Barcelona City Council. (2019). *Statistics of enterprises and establishments (DIRCE)*. https://www.bcn.cat/estadistica/angles/dades/economia/dirce/a2019/t111.htm

Barker, M., Page, S. J., & Meyer, D. (2003). Urban visitor perceptions of safety during a special event. *Journal of Travel Research*, *41*(4), 355–361. doi:10.1177/0047287503041004004

Barth, J. R., Li, T., McCarthy, D., Phumiwasana, T., & Yago, G. (2006). Economic Impacts of Global Terrorism: From Munich to Bali. Capital Studies, Milken Institute.

Bartik, A. W., Bertrand, M., Cullen, Z. B., Glaeser, E. L., Luca, M., & Stanton, C. T. (2020). *How are small businesses adjusting to COVID-19? Early evidence from a survey (No. w26989)*. National Bureau of Economic Research. doi:10.3386/w26989

Bassil, C., Saleh, A. S., & Anwar, S. (2019). Terrorism and tourism demand: A case study of Lebanon, Turkey and Israel. *Current Issues in Tourism*, *22*(1), 50–70. doi:10.1080/13683500.2017.1397609

BBC. (2009). https://www.bbc.com/mundo/lg/economia/2009/07/090728_1226_agencias_viajes_pea

Becken, S. (2013). Developing a framework for assessing resilience of tourism sub-systems to climatic factors. *Annals of Tourism Research*, *43*, 506–528. doi:10.1016/j.annals.2013.06.002

Becken, S., & Carmignani, F. (2016). Does tourism lead to peace? *Annals of Tourism Research*, *61*, 63–79. doi:10.1016/j.annals.2016.09.002

Becken, S., & Hay, J. E. (2007). *Tourism and climate change: Risks and opportunities*. Channel View Publications.

Behera, N. C. (2000). State, Identity & Violence: Jammu, Kashmir, and Ladakh. Manohar Publishers.

Beigel, S. (2015, July 11). *Inspirational ideas: "We don't just sell products; we sell the experience!"* EIP-AGRI - European Commission. https://ec.europa.eu/eip/agriculture/en/news/inspirational-ideas-%E2%80%9Cwe-don%E2%80%99t-just-sell-products-we-sell-experience%E2%80%9D-0

Belzunegui-Eraso, A., & Erro-Garcés, A. (2020). Teleworking in the context of the Covid-19 crisis. *Sustainability*, *12*(9), 2–18. doi:10.3390u12093662

Bennasar, T., Carbajo, M., Gallego, A., & Tárraga, J. C. (2020). *XV edición de la Jornada sobre Gestión de Riesgos en el sector turístico*. Academic Press.

Benur, A. M., & Bramwell, B. (2015). Tourism product development and product diversification in destinations. *Tourism Management*, *50*, 213–224. doi:10.1016/j.tourman.2015.02.005

Berbel Giménez, G., & Reyes Gómez, J. D. (2020). Impacto de la COVID-19 en la restauración. Estadísticas, visión de expertos y de clientes. In *F.J. Bauzá Martorell y F.J. Melgosa Arcos (Dirs.), Turismo Post COVID-19. El turismo después de la pandemia global. Análisis, perspectivas y vías de recuperación* (pp. 127–146). Aquilafuente. Ediciones Universidad de Salamanca.

Bernard, H. (2006). *Research methods in anthropology: Qualitative and quantitative approaches*. Alta Mira Press.

Bethapudi, A. (2015). Role of ICT in promoting rural tourism product. *Journal of Tourism & Hospitality (Los Angeles, Calif.)*, *4*(3), 1–3.

Bhat, I. (2019). New Delhi's demographic designs in Kashmir. *Foreign Policy*. https://foreignpolicy.com/2019/08/16/new-delhis-demographic-designs-in-kashmir

Bhattacharya, M., & Basu, K. (2010). Impact of terror incidents on the foreign tourist arrivals in India: An econometric exploration. *International Journal of Tourism Policy*, *3*(3), 213–222. doi:10.1504/IJTP.2010.037993

Bhuiyan, A. M., Hassan, S., Darda, M. A., & Habib, M. W. (in press). *Aspects of Sustainable Tourism Development and COVID-19 Pandemic*. https://www.preprints.org/manuscript/202008.0418/v1

Bianchi, R. (2006). Tourism and the globalisation of fear: Analysing the politics of risk and (in) security in global travel. *Tourism and Hospitality Research*, *7*(1), 64–74. doi:10.1057/palgrave.thr.6050028

Bigelow, L., & Barney, J. (2021). What can strategy learn from the business model approach? *Journal of Management Studies*, *58*(2), 528–539. doi:10.1111/joms.12579

Bigné, E., Caplliure, E. M., & Miquel, M. J. (2016). eWOM on travel agency selection: Specialized versus private label. *Psychology and Marketing*, *33*(12), 1046–1053. doi:10.1002/mar.20938

Binoy, T. A. (2009). The Media Strategy for Destination Image Building in Crisis. *Atna Journal of Tourism Studies*, *4*(1), 138–150. doi:10.12727/ajts.4.11

Blake, A., & Sinclair, M. T. (2003). Tourism crisis management: US response to September 11. *Annals of Tourism Research*, *30*(4), 813–832. doi:10.1016/S0160-7383(03)00056-2

Bo Liu, H., McCarthy, B., Chen, T., Guo, S., & Song, X. (2014). The Chinese wine market: A market segmentation study. *Asia Pacific Journal of Marketing and Logistics*, *26*(3), 450–471. doi:10.1108/APJML-07-2013-0089

Bocken, N. M., & Samuel, W. (2014). A Literature and Practice Review to Develop Sustainable Business Model Archetypes. *Journal of Cleaner Production*, *65*, 42–56. doi:10.1016/j.jclepro.2013.11.039

Bodosca, S. (2015). Sustainable decision making under the crisis for tourism businesses. *Ecoforum*, *4*(1), 127–131.

Bogers, M., & Jensen, J. D. (2017). Open for business? An integrative framework and empirical assessment for business model innovation in the gastronomic sector. *British Food Journal*, *119*(11), 2325–2339. doi:10.1108/BFJ-07-2017-0394

Bose, S. (2009). *Kashmir: Roots of conflict, paths to peace*. Harvard University Press.

Bradyand, J., & Widdows, R. (1988). The impact of world events on travel to Europe during the summer of 1986. *Journal of Travel Research*, *26*(3), 8–10. doi:10.1177/004728758802600304

Braun, P., & Hollick, M. (2008). Capacity building in SME Tourism networks. Encyclopedia of Networked and Virtual Organizations, 136-141. doi:10.4018/978-1-59904-885-7.ch018

Brau, R., Lanza, A., & Usai, S. (2008). *Tourism and sustainable economic development: Macroeconomic models and empirical methods*. Edward Elgar.

Breier, M., Kallmuenzer, A., Clauss, T., Gast, J., Kraus, S., & Tiberius, V. (2021). The role of business model innovation in the hospitality industry during the COVID-19 crisis. *International Journal of Hospitality Management*, *92*, 102723. doi:10.1016/j.ijhm.2020.102723

Brewer, P., & Sebby, A. G. (2021). The effect of online restaurant menus on consumers' purchase intentions during the COVID-19 pandemic. *International Journal of Hospitality Management*, *94*, 102777. Advance online publication. doi:10.1016/j.ijhm.2020.102777

Briedenhann, J., & Butts, S. (2006). Application of the Delphi technique to rural tourism project evaluation. *Current Issues in Tourism*, *9*(2), 171–190. doi:10.1080/13683500608668246

Brouder, P., Teoh, S., Salazar, N., Mostafanezhad, M., Mei Pung, J., Lapointe, D., Higgins, F., Haywood, M., Hall, M., & Balslev, H. (2020). Reflections and discussions: Tourism matters in the new normal post-COVID-19. *Tourism Geographies*, *22*(3), 735–746. doi:10.1080/14616688.2020.1770325

Buhalis, D. (2019). Technology in tourism-from information communication technologies to eTourism and smart tourism towards ambient intelligence tourism: A perspective article. *Tourism Review*, *75*(1), 267–272. doi:10.1108/TR-06-2019-0258

Buigut, S., & Amendah, D. D. (2016). Effect of terrorism on demand for tourism in Kenya. *Tourism Economics*, *22*(5), 928–938. doi:10.5367/te.2015.0467

Burnett, J. J. (1998). A strategic approach to managing crises. *Public Relations Review*, *24*(4), 475–488. doi:10.1016/S0363-8111(99)80112-X

Byrd, K., Her, E., Fan, A., Almanza, B., Liu, Y., & Leitch, S. (2021). Restaurants and COVID-19: What are consumers' risk perceptions about restaurant food and its packaging during the pandemic? *International Journal of Hospitality Management*, *94*, 102821. doi:10.1016/j.ijhm.2020.102821

Cadena, A., Child, F., Craven, M., Ferrari, F., Fine, D., Franco, J., & Wilson, M. (2020). *How to restart national economies during the coronavirus crisis*. McKinsey. https://tec.mx/sites/default/files/inline-files/How-to-restart-national-economies-during-the-coronavirus-crisis-vF_1.pdf

Çakar, K., & Aykol, S. (2020). Case Study as a Research Method in Hospitality and Tourism Research: A Systematic Literature Review (1974–2020). *Cornell Hospitality Quarterly*, *62*(1), 21–31. doi:10.1177/1938965520971281

Caloffi, A., Freo, M., Ghinoi, S., Rossi, F., & Russo, M. (2019). Is a policy mix more effective than individual policies for SME innovation? An exploratory analysis. *Fteval Journal for Research and Technology Policy Evaluation,* (47), 72-77.

Canalis, X. (2020, January 23). *La resiliencia de las agencias frente a internet, en una infografía*. https://www.hosteltur.com/126352_la-resiliencia-de-las-agencias-frente-a-internet-en-una-infografia.html

Candela, G., & Figini, P. (2012). *The economics of tourism destinations*. Springer. doi:10.1007/978-3-642-20874-4

Carr, A. (2020). COVID-19, indigenous peoples and tourism: A view from New Zealand. *Tourism Geographies*, 1–12.

Carter, W. N. (1991). *Disaster management: A disaster manager's handbook*. Asian Development Bank.

Castellanos-Verdugo, M., Oviedo, M. A., & Vega-Vázquez, M. (2014). CRM and RQ as key factors in retail setting services in an economic crisis context. The case of travel agencies. Springer-Verlag.

Cates, J. N., & Sussman, M. B. (1992). Family systems and inheritance. *Family Business Review*, *5*(2), 205–226. doi:10.1111/j.1741-6248.1992.00205.x

CEHE. (2021). *Confederación Empresarial de Hostelería de España*. Retrieved from https://www.cehe.es/index.html

Chahal, H., & Devi, A. (2015). Destination attributes and destination image relationship in volatile tourist destination: Role of perceived risk. *Metamorphosis*, *14*(2), 1–19. doi:10.1177/0972622520150203

Chang, Y. W., Hsu, P. Y., & Lan, Y. C. (2019). Cooperation and competition between online travel agencies and hotels. *Tourism Management*, *71*, 187–196. doi:10.1016/j.tourman.2018.08.026

Chaudary, M., Sharma, A., & Sharma, S. (2020). Are small travel agencies ready for digital marketing? Views of travel agency managers. *Tourism Management*, (79), 1–10.

Chauhan, V., & Khanna, S. (2008). Tourism: a tool for crafting peace process in Kashmir, J&K, India. Academic Press.

Cheah, S., Ho, Y. P., & Li, S. (2018). Business model innovation for sustainable performance in retail and hospitality industries. *Sustainability*, *10*(11), 3952. doi:10.3390u10113952

Chebli, A., & Ben Said, F. (2020). The Impact of Covid-19 on Tourist Consumption Behaviour: A Perspective Article. *Journal of Tourism Management Research*, *7*(2), 196–207. doi:10.18488/journal.31.2020.72.196.207

Chen, S., Law, R., & Zhang, M. (2020). Review of research on tourism-related diseases. *Asia Pacific Journal of Tourism Research*, 1-15.

Chesbrough, H. (2010). Business model innovation: Opportunities and barriers. *Long Range Planning*, *43*(2-3), 354–363. doi:10.1016/j.lrp.2009.07.010

Chiao, H., Chen, Y., & Huang, W. (2018, November). Examining the usability of an online virtual tour-guiding platform for cultural tourism education. *Journal of Hospitality, Leisure, Sport and Tourism Education*, *23*, 29–38. doi:10.1016/j.jhlste.2018.05.002

Christensen, C. M., & Raynor, M. E. (2003). *The innovator's solution: Creating and sustaining successful growth*. Harvard Business School Press.

Chung, J., Jin, B., Jeong, S. W., & Yang, H. (2019). Nie-based SME brand building in foreign markets: An exploratory study. *Journal of Product and Brand Management*, *28*(1), 63–79. doi:10.1108/JPBM-10-2016-1331

Cioccio, L., & Michael, E. J. (2007). Hazard or disaster: Tourism management for the inevitable in Northeast Victoria. *Tourism Management*, *28*(1), 1–11. doi:10.1016/j.tourman.2005.07.015

Cochrane, J. (2010). The sphere of tourism resilience. *Tourism Recreation Research*, *35*(2), 173–185. doi:10.1080/02508281.2010.11081632

Cohen, E. (1985). The tourist guide: The origins, structure and dynamics of a role. *Annals of Tourism Research*, *12*(1), 5–29. doi:10.1016/0160-7383(85)90037-4

Collier, P., & Hoeffler, A. (2004). Greed and grievance in civil war. *Oxford Economic Papers*, *56*(4), 563–595. doi:10.1093/oep/gpf064

Conyette, M. (2015). *21 century travel using websites, mobile and wearable technology devices.* Athens: ATINER'S Conference Paper Series.

Cooper, C., & Hall, C. M. (2019). *Contemporary tourism: An international approach.* Goodfellow Limited.

Correia, A., Kozak, M., & Reis, H. (2016). Conspicuous Consumption of the Elite: Social and Self-Congruity in Tourism Choices. *Journal of Travel Research, 55*(6), 738–750. doi:10.1177/0047287514563337

Cowling, M., Liu, W., & Zhang, N. (2018). Did firm age, experience, and access to finance count? SME performance after the global financial crisis. *Journal of Evolutionary Economics, 28*(1), 77–100. doi:10.100700191-017-0502-z

Cranmer, E. E., Tom Dieck, M. C., & Fountoulaki, P. (2020). Exploring the value of augmented reality for tourism. *Tourism Management Perspectives, 35*, 100672. doi:10.1016/j.tmp.2020.100672

Creswell, J. W. (1998). *Qualitative inquiry and research design: Choosing among five approaches.* SAGE.

Crick, D., Chaudhry, S., & Crick, J. M. (2018). Risks/rewards and an evolving business model. *Qualitative Market Research, 21*(2), 143–165. doi:10.1108/QMR-01-2017-0001

Cró, S., & Martins, A. M. (2017). Structural breaks in international tourism demand: Are they caused by crises or disasters? *Tourism Management, 63*, 3–9. doi:10.1016/j.tourman.2017.05.009 PMID:32287750

Crossley, É. (2020). Ecological grief generates desire for environmental healing in tourism after COVID-19. *Tourism Geographies*, 1–10.

Czinkota, M., Knight, G. A., & Liesch, P. W. (2004). Terrorism and international business: conceptual foundations. *Terrorism and the international business environment: The security-business nexus*, 43-57.

Dartford, K. (2021, abril 16). *What is vaccine tourism and is it ethical*? Euronews. https://www.euronews.com/travel/2021/04/16/what-is-vaccine-tourism-and-which-european-travel-companies-are-offering-trips

Davies, A. (2021). COVID-19 and ICT-Supported RemoteWorking: Opportunities for Rural Economies. *WORLD (Oakland, Calif.), 2*, 139–152.

De Montis, A., Ledda, A., Ganciu, A., Serra, V., & De Montis, S. (2015). Recovery of rural centres and "albergo diffuso": A case study in Sardinia, Italy. *Land Use Policy, 47*, 12–28. doi:10.1016/j.landusepol.2015.03.003

De Sousa Santos, B. (2020). *La cruel pedagogía del vírus* [The cruel pedagogy of the virus]. CLACSO. https://www.clacso.org/la-cruel-pedagogia-del-virus/

Dediu, L. (2018). Travel agency staff: The need for professional training in the information technologies context of use. *Cactus Tourism Journal, 17*(1), 9.

Del Alonso-Almeida, M. M., Bagur-Femenías, L., & Llach, J. (2015). The adoption of quality management practices and their impact on business performance in small service companies: The case of Spanish travel agencies. *Service Business, 9*(1), 57–75. doi:10.100711628-013-0218-6

Demil, B., & Lecocq, X. (2010). Business model evolution: In search of dynamic consistency. *Long Range Planning, 43*(2-3), 227–246. doi:10.1016/j.lrp.2010.02.004

Denchev, S., Trencheva, T., & Trenchev, I. (2020). Application of the Game Engine in the Promotion and Training of the Cultural Heritage. *Conference Proceedings: 14th International Technology, Education and Development Conference*, 2700-2706. 10.21125/inted.2020.0806

Desyllas, P., Salter, A., & Alexy, O. (2020). The breadth of business model reconfiguration and firm performance. *Strategic Organization*, 1–39.

Dhariwal, R. (2005). Tourist arrivals in India: How important are domestic disorders? *Tourism Economics, 11*(2), 185–205. doi:10.5367/0000000054183522

Díaz-Bravo, L., Torruco-García, U., Martínez-Hernández, M., & Varela-Ruiz, M. (2013). La entrevista, recurso flexible y dinámico. *Investigación en Educación Médica, 2*(7), 162–167. doi:10.1016/S2007-5057(13)72706-6

Dickinson, J., Hibbert, J., & Filimonau, V. (2016). Mobile technology and the tourist experience: (Dis)connection at the campsite. *Tourism Management, 57*, 193–201. doi:10.1016/j.tourman.2016.06.005

Dimitrov, P. (2009). Measuring Occupational Health and Safety Risks in Tourism Companies. *Ege Academic Review, 9*(3), 889-902.

Directive (EU) 2015/2302 of the European Parliament and of the Council of 25 November 2015 on package travel and linked travel arrangements, amending Regulation (EC) No 2006/2004 and Directive 2011/83/EU of the European Parliament and of the Council and repealing Council Directive 90/314/EEC, 32015L2302, CONSIL, EP, OJ L 326 (2015). https://data.europa.eu/eli/dir/2015/2302/oj/eng

Directorate of Economics & Statistics J&K Government. (2020). *Economic Survey 2017*. Srinagar, Jammu: Directorate of Economics & Statistics Jammu & Kashmir Government.

Divisekera, S., & Nguyen, V. K. (2018). Determinants of innovation in tourism evidence from Australia. *Tourism Management, 67*, 157–167. doi:10.1016/j.tourman.2018.01.010

Djanibekov, N., Herzfeld, T., & Weingarten, P. (2014). Rural areas in transition: An introduction. *Appl Agric Forestry Res, 64*, 125–126.

Dolnicar, S., & Laesser, C. (2007). Travel Agency Marketing Strategy: Insights from Switzerland. *Journal of Travel Research, 46*(2), 133–146. doi:10.1177/0047287507299573

Dorcic, J., Komsic, J., & Markovic, S. (2019). Mobile technologies and applications towards smart tourism–state of the art. *Tourism Review*, *74*(1), 82–103. doi:10.1108/TR-07-2017-0121

Dotti Sani, G. M. (2018). The economic crisis and changes in work–family arrangements in six European countries. *Journal of European Social Policy*, *28*(2), 177–193. doi:10.1177/0958928717700566

Drakos, K., & Kutan, A. M. (2003). Regional effects of terrorism on tourism in three Mediterranean countries. *The Journal of Conflict Resolution*, *47*(5), 621–641. doi:10.1177/0022002703258198

Druker, P. F. (1954). *The Practice of Management*. Harper and Row.

Duarte Alonso, A., Sakellarios, N., Alexander, N., & O'Brien, S. (2018). Strengths, innovation, and opportunities in a burgeoning industry: An exploratory study. *Asia Pacific Journal of Marketing and Logistics*, *30*(2), 276–296. doi:10.1108/APJML-05-2017-0105

Dube, K., Nhamo, G., & Chikodzi, D. (2020). COVID-19 cripples global restaurant and hospitality industry. *Current Issues in Tourism, 1-4*. Advance online publication. doi:10.1080/13683500.2020.1773416

Dutta, P. K. (2019). Kashmir: Curious case of demographic realities and perceptions. *India Today*.

Ebojoh, S., & Verma, N. (2020). *The influence of digitalization on the innovation strategy of small and medium-sized enterprises*. University of Gävle. http://urn.kb.se/resolve?urn=urn:nbn:se:hig:diva-31530

Egger, I., Lei, S. I., & Wassler, P. (2020). Digital free tourism – an exploratory study of tourist motivations. *Tourism Management*, *79*, 104098. doi:10.1016/j.tourman.2020.104098

EIM Business & Policy Research. (2010). *European SMEs under pressure. Annual report on EU SMEs 2009*. https://ec.europa.eu/docsroom/documents/15767/attachments/1/translations/en/renditions/native

Elhadary, T. (2018). Sufi tourism: The impact of Sufi heritage on Islamic religious tourism. *Islamic Tourism: Management of Travel Destinations*, 110-123. doi:10.1079/9781786394132.0110

Elwin, J. (1989). Site selection in tourism. In S. F. Witt & L. Moutinho (Eds.), *Tourism Marketing and Management Handbook* (pp. 403–406). Prentice Hall.

Enders, W., & Sandler, T. (1991). Causality between transnational terrorism and tourism: The case of Spain. *Studies in Conflict and Terrorism*, *14*(1), 49–58.

Enders, W., Sandler, T., & Parise, G. F. (1992). An econometric analysis of the impact of terrorism on tourism. *Kyklos*, *45*(4), 531–554. doi:10.1111/j.1467-6435.1992.tb02758.x

ESADE. (2021). *Informe económico y financiero*. 1er Semestre 2021, 28. Esade Center for Economy Policy. Retrieved from http://itemsweb.esade.es/wi/Prensa/InformeEconomicoFinancieroEsade28.pdf

Estrada, M. A. R., & Park, D. (2008). Korean unification: How painful and how costly? *Journal of Policy Modeling*, *30*(1), 87–100. doi:10.1016/j.jpolmod.2007.07.001

European Commission. (2013). *Commission recommendation concerning the definition of micro, small, and medium-sized enterprises*. https://eur-lex.europa.eu/LexUriServ/LexUriServ.do?uri=OJ:L:2003:124:0036:0041:en:PDF#:~:text=The%20category%20of%20micro%2C%20small,not%20exceeding%20EUR%2043%20million

European Commission. (2019). *Digitalisation support to SMEs*. https://ec.europa.eu/easme/sites/easmesite/files/dop_p2p_digital.pdf

European Commission. (2020). *Science for Policy Briefs. Telework in the EU before and after the COVID-19: where we were, where we head to*. https://ec.europa.eu/jrc/sites/jrcsh/files/jrc120945_policy_brief_-_covid_and_telework_final.pdf

Eurostat. (2017). *Services by employment size class*. https://appsso.eurostat.ec.europa.eu/nui/show.do?dataset=sbs_sc_1b_se_r2&lang=en

Eurostat. (2020). *Data on tourism*. https://ec.europa.eu/eurostat/web/tourism/data/database

Evans, A. (2000). The Kashmir insurgency: As bad as it gets. *Small Wars & Insurgencies*, *11*(1), 69–81. doi:10.1080/09592310008423261

Expedia Group. (n.d.). *History of the online travel industry pioneer*. https://www.expediagroup.com/about/history/

Express, F. (2019). *Article 370 removal will result in end of terrorism; bring development in Jammu and Kashmir, reaffirms: Amit Shah*. Retrieved on November 11, 2020 from https://www.financialexpress.com/india-news/article-370-removal-will-result-in-end-of terrorism-bring-development-in-jammu-and-kashmir-reaffirms-amit-shah/1672413/

Fakhimi, A. H., Khani, A. H., & Sardroud, J. M. (2021). Smart-city infrastructure components. In J. R. Vacca (Ed.), *Solving Urban Infrastructure Problems Using Smart City Technologies* (pp. 17–54). Elsevier. doi:10.1016/B978-0-12-816816-5.00002-4

Fakhrutdinova, L. R. (2018). Establishment of an effective enterprise in the touristic market. *Revista San Gregorio. Extra*, *25*, 131–137.

Fallon, F. (2004). After the LOMBOK riots, is sustainable Tourism Achievable? *Journal of Travel & Tourism Marketing*, *15*(2-3), 139–158. doi:10.1300/J073v15n02_08

Faulkner, B. (2001). Towards a framework for tourism disaster management. *Tourism Management*, *22*(2), 135–147. doi:10.1016/S0261-5177(00)00048-0

Femenia-Serra, F., Neuhofer, B., & Ivars-Baidal, J. A. (2019). Towards a conceptualisation of smart tourists and their role within the smart destination scenario. *Service Industries Journal*, *39*(2), 109–133. doi:10.1080/02642069.2018.1508458

Feridun, M., & Shahbaz, M. (2010). Fighting terrorism: Are military measures effective? Empirical evidence from Turkey. *Defence and Peace Economics*, *21*(2), 193–205. doi:10.1080/10242690903568884

Ferrario, F. F. (1979). The evaluation of tourist resources: An applied methodology. *Journal of Travel Research*, *17*(4), 24–30. doi:10.1177/004728757901700405

Fissi, S., Romolini, A., & Gori, E. (2020). Building a business model for a new form of hospitality: The albergo diffuso. *International Journal of Contemporary Hospitality Management*, *32*(1), 307–323. doi:10.1108/IJCHM-01-2019-0047

Fitzgerald, M. (2012). *O2O: O2 for local business.* Retrieved from the website: http://www. onlineeconomy. org/tag/online-to-offline

Fleischer, A., & Buccola, S. (2002). War, terror, and the tourism market in Israel. *Applied Economics*, *34*(11), 1335–1343. doi:10.1080/00036840110099252

Floyd, M. F., & Pennington-Gray, L. (2004). Profiling risk perceptions of tourists. *Annals of Tourism Research*, *31*(4), 1051–1054. doi:10.1016/j.annals.2004.03.011

Flyvbjerg, B. (2006). Five misunderstandings about case-study research. *Qualitative Inquiry*, *12*(2), 219–245. doi:10.1177/1077800405284363

Foss, N. J., & Saebi, T. (2017). Fifteen Years of Research on Business Model Innovation: How far have we come, and where should we go? *Journal of Management*, *43*(1), 200–227. doi:10.1177/0149206316675927

Foss, N. J., & Saebi, T. (2018). Business models and business model innovation: Between wicked and paradigmatic problems. *Long Range Planning*, *51*(1), 9–21. doi:10.1016/j.lrp.2017.07.006

Fraiz Brea, J. A., & Rodríguez-Toubes, D. (2012). Desarrollo de una política de gestión de crisis para desastres en el turismo. *Tourism & Management Studies*, (8), 10–18.

Franklin, A. (2007). The problem with tourism theory. In The critical turn in tourism studies (pp. 153-170). Routledge. doi:10.1016/B978-0-08-045098-8.50014-3

Frechtling, D. (2001). The tourism forecasting process. *Forecasting Tourism Demand*, 45-57. doi:10.1016/B978-0-7506-5170-7.50007-6

Freitas, A., Muritiba, P., Muritiba, S. N., Lima, Y. O., & Riascos, L. A. (2018). Measurement of innovations. In *SME: Exploratory study with emphasis on manufacturing processes in the south of São Paulo.* doi:10.14488/ENEGEP2018_TI_ST_265_520_36054

Frey, B. S., Luechinger, S., & Stutzer, A. (2007). Calculating tragedy: Assessing the costs of terrorism. *Journal of Economic Surveys*, *21*(1), 1–24. doi:10.1111/j.1467-6419.2007.00505.x

Fuchs, G., & Reichel, A. (2006). Tourist destination risk perception: The case of Israel. *Journal of Hospitality & Leisure Marketing*, *14*(2), 83–108. doi:10.1300/J150v14n02_06

Fusch, P. I., & Ness, L. R. (2015). Are we there yet? Data saturation in qualitative research. *Qualitative Report*, *20*(9), 1408–1416.

Gaffney, C., & Eeckels, B. (2020). Covid-19 and tourism risk in the Americas. *Journal of Latin American Geography*, *19*(3), 308–313. doi:10.1353/lag.2020.0054

Gallego, I., & Font, X. (2020). Changes in air passenger demand as a result of the COVID-19 crisis: Using Big Data to inform tourism policy. *Journal of Sustainable Tourism*, 1–20. doi:10.1080/09669582.2020.1773476

Galvani, A., Lew, A. A., & Perez, M. S. (2020). COVID-19 is expanding global consciousness and the sustainability of travel and tourism. *Tourism Geographies*, 1–10.

Ganguly, Š. (2002). *Conflict unending: India-Pakistan tensions since 1947*. Columbia University Press.

Garg, A. (2015). Travel risks vs tourist decision making: A tourist perspective. *International Journal of Hospitality and Tourism Systems*, *8*(1), 1–9. doi:10.21863/ijhts/2015.8.1.004

Garrod, B., Wornell, R., & Youell, R. (2006). Re-conceptualising rural resources as countryside capital: The case of rural tourism. *Journal of Rural Studies*, *22*(1), 117–128. doi:10.1016/j.jrurstud.2005.08.001

Gavinolla, M. R., Kaushal, V., Livina, A., Swain, S. K., & Kumar, H. (2021). (in press). Sustainable consumption and production of wildlife tourism in Indian tiger reserves: A critical analysis. *Worldwide Hospitality and Tourism Themes*. Advance online publication. doi:10.1108/WHATT-08-2020-0091

Gearing, C. E., Swart, W. W., & Var, T. (1974). Establishing a measure of touristic attractiveness. *Journal of Travel Research*, *12*(4), 1–8. doi:10.1177/004728757401200401

Gebicki, M. (2021, January 18). *The death of the travel agent: Will 2021 be the year it happens?* Traveller. https://www.traveller.com.au/the-death-of-the-travel-agent-will-2021-be-the-year-it-happens-h1tf3a

George, R. (2010). Visitor perceptions of crime-safety and attitudes towards risk: The case of Table Mountain National Park, Cape Town. *Tourism Management*, *31*(6), 806–815. doi:10.1016/j.tourman.2009.08.011

Gerritsen, P. (2014). Working with indigenous women on multifunctionality and sustainable rural tourism in western Mexico. *Journal of Rural and Community Development*, *9*(3), 243–257.

Getz, D., & Carlsen, J. (2005). Family business in tourism. *Annals of Tourism Research*, *32*(1), 237–258. doi:10.1016/j.annals.2004.07.006

Ghafur, A. (2020). India's concern for Kashmir. *Lancet*, *396*(10255), e47. doi:10.1016/S0140-6736(19)32076-8 PMID:31493894

Golechha, M. (2020). Integration of Kashmir for peace and prosperity. *Lancet*, *396*(10255), e48. doi:10.1016/S0140-6736(19)32077-X PMID:31493897

González, T. (2019, September 12). *Crean Asociación de Agentes y Agencias de Viajes Pymes de Argentina*. Hosteltur: Toda la información de turismo. https://www.hosteltur.com/lat/131228_crean-asociacion-de-agentes-y-agencias-de-viajes- pymes-de-argentina.html

González, T. (2020a, September 10). *Las agencias con ventas al 10% y ni el 25% de la plantilla reincorporada*. https://www.hosteltur.com/139139_las-agencias-con-ventas-al-10-y-ni-el-25-de-la-plantilla-reincorporada.html

González, T. (2020b, October 22*). La estrategia de las tres R para afrontar la recesión pandémica*. https://www.hosteltur.com/140026_la-estrategia-de-las-tres-r-para-afrontar-la-recesion-pandemica.html?code=home-page%7B2020-10-22%7D&utm_source=newsletter-es&utm_medium=email&utm_campaign=demanda-latente-explosiva-las-pernoctaciones-subieron-un-544-en-el

Gössling, S., & Michael Hall, C. (2019). Sharing versus collaborative economy: How to align ICT developments and the SDGs in tourism? *Journal of Sustainable Tourism*, *27*(1), 74–96. do i:10.1080/09669582.2018.1560455

Gössling, S., Scott, D., & Hall, C. M. (2020). Pandemics, tourism and global change: A rapid assessment of COVID-19. *Journal of Sustainable Tourism*, *29*(1), 1–20. doi:10.1080/0966958 2.2020.1758708

Gowreesunkar, G. V., & Gavinolla, M. R. (2020). Urbanism and Overtourism: Impacts and implications for the city of Hyderabad. In A. M. Morrison & J. A. Coca-Stefaniak (Eds.), *Routledge Handbook of Tourism Cities* (pp. 101–120). Routledge. doi:10.4324/9780429244605-7

Gowreesunkar, V., Cooper, C. P., & Durbarry, R. (2009). The role of internal marketing in sustainable destination management: Grand Bay, Mauritius, as a case study. *International Journal of Environmental, Cultural, Economic and Social Sustainability*, *5*(5), 21–24. doi:10.18848/1832-2077/CGP/v05i05/54684

Gretzel, U., Fuchs, M., Baggio, R., Hoepken, W., Law, R., Neidhardt, J., ... Xiang, Z. (2020). e-Tourism beyond COVID-19: A call for transformative research. *Information Technology & Tourism*, 1.

Gretzel, U., Sigala, M., Xiang, Z., & Koo, C. (2015). Smart Tourism, foundations and developments. *Electronic Markets*, *25*(3), 179–188. doi:10.100712525-015-0196-8

Group, I. D. M. (2019, September 16). *Cómo afrontan las pymes la resistencia al cambio en su digitalización*. IT Reseller; IT Digital Media Group. https://www.itreseller.es/pyme/2019/09/como-afrontan-las-pymes-la-resistencia-al-cambio- en-su-digitalizacion

Guerra, R. P. (2020). Algunas notas sobre el derecho administrativo del turismo: COVID-19. *Revista General de Derecho Administrativo, 54*(4).

Gunay, S., & Kurtulmuş, B. E. (2021). COVID-19 social distancing and the US service sector: What do we learn? *Research in International Business and Finance*, *56*, 101361. Advance online publication. doi:10.1016/j.ribaf.2020.101361

Gunts, E. (2020). Design for Distancing aims to bring life back to public spaces left dormant in lockdown. *The Architect's Newspaper*. Retrieved from https://www.archpaper.com/2020/07/design-for-distancing-initiative-aims-to-bring-life-back-to-public-baltimore-spaces/

Gustafsson, J. (2017). *Single case studies vs multiple case studies: A comparative study*. Academy of Business, Engineering and Science, Halmstad University.

Guttentag, D. (2015). Airbnb: Disruptive innovation and the rise of an informal tourism accommodation sector. *Current Issues in Tourism*, *18*(12), 1192–1217. doi:10.1080/13683500.2013.827159

Haleem, A., Javaid, M., & Vaishya, R. (2002). Effects of COVID-19 pandemic in daily life. *Current Medicine Research and Practice*, *10*(2), 78–19. doi:10.1016/j.cmrp.2020.03.011 PMID:32292804

Hall, C. M., Scott, D., & Gössling, S. (2020). Pandemics, transformations and tourism: Be careful what you wish for. *Tourism Geographies*, *22*(3), 577–598. doi:10.1080/14616688.2020.1759131

Hall, C. M., Williams, A. M., & Lew, A. A. (2004). *Tourism: Conceptualizations, institutions, and issues. In A companion to tourism*. Blackwell Publishing.

Hall, M. (2010). Crisis events in tourism: Subjects of crisis in tourism. *Current Issues in Tourism*, *13*(5), 401–417. doi:10.1080/13683500.2010.491900

Han, D. I., Jung, T., & Gibson, A. (2013). Dublin AR: implementing augmented reality in tourism. In *Information and communication technologies in tourism 2014* (pp. 511–523). Springer. doi:10.1007/978-3-319-03973-2_37

Hannam, K., Sheller, M., & Urry, J. (2006). Mobilities, immobilities and moorings. *Mobilities*, *1*(1), 1–22. doi:10.1080/17450100500489189

Hao, F., Xiao, Q., & Chon, K. (2020). COVID-19 and China's Hotel Industry: Impacts, a Disaster Management Framework, and Post-Pandemic Agenda. *International Journal of Hospitality Management*, *90*, 102636. doi:10.1016/j.ijhm.2020.102636 PMID:32834356

Harrison, D. (1993). Bulgarian tourism. *Annals of Tourism Research*, *20*(3), 519–534. doi:10.1016/0160-7383(93)90007-P

Hassan, F. (2020). *Kashmir's tourism sector faces losses over Rs 9,000 cr; 1.4 lakh people lose jobs*. Retrieved on November 11, 2020 from https://www.forbesindia.com/article/special/kashmir039s-tourism-sector-faces-losses-over-rs-9000-cr-14-lakh-people-lose-jobs/57275/1

Haywood, K. (2020). A post-COVID future: Tourism community reimagined and enabled. *Tourism Geographies*, *22*(3), 599–609. doi:10.1080/14616688.2020.1762120

Henderson, J. C. (2013). The great east japan earthquake and tourism: A preliminary case study. *Tourism Recreation Research*, *38*(1), 93–98. doi:10.1080/02508281.2013.11081731

Hernàndez, C. A. (2021). The role of citizens in smart cities and urban infrastructures. In J. R. Vacca (Ed.), *Solving Urban Infrastructure Problems Using Smart City Technologies* (pp. 213–234). Elsevier. doi:10.1016/B978-0-12-816816-5.00010-3

Higgins-Desbiolles, F. (2020). Socialising tourism for social and ecological justice after COVID-19. *Tourism Geographies*, 1–14.

Hinojosa, V. (2020a, September 11). *¿En qué está trabajando la intermediación para asegurar su futuro?* https://www.hosteltur.com/139179_en-que-esta-trabajando-la-intermediacion-para-asegurar-su-futuro.html

Hinojosa, V. (2020b, September 15). *Flexibilidad y adaptación, la vía que marca Asia para recuperar el turismo.* https://www.hosteltur.com/139220_flexibilidad-y-adaptacion-mantra-para-la-recuperacion-del-turismo-en-asia.html?code=home-page%7B2020-09- 15%7D&utm_source=newsletter-es&utm_medium=email&utm_campaign=flexibilidad-y- adaptacion-la-via-que-marca-asia-para-recu

Hjalager, A. M., & Madsen, E. L. (2018). Business Model Innovation in Tourism: Opportunities and Challenges. The Sage Handbook of Tourism Management, 373.

Hjalager, A. M. (2010). A review of innovation research in tourism. *Tourism Management*, *31*(1), 1–12. doi:10.1016/j.tourman.2009.08.012

Hjalager, A. M., Kwiatkowski, G., & Østervig Larsen, M. (2017). Innovation gaps in Scandinavian rural tourism. *Scandinavian Journal of Hospitality and Tourism*, *33*(2), 1–17.

Hojeghan, S., & Esfangareh, N. (2011). Digital economy and tourism impacts, influences and challenges. *Procedia: Social and Behavioral Sciences*, *19*, 308–316. doi:10.1016/j.sbspro.2011.05.136

Hoorani, B. H., Nair, L. B., & Gibbert, M. (2019). Designing for impact: The effect of rigor and case study design on citations of qualitative case studies in management. *Scientometrics*, *121*(1), 285–306. doi:10.100711192-019-03178-w

Hosteltur. (2020). *La COVID puede abocar al cierre del 20% de las agencias de viaje*. Edición España. Available on https://www.hosteltur.com/140017_la-covid-puede-abocar-al-cierre-del-20-de-las-agencias-de-viaje.html

Hosteltur. (2020, September 22). *Turismo lanza un programa de digitalización para pymes de la mano de Google | Economía*. Hosteltur: Toda la información de turismo. https://www.hosteltur.com/139380_turismo-lanza-un-programa-de-digitalizacion-para- pymes-de-la-mano-de-google.html

Houge, S., & Raymond, E. (2020). A conceptual model of adventure tour guide well-being. *Annals of Tourism Research*, *84*, 1–4.

Hui, T. K., & Wan, D. (2006). Factors affecting consumers' choice of a travel agency: The case of Singapore. *Journal of Travel & Tourism Marketing*, *19*(4), 1–12. doi:10.1300/J073v19n04_01

Hu, Y., & Ritchie, J. B. (1993). Measuring destination attractiveness: A contextual approach. *Journal of Travel Research*, *32*(2), 25–34. doi:10.1177/004728759303200204

Hyett, N., Kenny, A., & Dickson-Swift, V. (2014). Methodology or method? A critical review of qualitative case study reports. *International Journal of Qualitative Studies on Health and Well-being*, *9*(1), 23606. doi:10.3402/qhw.v9.23606 PMID:24809980

Hystad, P. W., & Keller, P. C. (2008). Towards a destination tourism disaster management framework: Long-term lessons from a forest fire disaster. *Tourism Management*, *29*(1), 151–162. doi:10.1016/j.tourman.2007.02.017

Ibanescu, B.-C., Eva, M., & Gheorghiu, A. (2020). Questioning the Role of Tourism as an Engine for Resilience: The Role of Accessibility and Economic Performance. *Sustainability*, *12*(14), 5527. doi:10.3390u12145527

IGE. (2021). *Padrón municipal de habitantes*. Statistics Galician Institute. Retrieved from https://www.ige.eu/web/mostrar_actividade_estatistica.jsp?idioma=es&codigo=0201001002

Ilnitsky, A., & Chikurova, A. (2015). Diversification as a tool of anti-crisis strategy of development of rural territories in Ukraine. *Baltic Journal of Economic Studies*, *1*(2), 51–55. doi:10.30525/2256-0742/2015-1-2-51-55

INE. (2020). *Edad media por comunidad autónoma y provincia, sexo y tamaño de municipio*. Statistics National Institute. Retrieved from https://www.ine.es/jaxi/Datos.htm?path=/t20/e245/p04/provi/l0/&file=0tamu005.px

INE. (2020). *Resultados mensuales detallados*. https://www.ine.es/jaxiT3/Tabla.htm?t=13884

INE. (2021). *Empresas por CCAA, actividad principal (grupos CNAE 2009) y estrato de asalariados*. Statistics National Institute. Retrieved from https://www.ine.es/jaxiT3/Datos.htm?t=39372

International Labour Office. (2017). *Pautas de la OIT sobre trabajo decente y turismo socialmente responsable*. ILO.

Irvine, W., & Anderson, A. R. (2005). The Impacts of Foot and Mouth Disease on a Peripheral Tourism Area: The Role and Effect of Crisis Management. *Journal of Travel & Tourism Marketing*, *19*(2-3), 47–60. doi:10.1300/J073v19n02_05

Israeli, A., & Perl, Y. (2011). Crisis management in the travel agency sector: A case of study. *Journal of Vacation Marketing*, (17), 115–125.

Ivanov, S., & Webster, C. (2011). Tourism's contribution to economic growth: A global analysis for the first decade of the Millenium. SSRN *Electronic Journal*. doi:10.2139srn.1962510

Ivanov, S.H., Webster, C., Stoilova, E., & Slobodskoy, D. (2020). Biosecurity, crisis management, automation technologies and economic performance of travel, tourism and hospitality companies – A conceptual framework. *Tourism Economics*, 1-24.

Ivanov, S. H., Webster, C., Stoilova, E., & Slobodskoy, D. (2020). Biosecurity, crisis management, automation technologies and economic performance of travel, tourism and hospitality companies–A conceptual framework. *Tourism Economics*.

Jackiewicz, E. L. (2005). Tourism without threat? Excerpts from rural Costa Rica. *Annals of Tourism Research*, *32*(1), 266–268. doi:10.1016/j.annals.2004.05.004

Jalilvand, M. R., Ebrahimi, A., & Samiei, N. (2013). Electronic word of mouth effects on tourists' attitudes toward Islamic destinations and travel intention: An empirical study in Iran. *Procedia: Social and Behavioral Sciences*, *81*, 484–489. doi:10.1016/j.sbspro.2013.06.465

Jensen, & Prebensen, N. (2015). Innovation and value creation in experience-based tourism. *Scandinavian Journal of Hospitality and Tourism*, *15*(1), 1-8,

Jha, S. (2020). Impact of Home Country Terrorism on Medical Tourism in India January 2020. *International Journal of Asian Social Science*, *10*(8), 407–416. doi:10.18488/journal.1.2020.108.407.416

Johnson, M., & Suskewicz, J. (2020). *Lead from the future*. HBR Press.

Jonas, A., Mansfeld, Y., Paz, S., & Potasman, I. (2011). Determinants of health risk perception among low-risk-taking tourists traveling to developing countries. *Journal of Travel Research*, *50*(1), 87–99. doi:10.1177/0047287509355323

Jones, P., & Comfort, D. (2020). The COVID-19 crisis and sustainability in the hospitality industry. *International Journal of Contemporary Hospitality Management*, *32*(10), 3037–3050. doi:10.1108/IJCHM-04-2020-0357

Jones, P., & Comfort, D. (2020). The COVID-19 Crisis, Tourism and Sustainable Development. *Athens Journal of Tourism*, *7*(2), 75–86. doi:10.30958/ajt.7-2-1

Jung, J. G. (2019). Megacities, utopian cities and planning our cities for the future. In S. Albert (Ed.), *Innovative Solutions for Creating Sustainable Cities* (pp. 269–289). Cambridge Scholars Publishing.

Jun, S.-H. (2020). The Effects of Perceived Risk, Brand Credibility and Past Experience on Purchase Intention in the Airbnb Context. *Sustainability*, *12*(12), 5212. doi:10.3390u12125212

Jurado-Almonte, J. M., & Pazos-García, F. J. (2016). Population and rural tourism in low demographic density territories in Spain. *Boletín de la Asociación de Geógrafos Españoles*, *71*, 511–514.

Kansakar, P., Munir, A., & Shabani, N. (2017). Technology in hospitality industry: Prospects and challenges. *IEEE Consum Electron Mag*, *8*(3), 60–65. doi:10.1109/MCE.2019.2892245

Kapur, S. P. (2006). The India-Pakistan Conflict: An Enduring Rivalry. *Canadian Journal of Political Science*, *39*(4), 966–967. doi:10.1017/S0008423906339960

Karabulut, G., Bilgin, M. H., Demir, E., & Doker, A. C. (2020). How pandemics affect tourism: International evidence. *Annals of Tourism Research*, *84*, 102991. doi:10.1016/j.annals.2020.102991 PMID:32834226

Kellens, W., Neutens, T., Deckers, P., Reyns, J., & De Maeyer, P. (2011). Coastal flood risks and seasonal tourism: Analysing the effects of tourism dynamics on casualty calculations. *Natural Hazards*, *60*(3), 1211–1229. doi:10.100711069-011-9905-6

Khan, M. J., Chelliah, S., & Ahmed, S. (2017). Factors influencing destination image and visit intention among young women travellers: Role of travel motivation, perceived risks, and travel constraints. *Asia Pacific Journal of Tourism Research*, *22*(11), 1139–1155. doi:10.1080/10941665.2017.1374985

Kim, H. M. (2021). Smart cities beyond COVID-19. In H. M. Kim, S. Sabri, & A. Kent (Eds.), *Smart Cities for Technological and Social Innovation* (pp. 299–308). Academic Press., doi:10.1016/B978-0-12-818886-6.00016-2

Kim, H., Kim, J., & Gu, Z. (2012). An examination of US hotel companies' risk features and their determinants of systematic risk. *International Journal of Tourism Research*, *14*(1), 28–39. doi:10.1002/jtr.828

Kim, J., Kim, J., & Wang, Y. (2021). Uncertainty risks and strategic reaction of restaurant firms amid COVID-19: Evidence from China. *International Journal of Hospitality Management*, *92*, 102752. Advance online publication. doi:10.1016/j.ijhm.2020.102752 PMID:33223596

Kim, J., & Lee, J. C. (2020). Effects of COVID-19 on preferences for private dining facilities in restaurants. *Journal of Hospitality and Tourism Management*, *45*, 67–70. doi:10.1016/j.jhtm.2020.07.008

Kim, K., Bonn, M. A., & Cho, M. (2021). Clean safety message framing as survival strategies for small independent restaurants during the COVID-19 pandemic. *Journal of Hospitality and Tourism Management*, *46*, 423–431. doi:10.1016/j.jhtm.2021.01.016

Kim, W. G., & Lee, H. Y. (2005). Comparison of web service quality between online travel agencies and online travel suppliers. *Journal of Travel & Tourism Marketing*, *17*(2-3), 105–116. doi:10.1300/J073v17n02_09

Klempert, R. (2020, September 29). *Agencias y pospandemia: Así será el nuevo modelo de distribución*. https://www.hosteltur.com/lat/139521_agencias-y-pospandemia-asi-sera-el-nuevo-modelo-de-distribucion.html

Komninos, N. (2006). The architecture of intelligent cities. *Intelligent Environments*, *6*, 53–61.

Komppula, R. (2014). The role of individual entrepreneurs in the development of competitiveness for a rural tourism destination – a case study. *Tourism Management*, *40*, 361–371. doi:10.1016/j.tourman.2013.07.007

Korstanje, M. E. (2011). Rent-a-car industry: A case study in Argentina. *Tourismos*, *6*(1), 1–15.

Korstanje, M. E. (2018). *The mobilities paradox: a critical analysis*. Edward Elgar Publishing. doi:10.4337/9781788113311

Korstanje, M. E., & Tarlow, P. (2012). Being lost: Tourism, risk and vulnerability in the post-'9/11'entertainment industry. *Journal of Tourism and Cultural Change*, *10*(1), 22–33. doi:10.1080/14766825.2011.639455

Kounavis, C. D., Kasimati, A. E., & Zamani, E. D. (2012). Enhancing the tourism experience through mobile augmented reality: Challenges and prospects. *International Journal of Engineering Business Management*, *4*, 10. doi:10.5772/51644

Kourtesopoulou, A., Theodorou, S. D., Kriemadis, A., & Papaioannou, A. (2019). The impact of online travel agencies web service quality on customer satisfaction and purchase intentions. In *Smart Tourism as a Driver for Culture and Sustainability* (pp. 343–356). Springer. doi:10.1007/978-3-030-03910-3_24

KPMG. (2019). *La gastronomía en la economía española. Impacto económico de los sectores asociados*. KPMG. Retrieved from https://assets.kpmg/content/dam/kpmg/es/pdf/2019/01/gastronomia-en-economia-espanola.pdf

Krajcik, V., Kljucnikov, A., & Rihova, E. (2019). Innovative Sharing Economy's Business Models in Tourism: Case of Airbnb in Prague. *Marketing and Management of Innovations*, *2*, 108–117. doi:10.21272/mmi.2019.2-10

Krakover, S. (2005). Estimating the effect of atrocious events on the flow of tourists to Israel. *Horror and human tragedy revisited: The management of sites of atrocities for tourism*, 183-194.

Krasteva, R., & Pantelis, E. (2017). Bridging relationships between education, refugee crisis and tourism. *EDULEARN17 Proceedings*. 10.21125/edulearn.2017.1063

Kraus, S., Clauss, T., Breier, M., Gast, J., Zardini, A., & Tiberius, V. (2020). The economics of COVID-19: Initial empirical evidence on how family firms in five European countries cope with the corona crisis. *International Journal of Entrepreneurial Behaviour & Research*, *26*(5), 1067–1092. doi:10.1108/IJEBR-04-2020-0214

Kristianto, D. A., Syamsu, M. N., & Wisnumurti, A. (2018). Online Travel Agents Use for Sales and Promotion of Hotel in Kaliurang Yogyakarta. *Kepariwisataan: Jurnal Ilmiah*, *12*(01), 15–28.

Krivý, M. (2018). Towards a critique of cybernetic urbanism: The smart city and the society of control. *Planning Theory*, *17*(1), 8–30. doi:10.1177/1473095216645631

Kumar, S., & Shekhar. (2020). Digitalization: A Strategic Approach for Development of Tourism Industry in India. *Paradigm*, *24*(1), 93–108. doi:10.1177/0971890720914111

Kumar, S., & Shekhar. (2020). Technology and innovation: Changing concept of rural tourism - A systematic review. *Open Geosciences*, *12*(1), 737–752. doi:10.1515/geo-2020-0183

Kunzmann, K. R. (2020). Smart Cities After Covid-19: Ten Narratives. *The Planning Review, 56*(2), 20–31. doi:10.1080/02513625.2020.1794120

Kyurova, V., & Koyundzhiyska-Davidkova, B. (2017). Current issues in family business management. *European Journal of Service Management, 22*, 47–52. doi:10.18276/ejsm.2017.22-06

Lage, E., & Alturas, B. (2013). Information sharing among innovative SME: An exploratory study within the Portuguese SME innovation network. *Innovation through Knowledge Transfer*, 159-168. doi:10.1007/978-3-642-34219-6_18

Lai, I. K. W. (2014). The role of service quality, perceived value, and relationship quality in enhancing customer loyalty in the travel agency sector. *Journal of Travel & Tourism Marketing, 31*(3), 417–442. doi:10.1080/10548408.2014.883346

Langford, G., & Weissenberg, A. (2018). *2018 Travel and Hospitality Industry Outlook*. Deloitte Center for Industry Insights. Retrieved from www.deloitte.com/us/travel-hospitality-trends

Lanzolla, G., & Markides, C. (2021). A Business Model View of Strategy. *Journal of Management Studies, 58*(2), 540–553. doi:10.1111/joms.12580

Larsen, S., Doran, R., & Wolff, K. (2017). How psychology can stimulate tourist experience studies. In N. Scott & J. Gao (Eds.), *Visitor experience design* (pp. 13–29). CABI. doi:10.1079/9781786391896.0013

Laukkanen, M., & Patala, S. (2014). Analysing barriers to sustainable business model innovations: Innovation systems approach. *International Journal of Innovation Management, 18*(6), 1440010. doi:10.1142/S1363919614400106

Law, R., Law, A., & Wai, E. (2002). The impact of the internet on travel agencies in Hong Kong. *Journal of Travel & Tourism Marketing, 11*(2-3), 105–126. doi:10.1300/J073v11n02_06

Laws, E., Prideaux, B., & Chon, K. S. (Eds.). (2007). *Crisis management in tourism*. Cabi. doi:10.1079/9781845930479.0000

Lee, S. J. (2017). A review of audio guides in the era of smart tourism. *Information Systems Frontiers, 19*(4), 705–715. doi:10.100710796-016-9666-6

Leitch, S., & Palmer, I. (2010). Analyzing texts in context: Current practices and new protocols for critical discourse analysis in organization studies. *Journal of Management Studies, 47*(6), 1194–1212. doi:10.1111/j.1467-6486.2009.00884.x

Leone, L., Fleischhacker, S., Anderson-Steeves, B., Harper, K., Winkler, M., Racine, E., Baquero, B., & Gittelsohn, J. (2020). Healthy Food Retail during the COVID-19 Pandemic: Challenges and Future Directions. *International Journal of Environmental Research and Public Health, 17*(20), 1–14. doi:10.3390/ijerph17207397 PMID:33050600

Lepp, A., & Gibson, H. (2008). Sensation seeking and tourism: Tourist role, perception of risk and destination choice. *Tourism Management, 29*(4), 740–750. doi:10.1016/j.tourman.2007.08.002

Leslie, D., & Black, L. (2005). Tourism and the Impact of the Foot and Mouth Epidemic in the UK. *Journal of Travel & Tourism Marketing*, *19*(2-3), 35–46. doi:10.1300/J073v19n02_04

Lew, A. (2020). *How to create a better post-COVID-19 World.* https://medium.com/new-earth-consciousness/creating-a-better-post-covid-19-world-36b2b3e8a7ae

Li, G. (2006). Forecasting tourism demand using econometric models. *Tourism Management Dynamics*, 219-228. doi:10.1016/B978-0-7506-6378-6.50033-0

Liedtka, J. (2021, April 3). *Why design thinking works.* Harvard Business Review. https://hbr.org/2018/09/why-design-thinking-works

Li, J., Feng, Y., Li, G., & Sun, X. (2020). Tourism companies' risk exposures on text disclosure. *Annals of Tourism Research*, *84*, 1–15. doi:10.1016/j.annals.2020.102986 PMID:32834224

Li, J., Nguyen, T., & Coca-Stefaniak, J. A. (2021, January). Coronavirus impacts on post-pandemic planned travel behaviours. *Annals of Tourism Research*, *86*, 102964. doi:10.1016/j.annals.2020.102964 PMID:32836567

Lisle, D. (2016). *Holidays in the danger zone: Entanglements of war and tourism.* University of Minnesota Press. doi:10.5749/minnesota/9780816698554.001.0001

Liugailaitė-Radzvickienė, L., & Jucevičius, R. (2014). Going to be an intelligent city. *Procedia: Social and Behavioral Sciences*, *156*, 116–120. doi:10.1016/j.sbspro.2014.11.131

Livina, A., & Reddy, M. (2017). Nature Park as a Resource for Nature Based Tourism. In *Proceedings of the 11th International Scientific and Practical Conference* (Vol. 179, p. 183). 10.17770/etr2017vol1.2590

Llorca-Vivero, R. (2008). Terrorism and international tourism: New evidence. *Defence and Peace Economics*, *19*(2), 169–188. doi:10.1080/10242690701453917

López, P. (2020, June 30). *La prohibición de la llegada de cruceros pone en jaque a las agencias de viajes.* https://tur43.es/transportes/la-prohibicion-de-la-llegada-de-cruceros-pone-en-jaque-a-las-agencias-de-viajes.html

López-Medel Báscones, J. (2020). *La cláusula rebus sic stantibus. Modificación o resolución de contratos ante la crisis del coronavirus.* Retrieved from: https://www.abogacia.es/actualidad/opinion-y-analisis/la-clausula-rebus-sic-stantibus-modificacion-o-resolucion-de-contratos-ante-la-crisis-del-coronavirus/

Loureiro, S. M. C., Guerreiro, J., & Ali, F. (2020). 20 years of research on virtual reality and augmented reality in tourism context: A text-mining approach. *Tourism Management*, *77*, 104028. doi:10.1016/j.tourman.2019.104028

Loureiro, S. M. C., & Jesus, S. (2019). How perceived risk and animosity towards a destination may influence destination image and intention to revisit: The case of Rio de Janeiro. *Anatolia*, *30*(4), 497–512. doi:10.1080/13032917.2019.1632910

Lyonnet, J. (2020, September 9). *Las tendencias que observa Amadeus para el reinicio del turismo.* https://www.hosteltur.com/lat/139122_las-tendencias-que-observa-amadeus-para-el-reinicio-del-turismo.html

Lyon, P. (2008). *Conflict between India and Pakistan: an encyclopedia.* Abc-Clio.

Madeira, A., Palrão, T., & Mendes, A. S. (2020). The Impact of Pandemic Crisis on the Restaurant Business. *Sustainability, 13*(1), 40. . doi:10.3390u13010040

Mair, J., Ritchie, B. W., & Walters, G. (2016). Towards a research agenda for post-disaster and post-crisis recovery strategies for tourist destinations: A narrative review. *Current Issues in Tourism, 19*(1), 1–26. doi:10.1080/13683500.2014.932758

Major, J. A. (2002). Advanced techniques for modeling terrorism risk. *The Journal of Risk Finance, 4*(1), 15–24. doi:10.1108/eb022950

Malbec, T. (2020). *Restaurant Technology in the Post-COVID-19 World.* Hospitality Technology. Retrieved from https://hospitalitytech.com/restaurant-technology-post-Covid-19-world

Malbec, T. (2020a). *Restaurant Business Model Changes in a Post COVID-19 World.* Hospitality Technology. Retrieved from https://hospitalitytech.com/restaurant-business-model-changes-post-Covid-19-world

Malecki, K., Keating, J., & Safdar, N. (2020). Crisis Communication and Public Perception of COVID-19 Risk in the Era of Social Media. *Clinical Infectious Diseases*, ciaa758.

Mansfeld, Y. (1999). Cycles of war, terror, and peace: Determinants and management of crisis and recovery of the Israeli tourism industry. *Journal of Travel Research, 38*(1), 30–36. doi:10.1177/004728759903800107

Mansfeld, Y., & Pizam, A. (2006). *Tourism, terrorism, and civil unrest issues. Tourism, Security and Safety.* Butterworth-Heinemann.

Mao, C. K., Ding, C. G., & Lee, H. Y. (2010). Post-SARS tourist arrival recovery patterns: An analysis based on a catastrophe theory. *Tourism Management, 31*(6), 855–861. doi:10.1016/j.tourman.2009.09.003 PMID:32287733

Martens, H. M., Feldesz, K., & Merten, P. (2016). Crisis Management in Tourism – A Literature Based Approach on the Proactive Prediction of a Crisis and the Implementation of Prevention Measures. *Athens Journal of Tourism, 3*(2), 89–101. doi:10.30958/ajt.3-2-1

Mason, P., Augustyn, M., & Seakhoa-King, A. (2009). Exploratory study in tourism: Designing an initial, qualitative phase of sequenced, mixed methods research. *International Journal of Tourism Research, 12*(5), 432–448. doi:10.1002/jtr.763

Mason, P., Grabowski, P., & Du, W. (2005). Severe acute respiratory syndrome, tourism, and the media. *International Journal of Tourism Research, 7*(1), 11–21. doi:10.1002/jtr.519

Masri, N. W., You, J.-J., Ruangkanjanases, A., Chen, S.-C., & Pan, C.-I. (2020). Assessing the Effects of Information System Quality and Relationship Quality on Continuance Intention in E-Tourism. *International Journal of Environmental Research and Public Health*, *17*(1), 174. doi:10.3390/ijerph17010174 PMID:31881762

Massa, L., Tucci, C. L., & Afuah, A. (2017). A Critical Assessment of Business Model Research. *The Academy of Management Annals*, *11*(1), 73–104. doi:10.5465/annals.2014.0072

Matović, I. M., & Đurković, J. V. (2020). Contemporary trends in human resource management in travel agencies. *Tourism International Scientific Conference Vrnjačka Banja-TISC*, *5*(1), 248–263.

Matyas, C., Srinivasan, S., Cahyanto, I., Thapa, B., Pennington-Gray, L., & Villegas, J. (2011). Risk perception and evacuation decisions of Florida tourists under hurricane threats: A stated preference analysis. *Natural Hazards*, *59*(2), 871–890. doi:10.100711069-011-9801-0

Maximiliano, E. K. (2018). Terrorism, tourism and hospitality. *The Challenges of Democracy in the War on Terror*, 105-122. doi:10.4324/9780429466182-7

Mazilu, M., Marinescu, R., Bălă, D., & Dragomir, L. (2019). Crisis Management in Tourism. *Quaestus Multidisciplinary Research Journal*, *14*(April), 289–298.

Mazzanti, M. (2002). Tourism growth and sustainable economic development: A note on economic issues. *Tourism Economics*, *8*(4), 457–462. doi:10.5367/000000002101298223

Mazzocchi, M., & Montini, A. (2001). Earthquake effects on tourism in central Italy. *Annals of Tourism Research*, *28*(4), 1031–1046. doi:10.1016/S0160-7383(01)00008-1

McCartney, M. (2015). Economic growth and economic structure since 1750. *Economic Growth and Development*, 166-182. doi:10.1007/978-1-137-29031-1_9

McKercher, B., & Chon, K. (2004). The over-reaction to SARS and the collapse of Asian tourism. *Annals of Tourism Research*, *31*(3), 716–719. doi:10.1016/j.annals.2003.11.002 PMID:32572281

McKercher, B., & Hui, E. L. (2004). Terrorism, economic uncertainty and outbound travel from Hong Kong. *Journal of Travel & Tourism Marketing*, *15*(2-3), 99–115. doi:10.1300/J073v15n02_06

Mehraliyev, F., Chan, I. C. C., Choi, Y., Koseoglu, M. A., & Law, R. (2020). A state-of-the-art review of smart tourism research. *Journal of Travel & Tourism Marketing*, *37*(1), 78–91. doi:10.1080/10548408.2020.1712309

Meinung, A. (1989). Determinants of the attractiveness of a tourism region. *Determinants of the attractiveness of a tourism region*, 99-101.

Mendoza, J. M. F., Gallego-Schmid, A., & Azapagic, A. (2019). A methodological framework for the implementation of circular economy thinking in higher education institutions: Towards sustainable campus management. *Journal of Cleaner Production*, *226*, 831–844. doi:10.1016/j.jclepro.2019.04.060

Menor, D. (2021, abril 19). *Viaje a Serbia con vacuna incluida por 500 euros*. La Voz de Galicia. https://www.lavozdegalicia.es/noticia/sociedad/2021/04/19/viaje-serbia-vacuna-incluida-500-euros/0003161882692262983851 2.htm

Merriam, S. B. (2009). *Qualitative research: A guide to design and implementation*. Jossey-Bass.

Miao, L., Im, J., Fu, X., Kim, H., & Zhang, Y. E. (2021). Proximal and distal post-COVID travel behavior. *Annals of Tourism Research*, *88*, 103159. doi:10.1016/j.annals.2021.103159

Mihalic, T., Zabkar, V., & Knezevic Cvelbar, L. (2012). A hotel sustainability business model: Evidence from Slovenia. *Journal of Sustainable Tourism*, *20*(5), 701–709. doi:10.1080/09669582.2011.632092

Mihardjo, L. W., & Alamsjah, F. (2019). Mediating Role of Co-creation Strategy on the Relationship between Business Model Innovation and Corporate Reputation: A Case Study on Indonesian Telecommunication Firms. *Journal of Technical Education and Training*, *11*(4), 67–76.

Mikusová, M. (2013). Do Small Organizations Have an Effort to Survive? Survey from Small Czech Organizations. *Economic Research-Ekonomska Istraživanja*, *26*(4), 59–76. doi:10.1080/1331677X.2013.11517630

Miles, L., & Shipway, R. (2020). Exploring the COVID-19 pandemic as a catalyst for stimulating future research agendas for managing crises and disasters at international sport events. *Event Management*, *24*(4), 537–552. doi:10.3727/152599519X15506259856688

Millán, S. (2020). La cancelación del Mobile por el coronavirus provoca fuertes pérdidas económicas. *Cinco Días*. Retrieved from https://cincodias.elpais.com/cincodias/2020/02/12/companias/1581541189_908014.html

Ministry of Industry, Trade and Tourism (2020b). *Guides for reducing spread of coronavirus SARS-COV-2 in the tourism sector.* https://www.mincotur.gob.es/en-us/COVID-19/Paginas/Guias-sector-turistico.aspx

Ministry of Industry, Trade and Tourism. (2020a). *Good practice guide for commercial sector establishments*. https://www.mincotur.gob.es/en-us/COVID-19/Paginas/COVID-19.aspx

Ministry of Labour and Social Economy. (2021). *Resumen datos de empleo*. Retrieved from https://www.mites.gob.es/ficheros/ministerio/estadisticas/documentos/RUD.pdf

Mohanty, S. (2018). The New Wave of Mobilisation in Kashmir: Religious or Political? *Social Change*, *48*(1), 59–71. doi:10.1177/0049085717743838

Moneo, M. (2018, July 7). *Las mayoristas sacan partido del intrusismo en las agencias*. https://www.preferente.com/noticias-de-agencias-de-viajes/las-mayoristas-sacan-partido-del-intrusismo-en-las-agencias-277847.html

Monterrubio, J. C. (2010). Short-term economic impacts of influenza A (H1N1) and government reaction on the Mexican tourism industry: An analysis of the media. *International Journal of Tourism Policy*, *3*(1), 1–15. doi:10.1504/IJTP.2010.031599

Morachat, C. (2003). *A Study of Destination Attractiveness through Tourists' Perspectives: A Focus On Chiang Mai, Thailand* (Unpublished doctoral thesis). Edith Cowan University.

Morgan, T., Anokhin, S., Ofstein, L. F., & Friske, W. (2020). SME response to major exogenous shocks: The bright and dark sides of business model pivoting. *International Small Business Journal: Researching Entrepreneurship*, *38*(5), 369–379. doi:10.1177/0266242620936590

Morris, M., Schindehutte, M., & Allen, J. (2005). The entrepreneur's business model: Toward a unified perspective. *Journal of Business Research*, *58*(6), 726–735. doi:10.1016/j.jbusres.2003.11.001

Mpiti, K. (2014). The potential use of ICT to enhance agritourism in Lesotho. In Technology and innovation: Changing concept of rural tourism (pp. 129-136). 751 Editors.

Musa, S. M., & Mamun, S. A. (2020). Business Model Innovation as Strategic Renewal Process: A Case of China's Most Innovative Firm. *Global Business and Management Research*, *10*(2), 1–18.

Naciones Unidas. (2020). *Informe de políticas: La Covid-19 y la transformación del turismo.* Agosto 2020.

Narain, A. (2016). Revival of violence in Kashmir: The threat to India's security. *Counter Terrorist Trends and Analyses*, *8*(7), 15–20.

Neher, P. A. (1971). *Economic growth development: A mathematical introduction.* J. Wiley.

Nespereira, A. (2021). La hostelería de la provincia es la más tocada de Galicia. La Región, 6.

Neumayer, E. (2004). The impact of political violence on tourism: Dynamic cross-national estimation. *The Journal of Conflict Resolution*, *48*(2), 259–281. doi:10.1177/0022002703262358

News, A. B. C. (2020). *Coronavirus update: Morrison brings in travel restrictions, Australian cases pass 250, Donald trump's COVID-19 test is clear.* Available at: www.abc.net.au/news/2020-03-15/coronavirus-update-latest-news-us-travel-ban-extended-trump-test/12057094

Nexotur. (2020a, September 16). *Las agencias piden una campaña para potenciar las ventas en España*. Nexotur. https://www.nexotur.com/noticia/109183/nexotur/las-agencias-piden- una-campana-para-potenciar-las-ventas-en-espana.html

Nexotur. (2020b, September 14). *Coronavirus: Las agencias, en estado crítico*. Nexotur. https://www.nexotur.com/noticia/109136/nexotur/coronavirus:-las-agencias-en-estado-critico.html

Nexotur. (2020c, September 24). *El papel de los agentes de viajes será mucho más importante, según Amadeus*. Nexotur. https://www.nexotur.com/noticia/109296/nexotur/el- papel-de-los-agentes-de-viajes-sera-mucho-mas-importante-segun-amadeus.html

Novelli, M., Burgess, L. G., Jones, A., & Ritchie, B. W. (2018). 'No Ebola… still doomed'– The Ebola-induced tourism crisis. *Annals of Tourism Research*, *70*, 76–87. doi:10.1016/j.annals.2018.03.006 PMID:32572288

Nunn, J. (2020, April 14). Restaurants will never be the same after coronavirus – but that may be a good thing. *The Guardian*. Retrieved from https://www.theguardian.com/commentisfree/2020/apr/14/coronavirus-restaurants-pandemic-workers-communities-prices

Observatori de Turisme. (2019). *Barcelona, 2019. Tourism activity report.* https://www.observatoriturisme.barcelona/sites/default/files/IAT19.pdf

Observatori de Turisme. (2020). *Perspectives d'activitat turística a la Destinació Barcelona.* https://www.observatoriturisme.barcelona/sites/default/files/201218_Perspectives%20d%27activitat.pdf

OECD. (2020). *Rebuilding tourism for the future: COVID-19 policy response and recovery.* https://read.oecd-ilibrary.org/view/?ref=137_137392-qsvjt75vnh&title=Rebuilding-tourism-for-the-future-COVID-19-policy-response-and-recovery

OECD. (2020a). *Tourism Policy Responses to the coronavirus (COVID-19).* https://www.oecd.org/coronavirus/policy-responses/tourism-policy-responses-to-the-coronavirus-covid-19-6466aa20/

OECD. (2020b). *Productivity gains from teleworking in the post COVID-19 era: How can public policies make it happen?* https://www.oecd.org/coronavirus/policy-responses/productivity-gains-from-teleworking-in-the-post-covid-19-era-a5d52e99/#section-d1e675

Okumus, F., Altinay, M., & Arasli, H. (2005). The impact of Turkey's economic crisis of February 2001 on the tourism industry in Northern Cyprus. *Tourism Management*, *26*(1), 95–104. doi:10.1016/j.tourman.2003.08.013

Organisation for Economic Co-operation and Development. (2017). *Tourism Policy Review of Mexico*. Organisation for Economic Co-operation and Development. https://read.oecd-ilibrary.org/industry-and-services/tourism-policy-review-of-mexico_9789264266575-en#page2

Osiyevskyy, O., & Dewald, J. R. (2018). The pressure cooker: When crisis stimulates explorative business model change intentions. *Long Range Planning*, *51*(4), 540–560. doi:10.1016/j.lrp.2017.09.003

Pablo-Romero, M. D. P., & Molina, J. A. (2013). Tourism and economic growth: A review of empirical literature. *Tourism Management Perspectives*, *8*, 28–41. doi:10.1016/j.tmp.2013.05.006

Packer, J. (2008). *Mobility without mayhem: Safety, cars, and citizenship*. Duke University Press. doi:10.2307/j.ctv123x6xj

Pain, R. (2014). Everyday terrorism: Connecting domestic violence and global terrorism. *Progress in Human Geography*, *38*(4), 531–550. doi:10.1177/0309132513512231

Paniccia, P., & Leoni, L. (2017). Co-evolution in tourism: The case of Albergo Diffuso. *Current Issues in Tourism*, 1–28.

Panzone, L. A., Larcom, S., & She, P. (2021). Estimating the impact of the first COVID-19 lockdown on UK food retailers and the restaurant sector. *Global Food Security*, *28*, 100495. Advance online publication. doi:10.1016/j.gfs.2021.100495

Papatheodorou, A., & Pappas, N. (2017). Greek tourism, economic crisis and political turmoil, between Scylla and Charybdis. *Tourism (Zagreb)*. Advance online publication. doi:10.23912/9781910158814-3433

Paraskevas, A., & Arendell, B. (2007). A strategic framework for terrorism prevention and mitigation in tourism destinations. *Tourism Management*, *28*(6), 1560–1573. doi:10.1016/j.tourman.2007.02.012

Pardo, C., & Ladeiras, A. (2020). Covid-19 "tourism in flight mode": A lost opportunity to rethink tourism – towards a more sustainable and inclusive society. *Worldwide Hospitality and Tourism Themes*, *12*(6), 671–678. doi:10.1108/WHATT-07-2020-0064

Parida, Y., Bhardwaj, P., & Chowdhury, J. R. (2015). Impact of terrorism on tourism in India. *Economic Bulletin*, *35*(4), 2543–2557.

Park, K., & Reisinger, Y. (2010). Differences in the perceived influence of natural disasters and travel risk on international travel. *Tourism Geographies*, *12*(1), 1–24. doi:10.1080/14616680903493621

Park, Y. A., Gretzel, U., & Sirakaya-Turk, E. (2007). Measuring web site quality for online travel agencies. *Journal of Travel & Tourism Marketing*, *23*(1), 15–30. doi:10.1300/J073v23n01_02

Patil, R., Gada, N., & Gala, K. (2019, December). Twitter Data Visualization and Sentiment Analysis of Article 370. In *2019 International Conference on Advances in Computing, Communication and Control (ICAC3)* (pp. 1-4). IEEE. 10.1109/ICAC347590.2019.9036800

Pencarelli, T. (2020). The digital revolution in the travel and tourism industry. *Information Technology & Tourism*, *22*(3), 455–476. doi:10.100740558-019-00160-3

Peric, M., & Wise, N. (2015). Understanding the delivery of experience: Conceptualising business models and sports tourism, assessing two case studies in Istria, Croatia. *Local Economy*, *30*(8), 1000–1016. doi:10.1177/0269094215604131

Permuy, I. (2021, abril 9). *Turismo de vacunas: A qué países se puede viajar para obtener antes la inoculación*. abc. https://www.abc.es/sociedad/abci-turismo-vacuna-paises-nsv-202104091208_noticia.html

Perpiña, L., Camprubí, R., & Prats, L. (2019). Destination image versus risk perception. *Journal of Hospitality & Tourism Research (Washington, D.C.)*, *43*(1), 3–19. doi:10.1177/1096348017704497

Pett, T. L., & Wolff, J. A. (2009). SME opportunity for growth or profit: What is the role of product and process improvement? *International Journal of Entrepreneurial Venturing*, *1*(1), 5. doi:10.1504/IJEV.2009.023817

Phillipson, J., Bennett, K., Lowe, P., & Raley, M. (2003). Adaptive responses and asset strategies: The experience of rural micro-firms and Foot and Mouth Disease. *Journal of Rural Studies*, *20*(2), 227–243. doi:10.1016/j.jrurstud.2003.08.006

Phillipson, J., Gorton, M., Turner, R., Shucksmith, M., Aitken-McDermott, K., Areal, F., Cowie, P., Hubbard, C., Maioli, S., McAreavey, R., Souza-Monteiro, D., Newbery, R., Panzone, L., Rowe, F., & Shortall, S. (2020). The COVID-19 Pandemic and Its Implications for Rural Economies. *Sustainability*, *12*(10), 1–9. doi:10.3390u12103973

Pikkemaat, B., & Peters, M. (2006). Towards the measurement of innovation—A pilot study in the small and medium sized hotel industry. *Journal of Quality Assurance in Hospitality & Tourism*, *6*(3-4), 89–112. doi:10.1300/J162v06n03_06

Pizam, A. & Mansfeld, Y. (2006). Toward a theory of tourism security. *Tourism, Security and Safety*, 1-27. doi:10.1016/B978-0-7506-7898-8.50004-7

Pizam, A., & Mansfeld, Y. (1996). *Tourism, crime, and international security issues*. John Wiley & Son Ltd.

Pizam, A., & Fleischer, A. (2002). Severity versus frequency of acts of terrorism: Which has a larger impact on tourism demand? *Journal of Travel Research*, *40*(3), 337–339. doi:10.1177/004728750204000301l

Pizam, A., & Smith, G. (2000). Tourism and terrorism: A quantitative analysis of major terrorist acts and their impact on tourism destinations. *Tourism Economics*, *6*(2), 123–138. doi:10.5367/000000000101297523

Platje, J., Harvey, J., & Rayman-Bacchus, L. (2020). COVID-19 – reflections on the surprise of both an expected and unexpected event. *The Central European Review of Economics and Management*, *4*(1), 149–162. doi:10.29015/cerem.874

Porte, M., Saur-Amaral, I., & Pinho, C. (2018). Research in auditing: Main themes. *Revista Contabilidade & Finanças*, *29*(76), 41–59. doi:10.1590/1808-057x201804410

Pratt, S. (2014). Potential economic contribution of regional tourism development in China: A comparative analysis. *International Journal of Tourism Research*, *17*(3), 303–312. doi:10.1002/jtr.1990

Prayag, G. (2020). Time for Reset? COVID-19 and Tourism Resilience. *Tourism Review International*, *24*(2-3), 179–184. doi:10.3727/154427220X15926147793595

Prescott, J. E., & Filatotchev, I. (2021). The Business Model Phenomenon: Towards Theoretical Relevance. *Journal of Management Studies*, *58*(2), 517–527. doi:10.1111/joms.12610

Press, E. (2009, February 2). *Economía/Empresas - Las agencias de viajes presentan alegaciones al anteproyecto de ley sobre la directiva Bolkestein*. Europa Press. https://www.europapress.es/economia/noticia-economia-empresas-agencias-viajes-presentan- alegaciones-anteproyecto-ley-directiva-bolkestein-20090202151820.html

Punzón, C. (2021, abril 17). *Agencias de Galicia y el resto de España venden viajes a Moscú para vacunarse con Sputnik V.* La Voz de Galicia. https://www.lavozdegalicia.es/noticia/sociedad/2021/04/17/agencias-galicia-resto-espana-venden-viajes-moscu-vacunarse-sputnik-v/0003_202104G17P3991.htm

Puri, B. (1993). *Kashmir towards insurgency* (No. 4). UN.

Puschita, E., Constantinescu, A., Dobrab, R., Colda, I., Vermesana, A., & Tudor, M. (2014). Challenges for a broadband service strategy in rural areas: A Romanian case study. *Telecommunications Policy*, *38*(2), 147–156. doi:10.1016/j.telpol.2013.08.001

Qiu, R., Park, J., Li, Sh., & Song, H. (2020). Social costs of tourism during the COVID-19 pandemic. *Annals of Tourism Research*, *84*, 1–14. doi:10.1016/j.annals.2020.102994 PMID:32834228

Quarantelli, E. L. (1988). Disaster Crisis Management: A Summary of Research Findings. *Journal of Management Studies*, *25*(4), 373–385. doi:10.1111/j.1467-6486.1988.tb00043.x

Quintana, T. A., Gil, S. M., & Peral, P. P. (2016). How could traditional travel agencies improve their competitiveness and survive? A qualitative study in Spain. *Tourism Management Perspectives*, *20*, 98–108. doi:10.1016/j.tmp.2016.07.011

Raj, R., & Griffin, K. (2017). Introduction to Conflicts, religion and culture in tourism. *Conflicts, Religion and Culture in Tourism*, 1-9. doi:10.1079/9781786390646.0001

Raja, P., & Raghu, A. (2020). *A critical review of terrorism effects and their impacts on tourist destination countries. In Materials Today: Proceedings*. Elsevier. doi:10.1016/j.matpr.2020.09.332

Ramón, D. (2020a, September 27). *El turismo está al borde del colapso: eurodiputados piden a la UE que actúe*. https://www.hosteltur.com/139476_el-turismo-esta-al-borde-del-colapso-eurodiputados-piden-a-la-ue-que-actue.html

Ramón, D. (2020b, October 1). *Agencias ofrecen pruebas rápidas de COVID-19 para impulsar los viajes*. https://www.hosteltur.com/139576_agencias-de-viaje-ofrecen-pruebas-rapidas-de-covid-19-a-sus-clientes.html

Randelli, F., Romei, P., & Tortora, M. (2014). An evolutionary approach to the study of rural tourism: The case of Tuscany. *Land Use Policy*, *38*, 276–281. doi:10.1016/j.landusepol.2013.11.009

Rapp, A., Beitelspacher, L. S., Grewal, D., & Hughes, D. E. (2013). Understanding social media effects across seller, retailer, and consumer interactions. *Journal of the Academy of Marketing Science*, *41*(5), 547–566. doi:10.100711747-013-0326-9

Rascombe, P. (2020). Rural areas at risk during the COVID-19 pandemic. *Lancet*, *20*(5), 545. doi:10.1016/S1473-3099(20)30301-7 PMID:32311327

Ray, A. (1997). *Kashmir Diary: Psychology of Militancy*. Manas.

Raza, S. A., & Jawaid, S. T. (2013). Terrorism and tourism: A conjunction and ramification in Pakistan. *Economic Modelling*, *33*, 65–70. doi:10.1016/j.econmod.2013.03.008

Reddy, M., Kaushal, V., Livina, A., Swain, S., & Kumar, H. (2021). Sustainable Consumption and Production of Wildlife Tourism in Indian Tiger Reserves: A Critical Analysis. *Worldwide Hospitality and Tourism Themes.*

Reddy, M., Boyd, S., & Nica, M. (2020). Towards a post-conflict tourism recovery framework. *Annals of Tourism Research*, *84*, 15–30. doi:10.1016/j.annals.2020.102940 PMID:32367898

Reddy, V., & Shilpa, R. (2016). Rural tourism – a catalyst for rural economic growth. International. *Journal of Human Social Sciences and Humanities Invention*, *5*(5), 14–19.

Reig, E. (2002). La multifuncionalidad del mundo rural [The multifunctionality of the rural world]. *ICE Revista de Economía*, *803*, 33–44.

Reinhold, S., Zach, F. J., & Krizaj, D. (2018). Business models in tourism – state of the art. *Tourism Review*, *74*(6), 1120–1134. doi:10.1108/TR-02-2018-0027

Reisinger, Y., & Mavondo, F. (2005). Travel anxiety and intentions to travel internationally: Implications of travel risk perception. *Journal of Travel Research*, *43*(3), 212–225. doi:10.1177/0047287504272017

Reisinger, Y., & Mavondo, F. (2006). Cultural differences in travel risk perception. *Journal of Travel & Tourism Marketing*, *20*(1), 13–31. doi:10.1300/J073v20n01_02

Reza, M., Gholipour, H., Feizi, M., Nunkoo, R., & Andargoli, A. (2020). International Tourism and Outbreak of Coronavirus (COVID-19): A Cross-Country Analysis. *Journal of Travel Research*, *60*(3), 1–6.

Richter, L. K., & Waugh, W. L. Jr. (1986). Terrorism and tourism as logical companions. *Tourism Management*, *7*(4), 230–238. doi:10.1016/0261-5177(86)90033-6

Riege, A. M. (2003). Validity and reliability tests in case study research: A literature review with "hands-on" applications for each research phase. *Qualitative Market Research*, *6*(2), 75–86. doi:10.1108/13522750310470055

Ritchie, B. W. (2004). Chaos, crises and disasters: A strategic approach to crisis management in the tourism industry. *Tourism Management*, *25*(6), 669–683. doi:10.1016/j.tourman.2003.09.004

Ritchie, B. W., & Jiang, Y. (2019). A review of research on tourism risk, crisis and disaster management: Launching the annals of tourism research curated collection on tourism risk, crisis and disaster management. *Annals of Tourism Research*, *79*, 1–15. doi:10.1016/j.annals.2019.102812

Ritchie, J. B., & Zins, M. (1978). Culture as determinant of the attractiveness of a tourism region. *Annals of Tourism Research*, *5*(2), 252–267. doi:10.1016/0160-7383(78)90223-2

Roa, M. (2020, April 29). *Nace PDAV, la asociación que da voz a las agencias pequeñas*. https://www.agenttravel.es/noticia-037724_Nace-PDAV-la-asociacion-que-da-voz-a-las- agencias-pequenas-.html

Robinson, L., & Jarvie, J. K. (2008). Post-disaster community tourism recovery: The tsunami and Arugam Bay, Sri Lanka. *Disasters*, *32*(4), 631–645. doi:10.1111/j.1467-7717.2008.01058.x PMID:18479472

Romagosa, F. (2020). *The COVID-19 crisis: Opportunities for sustainable and proximity tourism.* Tourism Geographies An International Journal of Tourism Space, Place and Environment.

Romagosa, F. (2020). The COVID-19 crisis: Opportunities for sustainable and proximity tourism. *Tourism Geographies, 20*(3), 690–694. doi:10.1080/14616688.2020.1763447

Romao, J. (2020). Tourism, smart specialisation, growth, and resilience. *Annals of Tourism Research, 84*, 31–46. doi:10.1016/j.annals.2020.102995 PMID:32834229

Romero, S. (2019). *Hostelería y turismo.* 7 Editores.

Rosato, P. F., Caputo, A., Valente, D., & Pizzi, S. (2021). 2030 Agenda and Sustainable Business Models in Tourism: A bibliometric analysis. *Ecological Indicators, 121*, 106978. doi:10.1016/j.ecolind.2020.106978

Rundshagen, V. M. (2013). Post Industrial Society. In S. Idowu, N. Capaldi, L. Zu, & A. Gupta (Eds.), *Encyclopedia of Corporate Social Responsibility* (p. 68). Springer. doi:10.1007/978-3-642-28036-8_175

Runfola, A., Rosati, M., & Guercini, S. (2013). New business model in online hotel distribution: Emerging private sales versus leading IDS. *Service Business, 7*(2), 183–205. doi:10.100711628-012-0150-1

Rusu, B. (2016). The impact of innovations on the business model: Exploratory analysis of a small travel agency. *Procedia: Social and Behavioral Sciences, 221*, 166–175. doi:10.1016/j.sbspro.2016.05.103

Saebi, T., Lien, L. B., & Foss, N. (2017). What Drives Business Model Adaptation? The Impact of Opportunities, Threats and Strategic Orientation. *Long Range Planning, 50*(5), 567–581. doi:10.1016/j.lrp.2016.06.006

Saha, S., & Yap, G. (2014). The moderation effects of political instability and terrorism on tourism development: A cross-country panel analysis. *Journal of Travel Research, 53*(4), 509–521. doi:10.1177/0047287513496472

Salemink, K., Strijker, D., & Bosworth, G. (2017). Rural development in the digital age: A systematic literature review on unequal ICT availability, adoption, and use in rural areas. *Journal of Rural Studies, 54*, 360–371. doi:10.1016/j.jrurstud.2015.09.001

San Juan, J. C. (2020, March 6). *La dificultad de la propiedad hotelera. Mi experiencia en Casual Hoteles.* https://ahoraturismo.com/2020/03/06/del-bi-al-bi-08-abr/

Sánchez-Zamora, P., Gallardo-Cobos, R., & Ceña-Delgado, F. (2014). Rural areas face the economic crisis: Analyzing the determinants of successful territorial dynamics. *Journal of Rural Studies, 35*, 11–25. doi:10.1016/j.jrurstud.2014.03.007

Sanders, D., & Laing, J. (2009). In the line of fire: The challenges of Managing tourism operations in the Victorian Alps. *Tourism Review International, 13*(4), 223–236. doi:10.3727/15442721 0X12741079930470

Santana, M. (2020, June 23). *Un 60% de agencias de viajes podrían cerrar si no se prorrogan los ERTE.* https://www.hosteltur.com/137610_un-60-de-agencias-de-viajes-podrian-cerrar-si-no-se-prorrogan-los-erte.html

Santana, G. (2004). Crisis management and tourism: Beyond the rhetoric. *Journal of Travel & Tourism Marketing, 15*(4), 299–321. doi:10.1300/J073v15n04_05

Santana, H. (2020). Good practices for e-business in the travel agencies. Izvestia Journal of the Union of Scientists - Varna. *Economic Sciences Series, 9*(1), 71–79.

Santos Lacueva, R. (2020). Estrategia de transformación del turismo rural frente a la COVID-19. In *M. Simancas Cruz, R. Hernández Martín, & N. Padrón Fumero (Coord.), Turismo pos-COVID-19. Reflexiones, retos y oportunidades* (pp. 523–534). Cátedra de Turismo Caja Canarias-Ashotel de la Universidad de La Laguna.

Sarrate, M. (2020). *La realidad de las empresas turísticas ante la COVID-19. XX Foro ACAVe.* El renacer del Turismo.

Saunders, B., Sim, J., Kingstone, T., Baker, S., Waterfield, J., Bartlam, B., Burroughs, H., & Jinks, C. (2018). Saturation in qualitative research: Exploring its conceptualization and operationalization. *Quality & Quantity-International Journal of Methodology, 52*(4), 1893–1907. doi:10.100711135-017-0574-8 PMID:29937585

Scott, N., & Laws, E. (2005). Tourism Crises and Disasters: Enhancing Understanding of System Effects. *Journal of Travel & Tourism Marketing, 19*(2-3), 149–158. doi:10.1300/J073v19n02_12

Scuri, S., Tesauro, M., & Petrelli, F. (2018). Implications of modified food choices and food-related lifestyles following the economic crisis in the Marche Region of Italy. *Annali Di Igiene Medicina Preventiva e Di Comunnità, 2*, 173–179. doi:10.7416/ai.2018.2208 PMID:29465154

Seabra, C., Abrantes, J. L., & Kastenholz, E. (2014). The influence of terrorism risk perception on purchase involvement and safety concern of international travellers. *Journal of Marketing Management, 30*(9-10), 874–903. doi:10.1080/0267257X.2014.934904

Séraphin, H., Butcher, J., & Korstanje, M. (2017). Challenging the negative images of Haiti at a pre-visit stage using visual online learning materials. *Journal of Policy Research in Tourism, Leisure & Events, 9*(2), 169–181. doi:10.1080/19407963.2016.1261146

Seraphin, H., & Dosquet, F. (2020). Mountain tourism and second home tourism as post-COVID 19 lockdown placebo? *Worldwide Hospitality and Tourism Themes, 12*(4), 485–500. doi:10.1108/WHATT-05-2020-0027

Seraphin, H., Gowreesunkar, V., Roselé-Chim, P., Duplan, Y. J. J., & Korstanje, M. (2018). Tourism planning and innovation: The Caribbean under the spotlight. *Journal of Destination Marketing & Management, 9*, 384–388. doi:10.1016/j.jdmm.2018.03.004

Séraphin, H., Korstanje, M., & Gowreesunkar, V. G. (Eds.). (2020). *Post-Disaster and Post-Conflict Tourism: Toward a New Management Approach.* CRC Press. doi:10.1201/9780429276224

Serravalle, F., Ferraris, A., Vrontis, D., Thrassou, A., & Christofi, M. (2019). Augmented reality in the tourism industry: A multi-stakeholder analysis of museums. *Tourism Management Perspectives*, *32*, 100549. doi:10.1016/j.tmp.2019.07.002

Setyawan, A. A., Isa, M., Wajdi, F., & Syamsudin, S. (2017). Disaster as business risk IN SME: An exploratory study. *International Journal of Research in Business and Social Science*, *6*(6), 52-63. doi:10.20525/ijrbs.v6i6.831

Shafi, M., Liu, J., & Ren, W. (2002). Impact of COVID-19 Pandemic on Micro, Small, and Medium-Sized Enterprises operating in Pakistan. *Research in Globalization*, *2*, 1–4.

Shanker, D. (2008). ICT and tourism: Challenges and opportunities. *J Manage Res Anal*, *5*(1), 276–283.

Shehawy, Y., & Ragab, A. (2017). Crisis Management Actions in the Egyptian Travel Agencies: A Managerial Perspective. *Journal of the Faculty of Tourism and Hotels- University of Sadat City*, *1*(1), 111–131. doi:10.21608/mfth.2017.26071

Sheller, M., & Urry, J. (2004). *Tourism mobilities: Places to play, places in play*. Routledge. doi:10.4324/9780203340332

Sheng-Hshiung, Gwo-Hshiung, T., & Kuo-Ching, W. (1997). T. Sheng-Hshiung, T. Gwo-Hshiung, W. Kuo-Ching (2019). Evaluating tourist risks from fuzzy perspectives. *Annals of Tourism Research*, *24*(4), 796–812. doi:10.1016/S0160-7383(97)00059-5

Shivakumar, K., & Agarwal, R. (2014). Significance of user-generated content in travel and tourism: An exploratory study. *Journal of Tourism & Hospitality (Los Angeles, Calif.)*, *03*(03). Advance online publication. doi:10.4172/2167-0269.1000137

Sigala, M. (2020). Tourism and COVID-19: Impacts and implications for advancing and resetting industry and research. *Journal of Business Research*, *117*, 312–321. doi:10.1016/j.jbusres.2020.06.015 PMID:32546875

Singh, S. (1998). War, terrorism & tourism. *Tourism Recreation Research*, *23*(2), 83–83. doi:10.1080/02508281.1998.11014846

Sinkovics, R. R., & Alfoldi, E. A. (2012). Progressive focusing and trustworthiness in qualitative research: The enabling role of computer-assisted qualitative data analysis software (CAQDAS). *Management International Review*, *52*(6), 817–845. doi:10.100711575-012-0140-5

Sloboda, B. W. (2003). Assessing the effects of terrorism on tourism by use of time series methods. *Tourism Economics*, *9*(2), 179–190. doi:10.5367/000000003101298349

Smartpoint. (2019). *Informe eApyme*. Asociación eApyme. https://www.eapyme.com/app/download/17019151/Informe+eAPyme+2019.pdf

SmartTravel. (2020). *Qué pueden hacer las agencias de viajes ante la crisis del coronavirus*. Available on https://www.smarttravel.news/pueden-las-agencias-viaje-ante-la-crisis-del-coronavirus/

Sobaih, A. E. E., Elshaer, I., Hasanein, A. M., & Abdelaziz, A. S. (2021). Responses to COVID-19: The role of performance in the relationship between small hospitality enterprises' resilience and sustainable tourism development. *International Journal of Hospitality Management*, *94*, 102824. Advance online publication. doi:10.1016/j.ijhm.2020.102824

Solarin, S. A. (2015). September 11 attacks, H1N1 influenza, global financial crisis and tourist arrivals in Sarawak. *Anatolia*, *26*(2), 298–300. doi:10.1080/13032917.2014.991933

Sönmez, S. F. (1998). Tourism, terrorism, and political instability. *Annals of Tourism Research*, *25*(2), 416–456. doi:10.1016/S0160-7383(97)00093-5

Sönmez, S. F., & Graefe, A. R. (1998). Influence of terrorism risk on foreign tourism decisions. *Annals of Tourism Research*, *25*(1), 112–144. doi:10.1016/S0160-7383(97)00072-8

Sosna, M., Trevinyo-Rodríguez, R. N., & Velamuri, S. R. (2010). Business model innovation through trial-and-error learning: The Naturhouse case. *Long Range Planning*, *43*(2-3), 383–407. doi:10.1016/j.lrp.2010.02.003

Souto, J. E. (2015). Business model innovation and business concept innovation as the context of incremental innovation and radical innovation. *Tourism Management*, *51*, 142–155. doi:10.1016/j.tourman.2015.05.017

Spence, M., & Hamzaoui Essoussi, L. (2010). SME brand building and management: An exploratory study. *European Journal of Marketing*, *44*(7/8), 1037–1054. doi:10.1108/03090561011047517

Spieth, P., Schneckenberg, D., & Ricart, J. E. (2014). Business model innovation–state of the art and future challenges for the field. *R & D Management*, *44*(3), 237–247. doi:10.1111/radm.12071

Stafford, G., Yu, L., & Armoo, A. K. (2002). Crisis management and recovery. *The Cornell Hotel and Restaurant Administration Quarterly*, *43*(5), 27–40.

Stake, R. E. (2005). Qualitative case studies. In N. K. Denzin & Y. S. Lincoln (Eds.), *The Sage handbook of qualitative research* (3rd ed., pp. 443–466). SAGE.

Stamboulis, Y., & Skayannis, P. (2003). Innovation strategies and technology for experience-based tourism. *Tourism Management*, *24*(1), 35–43. doi:10.1016/S0261-5177(02)00047-X

Steffen, W., Broadgate, W., Deutsch, L., Gaffney, O., & Ludwig, C. (2015). The trajectory of the Anthropocene: The Great Acceleration. *The Anthropocene Review*, *2*(1), 81–98. doi:10.1177/2053019614564785

Steiner, A., & Atterton, J. (2015). Exploring the contribution of rural enterprises to local resilience. *Journal of Rural Studies*, *40*, 30–45. doi:10.1016/j.jrurstud.2015.05.004

Steinerowski, A., & Steinerowska-Streb, I. (2012). Can social enterprise contribute to creating sustainable rural communities? Using the lens of structuration theory to analyze the emergence of rural social enterprise. *Local Economy*, *27*(2), 167–182. doi:10.1177/0269094211429650

Subrahmanyam, K. (1990). Kashmir. *Strategic Analysis*, *23*(11), 111–198.

Sundbo, J., Orfila-sintes, F., & Sorensen, F. (2007). The innovative behavior of tourism firmscomparative studies of Denmark and Spain. *Research Policy*, *36*(1), 88–106. doi:10.1016/j.respol.2006.08.004

Swain, A. (2009). The Indus II and Siachen Peace Park: Pushing the India–Pakistan Peace Process Forward. *The Round Table*, *98*(404), 569–582. doi:10.1080/00358530903151854

Swami, P. (2003). Terrorism in Jammu and Kashmir in theory and practice. *India Review*, *2*(3), 55–88. doi:10.1080/14736480412331307082

Szromek, A. R., & Naramski, M. (2019). A Business Model in Spa Tourism Enterprises: Case Study from Poland. *Sustainability*, *11*(10), 2880. doi:10.3390u11102880

Talwar, R. (2010). Scanning the horizon. *Trends and Issues in Global Tourism*, *2010*, 237–268. doi:10.1007/978-3-642-10829-7_26

Tang, J. C., & Rochananond, N. (1990). Attractiveness as a tourist destination: A comparative study of Thailand and selected countries. *Socio-Economic Planning Sciences*, *24*(3), 229–236. doi:10.1016/0038-0121(90)90005-R

Tarlow, P. (2014). *Tourism security: strategies for effectively managing travel risk and safety*. Elsevier.

Taylor, S. C., Torpy, D. J., & Das, D. K. (Eds.). (2012). *Policing global movement: Tourism, migration, human trafficking, and terrorism*. CRC Press. doi:10.1201/b12892

Taylor, S. Jr. (2020). The socially distant service scape: An investigation of consumer preference's during the re-opening phase. *International Journal of Hospitality Management*, *91*, 102692. Advance online publication. doi:10.1016/j.ijhm.2020.102692 PMID:33012948

Teece, D. J. (2010). Business models, business strategy and innovation. *Long Range Planning*, *43*(2-3), 172–194. doi:10.1016/j.lrp.2009.07.003

The European Travel Agents and Tour Operators Association. (2020). *Covid-19: Tourism Ministers and the Commission need to step up efforts for* [Press release]. Group of National Travel Agents' and Tour Operators' Associations within the EU. https://www.ectaa.org/Uploads/press-releases/PR-20200424-Tourism-Ministers-and- Commission-need-to-step-up-efforts.pdf

Thomas, K. C. (2020). *2 economic growth and economic development*. Competition Law in Developing Countries. doi:10.1093/law-ocl/9780198862697.003.0003

Thomas, R., & Wood, E. (2014). Innovation in tourism: Re-conceptualising and measuring the absorptive capacity of the hotel sector. *Tourism Management*, *45*, 39–48. doi:10.1016/j.tourman.2014.03.012

Thomé-Ortiz, H. (2020). O coronavírus reescreverá o turismo rural? Reinvenção, adaptação e ação no contexto latino-americano [Will the coronavirus rewrite rural tourism? Reinvention, adaptation and action in the Latin American context]. *Cenário*, *8*, 55–72.

Thompson, A. (2011). Terrorism and tourism in developed versus developing countries. *Tourism Economics*, *17*(3), 693–700. doi:10.5367/te.2011.0064

Toubes, D., Araújo-Vila, N., & Fraiz-Brea, J.A. (2020). Be water my friend: Building a liquid destination through collaborative networks. *Tourism Management Perspectives*, *33*, 100619. doi:10.1016/j.tmp.2019.100619

Tsaur, H., & Teng, H. (2017). Exploring tour guiding styles: The perspective of tour leader roles. *Tourism Management*, *59*, 438–448. doi:10.1016/j.tourman.2016.09.005

Tse, A. C. B., & Leo Sin, S. S. (2006). Crisis management and recovery: How restaurants in Hong Kong responded to SARS. *International Journal of Hospitality Management*, *25*(1), 3–11. doi:10.1016/j.ijhm.2004.12.001 PMID:32287848

Tuclea, C., Vasile, C., Schiopu, F., & Marin, M. (2014). Facets of economic and financial crisis impact on strategic planning of travel agencies. *Amfiteatru Economic Journal*, *16*(8), 1222–1237.

Turijobs. (2020). *Mirando hacia el futuro: La lucha del sector turístico frente al Covid-19.* L'Hospitalet de Llobregat.

UN. (2020). *Policy Brief: Covid-19 and Transforming Tourism.* https://www.un.org/sites/un2.un.org/files/sg_policy_brief_covid-19_tourism_august_2020.pdf

UNCTAD. (2020). *World Investment Report.* https://unctad.org/webflyer/world-investment-report-2020

UNWTO world Tourism barometer and Statistical Annex, November 2019. (2019). *UNWTO World Tourism Barometer, 17*(4), 1-44.

UNWTO. (2020a). *International tourist numbers could fall 60-80% in 2020, UNWTO reports.* https://www.unwto.org/news/covid-19-international-tourist-numbers-could-fall-60-80-in-2020

UNWTO. (2020b). *Tourism and COVID-19, Issue 1. How are countries supporting tourism recovery?* https://www.e-unwto.org/doi/book/10.18111/9789284421893

UNWTO. (2020c). *Covid-19 measures to support the travel and tourism sector.* https://webunwto.s3-eu-west-1.amazonaws.com/s3fs-public/2020-06/europe.pdf

UNWTO. (2020c). *International tourism and COVID-19.* https://www.unwto.org/international-tourism-and-covid-19

Urry, J. (2006). Inhabiting the car. *The Sociological Review*, *54*(1, suppl), 17–31. doi:10.1111/j.1467-954X.2006.00635.x

Vargas, A. (2020, September 30). *Oportunidades para el agente de viajes en un mundo transformado.* https://www.hosteltur.com/139520_oportunidades-para-el-agente-de-viajes-en- un-mundo-transformado.html?utm_content=buffer90ef0&utm_medium=social&utm_source=facebook&utm_campaign=noticia&fbclid=IwAR3DfaoVSO4bpVQAD6dvmE1TDB8_npqu56YRGmVn9scmmw-NauA6OtPlSZo

Vargas-Sánchez, A. (2018). Crises in tourist destinations: how can they be managed? *Enlightening Tourism. A Pathmaking Journal*, *8*(1), 47-69.

Vargas-Sánchez, A. (2020a). Entender el turismo post-coronavirus: posibles escenarios. In F.J. Bouzá Martorell, & F.J. Melgosa Arcos (Dir.), *El turismo después de la pandemia global: análisis, perspectivas y vías de recuperación*. Documento de Trabajo de AECIT. Retrieved from https://aecit.org/documento-de-trabajo-el-impacto-de-la-crisis-sanitaria-en-el-sector-t/news/193/

Vargas-Sánchez, A. (2020b). Estrategias de destinos y empresas en el mercado turístico pos-COVID-19. In M. Simancas Cruz, R. Hernández Martín, & N. Padrón Fumero (Eds.), *Turismo pos-COVID-19. Reflexiones, retos y oportunidades* (pp. 147–160). Cátedra de Turismo Caja Canarias-Ashotel de la Universidad de La Laguna.

Varma, A., Sprott, D. E., Silverman, S. N., & Stem, D. E. (2000). Adoption of Internet shopping: The role of consumer innovativeness. *Industrial Management & Data Systems*.

Vasenska, I., Dimitrov, P., Koyundzhiyska-Davidkova, B., Krastev, V., Durana, P., & Poulaki, I. (2021). Financial transactions USING Fintech during the Covid-19 crisis in Bulgaria. *Risks*, *9*(3), 48. doi:10.3390/risks9030048

Vaughan, D. R., & Edwards, J. R. (1999). Experiential perceptions of two winter sun destinations: The Algarve and Cyprus. *Journal of Vacation Marketing*, *5*(4), 356–368. doi:10.1177/135676679900500405

Vázquez, M. (2017). Ourense tiene más bares por habitante que las otras tres capitales gallegas. *La Voz de Galicia*. Retrieved from https://www.lavozdegalicia.es/noticia/ourense/a-gudina/2017/06/18/ourense-bares-habitante-tres-capitales-gallegas/0003_201706O18C5991.htm

Veal, A. J. (2011). *Research Methods for Leisure and Tourism: A Practical Guide* (4th ed.). Prentice Hall.

Veal, A. J. (2018). *Research methods for leisure and tourism*. Pearson.

Venkatesan, V. (2020). *Abrogation of Articles 370 and 35A: Assault on the Constitution*. Frontline.

Verma, S., & Gustafsson, A. (2020). Investigating the emerging COVID-19 research trends in the field of business and management: A bibliometric analysis approach. *Journal of Business Research*, *118*, 253–261. Advance online publication. doi:10.1016/j.jbusres.2020.06.057 PMID:32834211

Verreynne, M., Williams, A. M., Ritchie, B. W., Gronum, S., & Betts, K. S. (2019). Innovation diversity and uncertainty in small and medium sized tourism firms. *Tourism Management*, *72*, 257–269. doi:10.1016/j.tourman.2018.11.019

Visontay, E. (2020, July 18). Death of the travel agent: Covid-19 an existential threat for an industry without "any product to sell." *The Guardian*. http://www.theguardian.com/australia-news/2020/jul/19/death-of-the-travel-agent-covid-19-an-existential-threat-for-an-industry-without-any-product-to-sell

Wahab, S. (1996). Tourism and Terrorism: synthesis of the problem with emphasis on Egypt. In A. Pizam & Y. Mansfeld (Eds.), *Tourism, crime &International security Issues* (pp. 175–188). Wiley.

Wall, G. (1996). Terrorism and Tourism: An Overview and an Irish Example. In A. Pizam & Y. Mansfeld (Eds.), *Tourism, Crime and International Security Issues* (pp. 143–158). Wiley.

Wearing, S., & Neil, J. (1999). *Ecoturismo: Impacto, Tendencias y posibilidades*. Editorial Síntesis.

Weber, K. (1997). The assessment of tourist satisfaction using the expectancy disconfirmation theory: A study of the German travel market in Australia. *Pacific Tourism Review, 1*(1), 35–45.

Weber, S. (1998). War, terrorism, and tourism. *Annals of Tourism Research, 25*, 760–762.

Weiler, B., & Black, R. S. (2015). The changing face of the tour guide: One-way communicator to choreographer to co-creator of the tourist experience. *Tourism Recreation Research, 40*(3), 364–378. doi:10.1080/02508281.2015.1083742

Weiler, B., & Kim, A. K. (2011). Tour guides as agents of sustainability: Rhetoric, reality and implications for research. *Tourism Recreation Research, 36*(2), 113–125. doi:10.1080/02508281.2011.11081313

Weimann, G. (2008). The psychology of mass-mediated terrorism. *The American Behavioral Scientist, 52*(1), 69–86. doi:10.1177/0002764208321342

Wen, J., Kozak, M., Yang, S., & Liu, F. (2020). COVID-19: potential effects on Chinese citizens' lifestyle and travel. *Tourism Review*.

Wen, J., Wang, W., Kozak, M., Liu, X., & Hou, H. (2020). Many brains are better than one: The importance of interdisciplinary studies on COVID-19 in and beyond tourism. *Tourism Recreation Research*, 1–4. doi:10.1080/02508281.2020.1761120

Whitacre, B., & Mills, B. (2007). Infrastructure and the rural urban divide in high-speed residential internet access. *International Regional Science Review, 2007*(30), 249–273. doi:10.1177/0160017607301606

WHO. (2004). *WHO guidelines for the global surveillance of severe acute respiratory syndrome (SARS)*. Updated recommendations, October 2004. World Health Organization. Retrieved from https://www.who.int/csr/resources/publications/WHO_CDS_CSR_ARO_2004_1.pdf

Williams, A. M., & Shaw, G. (1998). *Tourism and economic development: European experiences*. J. Wiley.

Williams, C., & Ferguson, M. (2005). Recovering from crisis: Strategic alternatives for leisure and tourism providers based within a rural economy. *International Journal of Public Sector Management, 18*(4), 350–366. doi:10.1108/09513550510599265

Wilson, S., Fesenmaier, D. R., Fesenmaier, J., & Van Es, J. C. (2001). Factors for Success in Rural Tourism Development. *Journal of Travel Research, 40*(2), 132–138. doi:10.1177/004728750104000203

Wiprächtiger, D., Narayanamurthy, G., Moser, R., & Sengupta, T. (2019). Access-based business model innovation in frontier markets: Case study of shared mobility in Timor-Leste. *Technological Forecasting and Social Change*, *143*, 224–238. doi:10.1016/j.techfore.2019.02.004

Wolff, K., & Larsen, S. (2014). Can terrorism make us feel safer? Risk perceptions and worries before and after the July 22nd attacks. *Annals of Tourism Research*, *44*, 200–209. doi:10.1016/j.annals.2013.10.003

Wolff, K., Larsen, S., & Øgaard, T. (2019). How to define and measure risk perceptions. *Annals of Tourism Research*, *79*, 1–9. doi:10.1016/j.annals.2019.102759

Wolpert, S. (2010). India and Pakistan: continued conflict or cooperation? Univ of California Press. doi:10.1525/9780520948006

World Health Organization. (2020). *Ethical standards for research during public health emergencies: Distilling existing guidance to support COVID-19 R&D*. WHO. https://apps.who.int/iris/handle/10665/331507

World Tourism Organization. (2020). *Framework Convention on Tourism Ethics*. UNWTO. doi:10.18111/9789284421671

Wu, P. J. (2020). *A Study on the Impact of COVID-19 on Travel Agencies in Taiwan and their Response*. Academic Press.

Xue, L., & Kerstetter, D. (2019). Rural Tourism and Livelihood Change: An Emic Perspective. *Journal of Hospitality & Tourism Research (Washington, D.C.)*, *43*(3), 416–437. doi:10.1177/1096348018807289

Yang, Y., Zhang, H., & Chen, X. (2020). Coronavirus pandemic and tourism: Dynamic stochastic general equilibrium modeling of infectious disease outbreak. *Annals of Tourism Research*, *83*, 102913. doi:10.1016/j.annals.2020.102913 PMID:32292219

Yarkova, Y., Stoykova, B., Markov, N., & Bankova, M. (2019). Regional economy within the context of bioeconomy. *Trakia Journal of Sciences*, *17*(Suppl.1), 5–13. doi:10.15547/tjs.2019.s.01.002

Yigitcanlar, T., Kamruzzaman, M., Foth, M., Sabatini-Marques, J., da Costa, E., & Ioppolo, G. (2019). Can cities become smart without being sustainable? A systematic review of the literature. *Sustainable Cities and Society*, *45*, 348–365. doi:10.1016/j.scs.2018.11.033

Yin, R. K. (2003). *Case Study Research: Design and Methods* (3rd ed.). Sage.

Yin, R. K. (2018). *Case Study Research and Application: Design and Methods* (6th ed.). Sage.

Yin, Y., Zhang, X., Peng, D., & Li, X. (2009). Model validation and case study on internally cooled/heated dehumidifier/regenerator of liquid desiccant systems. *International Journal of Thermal Sciences*, *48*(8), 1664–1671. doi:10.1016/j.ijthermalsci.2008.12.017

Yüksel, A., & Yüksel, F. (2007). Shopping risk perceptions: Effects on tourists' emotions, satisfaction and expressed loyalty intentions. *Tourism Management*, *28*(3), 703–713. doi:10.1016/j.tourman.2006.04.025

Yu, M., Li, Z., Yu, Z., He, J., & Zhou, J. (2020). Communication-related health crisis on social media: A case of COVID-19 outbreak. *Current Issues in Tourism*, 1–7. doi:10.1080/13683500.2020.1752632

Yung, R., & Khoo-Lattimore, C. (2019). New realities: A systematic literature review on virtual reality and augmented reality in tourism research. *Current Issues in Tourism*, *22*(17), 2056–2081. doi:10.1080/13683500.2017.1417359

Zabalza, M., & Zabalza, A. (2012). *Innovación y cambio en las instituciones educativas*. Homo Sapiens.

Zapata, M., Hall, M., Lindo, P., & Vanderschaeghe, M. (2011). Can community-based tourism contribute to development and poverty alleviation? Lessons from Nicaragua. *Current Issues in Tourism*, *14*(8), 725–749. doi:10.1080/13683500.2011.559200

Zeng, B., Carter, R. W., & De Lacy, T. (2005). Short-term perturbations and tourism effects: The case of SARS in China. *Current Issues in Tourism*, *8*(4), 306–322. doi:10.1080/13683500508668220

Zeng, Z., Chen, P., & Lew, A. A. (2020). From high-touch to high-tech: COVID-19 drives robotics adoption. *Tourism Geographies*, *22*(3), 724–734. doi:10.1080/14616688.2020.1762118

Zenker, S., & Kock, F. (2020). The coronavirus pandemic – A critical discussion of a tourism research agenda. *Tourism Management*, *81*, 104164. doi:10.1016/j.tourman.2020.104164 PMID:32518437

Zhang, W., Daim, T., & Zhang, Q. (2019). Exploring the multi-phase driven process for disruptive business model innovation of e-business microcredit: A multiple case study from China. *Journal of the Knowledge Economy*, *10*(2), 590–617. doi:10.100713132-017-0483-z

Zott, C., & Amit, R. (2010). Business model design: An activity system perspective. *Long Range Planning*, *43*(2-3), 216–226. doi:10.1016/j.lrp.2009.07.004

Zott, C., Amit, R., & Massa, L. (2011). The business model: Recent developments and future research. *Journal of Management*, *37*(4), 1019–1042. doi:10.1177/0149206311406265

Zuvekas, C. (1979). Economic growth and economic development. *Economic Development*, 7-29. doi:10.1007/978-1-349-16275-8_2

About the Contributors

Diego R. Toubes is a PhD in Economic and Business Sciences. He is an Associate professor of Strategic Management in the Faculty of Administration and Tourism of Ourense, at the University de Vigo (Spain). He is a member of the research group Emitur of the University of Vigo. His research interests include risk management and organizational resilience. He has published academic papers and collaborations in the fields of entrepreneurship, risk perception and crisis communication in tourism.

Noelia Araújo Vila is PhD in Tourism Management and Planning (2013) and Bachelor in Business Administration and Management (2006). Since 2007 she collaborates in the coordination tasks of the University Master in Tourism Management and Planning. She teaches as professor in the Departments of Business Organization and Marketing and Financial Economics and Accounting (University of Vigo).

* * *

Ruska Bozhkova, PhD, is a lecturer at the Department of Tourism at the South-West University "Neofit Rilski" of Blagoevgrad. Her major line of inquiry is based on international tourism, cultural studies, strategies of tourism flows and cultural heritage.

Lucília Cardoso is a postdoctoral researcher at IP Leiria Portugal and in Switzerland she's studying Neurosciences applied to Tourism. She is a Researcher of CITUR - Portugal and her publications and empirical studies cover a large range of cognitive psychology applied to marketing and tourism. In recent years, her publications have also covered the management field and bibliometrics applied to tourism and hospitality research performance.

Sara Fidalgo holds a Tourism Bachelor Degree from the University of Vigo, Spain. She is a junior researcher at the University of Vigo and is a collaborator of the International Week at this University every year. Her research interests include primarily the MICE sector. She has taken an active part in international conferences related to tourism. Her professional experience includes business internships at international tourism fairs and event coordinator in the automotive industry.

Luis Felipe García-Rodea has a degree in tourism from the Universidad Autónoma del Estado de México (Mexico). He received a Master's degree in rural agribusiness, territorial development and agri-food tourism and is currently studying a PhD in agricultural sciences and natural resources. His research interests are in the areas of rural tourism, rural development, wine tourism and agri-food tourism.

Mahender Reddy Gavinolla is Guest Lecturer at Vidzeme university of Applied Sciences, Latvia and Assistant Professor & Coordinator-International Relations at National Institute of Tourism & Hospitality Management, Hyderabad, India. He has published several papers in reputed journals and edited book such as Sage, Routledge, Emerald, De Gruyter, in the area of responsible and sustainable tourism. ORCiD: 0000-0002-3476-3132.

Anubha Kaura is an Assistant Lecturer, School of Hospitality, National Institute of Tourism and Hospitality Management, Hyderabad, India. She teaches food production courses and specializes in Continental cuisine. She also regularly contributes writings on food in Indian English-language daily newspaper, Deccan Chronicle. Her area of research includes food tourism and disaster tourism.

Teodora Kiryakova-Dineva is an Associate Professor with an interdisciplinary profile. Her research interests lie in tourism studies, intercultural studies and all types of intangible assets, including management and theoretical and practical approaches to contemporary phenomena. She is an academic member of the South-West University "Neofit Rilski", Bulgaria.

Maximiliano E. Korstanje is editor in chief of International Journal of Safety and Security in Tourism (UP Argentina) and Editor in Chief Emeritus of International Journal of Cyber Warfare and Terrorism (IGI-Global US). Besides being Senior Researchers in the Department of Economics at University of Palermo, Argentina, he is a global affiliate of Tourism Crisis Management Institute (University of Florida US), Centre for Ethnicity and Racism Studies (University of Leeds), The Forge (University of Lancaster and University of Leeds UK) and The International Society for Philosopher, hosted in Sheffield UK. With more than 1200 published papers and 35 books. As author his recent works include, Strategic Tool and Methods for promoting Hospitality and Tourism Services, Virtual Traumascapes and exploring the roots of Dark Tourism, & Research Practices and innovations in Global Risk and Contingency Management by IGI Global, The Rise of Thana Capitalism and Tourism by Routledge, Terrorism, Tourism and the end of Hospitality in the West by Springer Nature, The Mobilities paradox, a critical analysis by Edward Elgar, Risk and Safety Challenges for Religious Tourism and Events by CABI, and Terrorism in a Global Village by Nova among others. Now he co-edits almost 10 specialized journals and takes part of almost 30 journals associated to themes as human rights, mobility, tourism and terrorism. Korstanje was nominated to 5 honorary doctorates for his contribution in the study of the effects of terrorism in tourism. In 2015 he was awarded as Visiting Research Fellow at School of Sociology and Social Policy, University of Leeds, UK and the University of La Habana Cuba. In 2017 is elected as Foreign Faculty Member of AMIT, Mexican Academy in the study of Tourism, which is the most prominent institutions dedicated to tourism research in Mexico. He had a vast experience in editorial projects working as advisory member of Elsevier, Routledge, Springer, IGI global and Cambridge Scholar publishing. Korstanje had visited and given seminars in many important universities worldwide. He has also recently been selected to take part of the 2018 Albert Nelson Marquis Lifetime Achievement Award. a great distinction given by Marquis Who´s Who in the world.

Elizabeth López Carré has a PhD in Tourism Studies from the Universidad Autónoma del Estado de México. She is a full-time professor at the school of Tourism and Gastronomy of which she has been principal.

María-Concepción López-Fernández, PhD, is a Professor of Management and Strategy at the Business Administration Department (University of Cantabria - UC). She colaborates with the Banco Santander Chair in Family Business. Her research interests include innovation management, entrepreneurship, family business, business strategy and structure, organisational flexibility, and tourism. She has co-authored a significant number of national and international conference papers, books and journal articles including those appearing in Family Business Review, Journal of Family Business Strategy, Journal of Small Business Management, Tourism Management, Organization & Environment, R&D Management, Cornell Hospitality Quarterly, Total Quality Management & Business Excellence, International Journal of Production Research or Journal of Manufacturing and Technology Management.

Almudena Otegui Carles is a graduate in Tourism and has Master's degrees in Tourism, Education, and Protocol and External Relations. She has worked for the last fifteen years as a travel agency manager. In recent years, she has combined her experience in the latter field with her academic work in the tourism sector, serving as a teacher-tutor for Master's students in Events Organisation, Ceremonial, and Protocol. She is currently pursuing a doctorate in Tourism at the University of Vigo.

Marta Perez-Perez is associate professor of Operations Management at the University of Cantabria, Spain. Her main research interests are: Operational flexibility, Tourism, Family firm, Bibliometrics tools and knowledge management.

Elena Puiggros is a Doctor of Sociology, coordinator of research group: Image, Satisfaction & Tourist Behaviour, professor and coordinator at EU Mediterrani (Tourism degree),

Ana-María Serrano-Bedia is associate professor of Operations Management, at the Department of Business Administration, University of Cantabria, Spain. Her main research interests are: Operational flexibility, Tourism, Operations strategy, Quality management, Family firm, Bibliometric techniques, Knowledge management, Innovation management.

Humberto Thomé-Ortiz has a degree in Social Communication from the Universidad Autónoma Metropolitana (México), Master of Science in Rural Development from the Colegio de Postgraduados (México), Doctor of Agricultural Sciences from the Universidad Autónoma Chapingo (México). Full-time Research Professor at the Institute of Agricultural and Rural Sciences (ICAR) of the Universidad Autónoma del Estado de México. Specialist in Agri-Food Studies and Agrifood Tourism, he is president of the Mexican Association of Rural Tourism. He has developed research on tourism and agri-food systems in different rural areas of Mexico, Brazil, Ecuador, Colombia and Chilean Polynesia. He has also worked on agri-food studies in different areas of central Mexico. Cultivate the disciplines of Sociology of Food, Gender and Food and Sustainable Development. He is a member of the National System of Researchers of Mexico Level 1. He has published more than 150 academic products including books, chapters and scientific articles in journals such as the Journal of Heritage Tourism, British Food Journal, Apettite, and Routledge Publishing. He currently works as Coordinator of the Master's Degree in Rural Agroindustry, Territorial Development and Agrifood Tourism at UAEMEX.

Najat Tort is a PhD student in Tourism, member of research group: Image, Satisfaction & Tourist Behaviour, and lecturer at EU Mediterrani.

Alfonso Vargas-Sánchez is Full Professor at the University of Huelva (Spain), in its Management and Marketing Department. For more details, please go to: http://www.uhu.es/alfonso_vargas/.

Index

A

B

C

D

E

F

www.ingramcontent.com/pod-product-compliance
Lightning Source LLC
LaVergne TN
LVHW061546290926
849162LV00004B/922